THE NEW NATURALIST
A SURVEY OF BRITISH NATURAL HISTORY

THE POLLINATION OF FLOWERS

The aim of this series is to interest the general reader in the
wild life of Britain by recapturing the inquiring spirit of the
old naturalists. The Editors believe that the natural pride of
the British public in the native fauna and flora, to which
must be added concern for their conservation, is best fostered
by maintaining a high standard of accuracy combined with
clarity of exposition in presenting the results of modern
scientific research.

THE
POLLINATION
OF FLOWERS

MICHAEL PROCTOR

M.A., PH.D., F.R.P.S.

SENIOR LECTURER
DEPARTMENT OF BIOLOGICAL SCIENCES
UNIVERSITY OF EXETER

and

PETER YEO

M.A., PH.D.

TAXONOMIST, UNIVERSITY BOTANIC GARDEN,
CAMBRIDGE

WITH 19 COLOUR PHOTOGRAPHS
181 PHOTOGRAPHS IN BLACK AND WHITE
AND 134 LINE DRAWINGS

TAPLINGER PUBLISHING COMPANY
NEW YORK

To
JANE AND AVRIL

First published in the United States in 1972 by
TAPLINGER PUBLISHING CO., INC.
New York, New York

Library of Congress Catalog Card Number: 70-185876
ISBN 0-8008-6408-5

CONTENTS

PLATES IN COLOUR

BLACK AND WHITE PLATES

EDITORS' PREFACE

The methods by which pollen grains reach a stigma, thus enabling fertilisation and seed production to take place, include some of the most varied and fascinating mechanisms in the whole world of living things. As the authors point out in their Historical Introduction, Theophrastus, in the fourth century B.C., had an inkling of the occurrence of sex in plants, but it was not until the end of the seventeenth century that the phenomenon began to be generally accepted, and a study of pollination mechanisms could begin. The knowledge thus accumulated was first gathered together by Hermann Müller in his classic work, translated into English in 1883 under the title *The Fertilisation of Flowers*, which was followed in 1906-8 by a translation of P. Knuth's *Handbook of Flower Pollination* in three volumes. Both of these are now, of course, long out of date, and although a number of books have appeared in recent years covering various aspects of the subject, the Editors felt that a New Naturalist volume, dealing primarily with British plants, but including considerable reference to the rest of the world, would be a very valuable addition to the Series, and biological literature in general.

We have been extremely fortunate in persuading Dr Michael Proctor and Dr Peter Yeo to co-operate as joint authors of such a volume. Both are professional botanists with a special interest in pollination, but both have other qualifications which make them particularly fitted for this task. Dr Proctor, as is shown by the illustrations to the volume, is an outstanding photographer, and both have a wide knowledge of insects – particularly, in the case of Dr Yeo, of the Hymenoptera, which play so important a part in flower pollination. The book they have written is the only one that includes, within the covers of a single volume, descriptions and illustrations both of pollination mechanisms and of the detailed structure of the insects concerned with so many of these mechanisms.

We hope that this volume will appeal not only to professional biologists but also to amateur naturalists, both botanists and zoologists. As pointed out by the authors in their Preface, observations on pollination mechanisms are eminently suited to amateurs, requiring, as they do, mainly time, patience, and accurate observation and recording. An outstanding example is the discovery described in Chapter 7 of the pseudocopulatory mechanism in various orchids of the genus *Ophrys* by M. Pouyanne, Counsellor of the Court of Appeal in Algiers, later confirmed and extended by the long and patient observations of Colonel Godfery on cut flowers on a hotel balcony in France! There is still much to be done on these lines, even on British plants, and we hope that the publication of this volume will stimulate such work by both professionals and amateurs.

The Editors

AUTHORS' PREFACE

THERE is a particular fascination about the reciprocal adaptations of flowers and flower-visiting insects, and few scientific topics have become so firmly rooted in our everyday consciousness as the pollination of flowers. Much of our knowledge goes back to the great floral biologists of the latter part of the nineteenth century. However, in the last sixty or seventy years floral biology has grown in many directions, and there have been great advances in related scientific fields. We have set out to produce an up-to-date book on flower pollination, in a context which is now very different from that in which the classical accounts were written. Inevitably we have been selective in our choice of topics and examples: the field is now so vast, and impinges so much on other subjects, that a comprehensive book on floral biology could probably not be written. However, we hope to have produced a reasonably rounded account, which will give something of interest, pleasure or instruction to a diversity of readers.

In recent years there has been a great revival of interest in pollination biology. It is an ideal study for the amateur naturalist, whether he is primarily interested in plants or insects, because even the simplest observation may add something significant to the fund of knowledge, while much useful scientific work can be carried out with comparatively simple equipment. That it is also an important and rewarding field for the professional research worker is amply demonstrated by the number and interest of the papers published. Pollination studies often require a combination of botanical and zoological knowledge and methods, and are thus in a particular position to benefit from the current emphasis on the unity of biology.

The aim of the *New Naturalist* '. . . is to interest the general reader in the wild life of Britain . . .', and we have therefore written this book with the native British flora and fauna uppermost in our minds. However, pollination biology is an obstinately international subject. A great deal of research relevant to the pollination of British plants has been carried out on the Continent or in North America, and we have drawn freely on this work. In addition, we have included a chapter on some of the more important or remarkable pollination relationships and mechanisms that are not represented in the British flora.

We should perhaps explain and justify the many references in the text giving the sources of our statements. They have been included for two important reasons. Firstly, the literature of pollination biology is very widely scattered, and the references will guide the reader to where he can find more detailed information about matters that we may have touched upon only briefly. Secondly, they allow the reader to check the origin and reliability of particular ideas and observations. By constant repetition in print, a statement can acquire an air of authority and generality that may not be justified, and sometimes that its original author never intended. We have tried to check to their original sources many statements that seemed to us interesting or important, though often, because of the size of the task, we have had to be content with secondary sources. The conventions we have used in citing references are explained at the beginning of the bibliography (p. 386).

For the Latin names of British plants we have followed the second edition of Clapham, Tutin and Warburg's *Flora of the British Isles* (Cambridge, 1962), with a few minor exceptions, notably the use of the generic name *Dactylorhiza* instead of *Dactylorchis* for the Spotted and Marsh Orchids. The names of insects in general follow Kloet and Hincks's *A Check List of British Insects* (1945), except for the *Syrphidae*, where we have used the names in the Royal Entomological Society's *Handbook* by R. L. Coe (1953). For English names of plants we have followed McClintock and Fitter's *The Pocket Guide to Wild Flowers* (Collins, 1956).

In writing this book, M.C.F.P. was primarily responsible for Chapters 1, 2, 6, 7, 8 and 9, and P.F.Y. for Chapters 3, 4, 5, 10 and 11; Chapter 12 was written jointly, and throughout we have discussed and revised one another's drafts. We have tried to keep technical terms to a minimum, and to define those that we have used. In particular, Chapter 2 was written with some of the functions of a glossary in mind, and parts of Chapters 4 and 5 will serve something of the same purpose. This is not intended as a book to be read without pause or deviation from beginning to end. We realise that many people may not always want to read the chapters in the order in which they appear here; we have tried to put as few difficulties in their way as possible. We have used photographs and line-drawings freely; we hope that both these and the text will stimulate the reader to observe flowers and their pollinators closely for himself in the field, confident that this will enhance his pleasure in reading as much as it enhanced ours in writing. Unacknowledged photographs are by M.C.F.P., insect drawings are by P.F.Y. and others mainly by M.C.F.P.

It is a pleasure to thank John Gilmour and Max Walters for encouraging us to write this book in the first place; Avril Yeo, for reading critically and editing many of the chapters, Jean Proctor, for preparing most of the line-drawings for the block-maker from our pencil originals, and both of them for much tolerant forbearance; those who so readily allowed us to use their drawings and photographs (acknowledged individually in the captions); Dr C. North for providing the cultivated flowers of *Cypripedium calceolus* illustrated in Fig. 75 and Plate 40a and Paignton Zoo for permission to photograph the flower-visiting birds illustrated in Plate 55a–c; and finally the editors and publishers of the *New Naturalist* series for their unfailing patience and willingness to help.

M.C.F.P.
P.F.Y.

CHAPTER I

HISTORICAL INTRODUCTION

THE history of familiar things is easily forgotten and often surprising. Flowers and insects are closely linked in our consciousness, and the role of insects in pollinating flowers is a commonplace: so commonplace that it is easy to forget that the discovery of the pollination of flowers by insects is little older than the invention of the steam engine.

Honey must be one of the oldest foods of mankind, and there have been references to it and to the bees that gather it in the literature of all ages. The Neolithic Jacob sent his sons down into Egypt with 'a little balm and a little honey, spices, and myrrh, nuts and almonds'; and Joshua led his descendants to a 'land that floweth with milk and honey'. Even if the details of the biblical story reflect the everyday experience of its writer rather than historical fact, there is no reason to suppose that the honey was an anachronism. The classical world was certainly familiar with the association of bees and flowers – Aristotle describes the habits of bees in his *History of Animals* and Virgil devotes the fourth book of his *Georgics* to honey and the ways of bees – but people were generally content to accept it as a fact of nature and, rather anthropocentrically, not to seek any significance in it for the flowers. This seems all the more curious, since the Greek and Roman writers were familiar with the need for pollination of the date palm (see p. 351). An excellent account of the way in which date palms were fertilised was given by Theophrastus (*c.* 373–287 B.C.), who says: 'With dates it is helpful to bring the male to the female; for it is the male which causes the fruit to persist and ripen . . . The process is thus performed; when the male palm is in flower, they at once cut off the spathe on which the flower is, just as it is, and shake the bloom with the flower and the dust over the fruit of the female, and, if this is done to it, it retains the fruit and does not shed it. In the case both of the fig and the date, it appears that the "male" renders aid to the "female" – for the fruit-bearing tree is called "female" – but while in the latter case there is a union of the two sexes, in the former the result is brought about somewhat differently.' Theophrastus repeatedly refers to plants as 'male' and 'female', and records that '. . . they say that in the citron those flowers which have a kind of distaff growing in the middle are fruitful, but those that have it not are sterile.' After this it is disappointing to read about the differences '. . . by which men distinguish the "male"

P.F.

B

and "female", the latter being fruit-bearing, the former barren in some kinds. In those kinds in which both forms are fruit-bearing the "female" has fairer and more abundant fruit; however, some call these the "male" trees – for there are those who actually thus invert the names. This difference is of the same character as that which distinguishes the cultivated from the wild tree . . .' It seems that the sex of a plant meant no more to Theophrastus than the possession of some characters associated with maleness or femaleness; he had little idea of a process in plants analogous to sexual union in animals. He seems to have rejected the idea of a real sexual union in the date palm because a similar state of affairs could not be seen in other plants. Theophrastus was certainly the greatest botanist of classical times, and his account of plants was not surpassed until the sixteenth century. The vague notion of sex in plants which was current in his day lingered on; a relic of it is preserved in the names of two of our commonest ferns, the Male Fern (*Dryopteris filix-mas*) and the Lady Fern (*Athyrium filix-femina*). Their only qualification for these names is that the Lady Fern is more delicate and lady-like than the other! It is scarcely surprising that some botanists, like the Italian Caesalpino (1519–1603), rejected the idea of sex in plants altogether.

Like many other great discoveries, the idea that a sexual fusion takes place in the reproduction of plants, and that the stamens are the male organs of the flower, seems to have developed independently in the minds of a number of botanists towards the end of the seventeenth century. In a paper on 'The Anatomy of Flowers' read before the Royal Society in 1676, and published in 1682 as a part of his *Anatomy of Plants*, the English botanist Nehemiah Grew said that he had discussed the connection of the stamens with the formation of seeds with Thomas Millington (at that time Sedleian Professor of Natural Philosophy at Oxford), who had suggested to him that 'the attire [stamens] doth serve, as the male, for the generation of the seed . . .', and that he, Grew, agreed with him.* In 1686 John Ray clarified and cautiously accepted Grew's opinion in his *Historia Plantarum*, adding 'This opinion of Grew, however, of the use of the pollen before mentioned wants yet more decided proofs; we can only admit the doctrine as extremely probable.' (Vol. 2, p. 18.)

The 'more decided proofs' were supplied a short time afterwards by Rudolph Jacob Camerarius (1665–1721), who was Professor of Physic at

---

* In 1671 Grew had considered 'The Use of the *Attire* . . . to be not only Ornament and Distinction to us, but also Food for a vast number of little Animals, who have their peculiar provisions stored up in these *Attires* of Flowers: each Flower becoming their Lodging and their Dining-room, both in one.' (Account of *The Anatomy of Vegetables begun* in *Phil. Trans.* No. 78, p. 3041 (1671).) Grew seems to have been thinking primarily of insects: his 'little animals' evidently have no connection with the 'animalcules' Leeuwenhoeck was to describe from water and from the semen of animals a few years later.

Tübingen in Germany. Camerarius carefully examined flowers, and carried out a number of experiments on pollination. He found, for instance, that when he removed the anthers from the male flowers of the true Castor Oil plant (*Ricinus communis*, not the so-called 'Castor Oil plant' of gardens, *Fatsia*), the female flowers failed to set seed; and Maize failed to set seed when he removed the stigmas from the female flowers. Similarly he found that female plants of Mulberry, Mercury (*Mercurialis*) and Spinach failed to produce viable seed in the absence of the male plants. Many of the earlier botanists seem to have been worried by the occurrence of the two sexes together in plants. Camerarius, like Grew, quotes Swammerdam's discovery of hermaphroditism in snails, and he suggests that what is the exception in animals is the rule in plants. Camerarius set out his observations and his conclusions together with a long discussion of previous writings on the subject, in a dissertation entitled *Epistola de Sexu Plantarum* addressed to Michael Bernard Valentini (1657–1729), who was Professor of Physic in Giessen, on 25th August, 1694.

Camerarius's conclusions were not everywhere accepted at once, or without controversy. Some of his experiments had appeared inconclusive or contradictory, and in 1700 the great French botanist, Tournefort, apparently in ignorance of Camerarius's work, still considered that the stamens served to excrete unwanted portions of the sap in the form of pollen, and he doubted the need for pollination of the date palm. For half a century little was added to Camerarius's experimental demonstration of the need for pollination, though sporadic experiments are recorded, of which the most interesting are those of Richard Bradley (Fellow of the Royal Society and Professor of Botany in Cambridge from 1724 until his death in 1732*), Philip Miller and James Logan. Bradley describes his experiments in his *New Improvements of Planting and Gardening*, published in 1717.

'I made my first experiment upon the *Tulip*, which I chose rather than any other *Plant*, because it seldom misses to produce *Seed*. Sveral years I had the Conveniency of a large Garden, wherein there was a considerable bed of *Tulips* in one Part, containing about 400 Roots; in another part of it, very remote from the former, were Twelve *Tulips* in perfect Health, at the first opening of the Twelve, which I was very careful to observe, I cautiously took out all of their *Apices*, before the *Farina Fecundens* was ripe or in any way appeared: these *Tulips* being thus *castrated*, bore no Seed that Summer, while on the other Hand, every one of the 400 Plants which I had let alone produced *Seed*.'

*Bradley was a prolific and popular writer, but in Cambridge he was evidently felt to be something of a charlatan, and his ignorance of Latin and Greek and his neglect of his teaching duties excited great scandal. In extenuation it must be said that he seems to have contributed more to his subject than some of his more respectable contemporaries.

This experiment seems to be the first on hermaphrodite flowers. Bradley then commends to his reader the experiment of removing the young male catkins of an isolated Hazel or Filbert, which will then not bear unless the female flowers are dusted with pollen from 'Catkins from another Tree, gather'd fresh every Morning for three or four Days successively, and dusted lightly over it, without bruising its tender *Fibres*'. He goes on to describe the production of an artificial hybrid between a Carnation and a Sweet William by Thomas Fairchild (1667–1729), and looks forward to the use of artificial pollination for the selective breeding of plants.

Philip Miller (1691–1771) performed an experiment on tulips similar to Bradley's in 1721, apparently at the suggestion of Patrick Blair, a medical man and Fellow of the Royal Society who was sentenced to death (but reprieved) for acting as surgeon with the Jacobite forces during the rebellion of 1715. In a letter to Blair, dated 11th November, 1721, Miller described how he had

'. . . experimented with twelve Tulips, which he set by themselves about six or seven Yards from any other, and as soon as they blew, he took out the *Stamina* so very carefully, that he scattered none of the Dust, and about two Days afterwards, he saw Bees working on Tulips, in a bed where he did not take out the *Stamina*, and when they came out, they were loaded with Dust on their Bodies and Legs; He saw them fly into the Tulips, where he had taken out the *Stamina*, and when they came out, he went and found they had left behind them sufficient to impregnate these Flowers, for they bore good ripe Seed: which persuades him that the *Farina* may be carried from Place to Place by Insects . . .' (Blair, 1721).

Miller included an account of his experiments in his *Gardener's and Florist's Dictionary* (1724) and in his *Gardener's Dictionary* (1731).* James Logan (1674–1751), who was born in County Armagh, went as William Penn's secretary to Pennsylvania in 1699, and was chief justice and president of the council of the province at the time of his experiments, described in a letter dated 20th November, 1735, to his fellow-Quaker, Peter Collinson, F.R.S. In each corner of his garden in Philadelphia, Logan had planted a hill of 'Mayze or *Indian Corn*'.

'. . . from one of these Hills, I cut off the whole Tassels, on others I carefully opened the Ends of the Ears, and from some of them I cut or pinched off all the silken Filaments: from others I took about half, and from others one fourth and three fourths &c. with some variety, noting the Heads, and the Quantity taken from each; Other Heads again I tied up at their Ends, just before the Silk was putting out, with fine Muslin, but the Fuzziest or most Nappy I could find, to prevent the passage of the *Farina*: but that would obstruct neither the Sun, Air or Rain. I fastened it also very loosely, as not to give the least Check to Vegetation.'

---

* The article on 'Generation', in which Miller's account of his experiments appears, was omitted from the 3rd–5th editions of the *Gardener's Dictionary*, but reinstated in the 6th edition (1752).

He found that the plants in the group from which the male panicles had been removed produced no good grains – apart from a single large cob which had its stigmas fully exposed in the direction of one of the other groups of plants, and produced 20 or 21 out of a possible total of some 480 grains; Logan plausibly attributes this to carriage of pollen by wind. The cobs wrapped in muslin again produced no seed. On cobs from which he had removed some of the stigmas, Logan found seed in proportion to the stigmas he had left.

Various other observations and accounts of the pollination of flowers were published during the first half of the eighteenth century. Samuel Morland (1703) discussed whether the pollen grains passed down the 'tubes' of the styles to fertilise the ovules. He was unable to discover whether the ovules contained an embryo before pollination, but suspected they did not, and recommended 'the inquiry to those gentlemen who are masters of the best microscopes, and address in using them.' However, he observed that the 'seminal plant always lies in that part of the seed which is nearest to the insertion of this stylus, or some propagation of it into the seed vessel', and continues 'I have discovered in beans and peas and phaseoli, just under the extremity of what is called the eye, a manifest perforation, discernible by the larger magnifying glasses, which leads directly to the seminal plant, and at which I suppose the seminal plant entered . . .'.* At this period pollen was frequently referred to as the Farina Fecundens, and striking instances of the transmission of characters by the pollen were thought worthy of comment. In 1721 Philip Miller described the motley progeny of a batch of seed saved from savoys which had grown close to red and white cabbages (Blair, 1721). Benjamin Cooke (1749) grew red and white Maize together, and writes '. . . you may with pleasure observe the filament in the white plant, which has been struck with the red farina, discovering its alien commerce by a conscious blush, and by counting the threads thus stained, foretell how many corresponding seeds will appear red, at the opening of the ear, when ripe.'† Certainly, by the middle of the century sexuality in plants was regarded as an established fact. The climate of opinion was evidently ready for the idea, and the theory had champions whose influence carried great weight even though they added little new evidence, notably Sebastien

*Grew had described the micropyle of the seed (see p. 42) in 1671, but it was not until 1830 that Amici was able to trace the path of the pollen-tube from the germinating pollen-grain to the ovule (Sachs, 1875, p. 467).
† The Hon. Paul Dudley had described the transmission of seed colour between rows of Maize plants in New England in 1724 – 'even at the distance of 4 or 5 rods: and particularly in one place where there was a broad ditch of water between them . . . Mr. D. is therefore of the opinion that the stamina, or principles of this wonderful copulation, or mixing of colours, are carried through the air by the wind . . .' (Phil. Trans., Vol. 33, p. 194).

Vaillant (1669–1722), whose *Discours sur la Structure des Fleurs* appeared in 1718, and that most influential of eighteenth-century botanists, Carl Linnaeus. In England, the account of the generation of plants in Patrick Blair's *Botanick Essays* (1720), which quotes Grew, Ray, Camerarius, Vaillant and Bradley, and which was reproduced in Philip Miller's *Gardener's Dictionary* (1731), was widely read.

In 1750, Arthur Dobbs observed bees around his home near Carrick-fergus in County Antrim, confirming Miller's observation made nearly thirty years earlier that flowers could be pollinated by insects, and Aristotle's brief comment on the flower-constancy of bees.* He communicated his observations to the Royal Society, and in the *Philosophical Transactions* we read:

'... I have frequently follow'd a Bee loading the *Farina*, Bee-Bread or crude Wax, upon its Legs, through a Part of a great Field in Flower: and upon whatsoever Flower I saw it first alight and gather the *Farina*, it continued gathering from that Kind of Flower: and has passed over many other Species of Flowers, tho' very numerous in the Field, without alighting upon or loading from them: tho' the Flower it chose was much scarcer in the field than the others; So that if it began to load from a Daisy, it continued loading from them, neglecting Clover, Honey-suckles, Violets, &c.; and if it began with any of the others, it continued loading from the same Kind, passing over the Daisy. So in a Garden upon my Wall-Trees, I have seen it load from a Peach, and pass over Apricots, Plums, Cherries, &c. yet made no distinction betwixt a Peach and an Almond.'

Dobbs points out that this observation is confirmed by examining the pollen-loads carried by bees returning to the hive, and continues:

'Now if the Facts are so, and my Observations true, I think that Providence has appointed the Bee to be very instrumental in promoting the Increase of Vegetables . . .

'Now if the Bee is appointed by Providence to go only, at each Loading, to Flowers of the same Species, as the abundant *Farina* often covers the whole Bee, as well as what it loads upon its Legs, it carries the *Farina* from Flower to Flower, and by its walking upon the *Pistillum* and Agitation of its Wings, it contributes greatly to the *Farina's* entering into the *Pistillum*, and at the same time prevents the heterogeneous Mixture of the *Farina* of different Flowers with it; which, if it stray'd from Flower to Flower at random, it would carry to Flowers of a different Species.'

The main credit for the demonstration of the significance of insects in flower pollination must go to Joseph Gottlieb Kölreuter, Professor of Natural History in the University of Karlsruhe. Kölreuter did experiments in hybridisation and made systematic observations on pollination, which

---

* 'On each expedition the bee does not fly from a flower of one kind to a flower of another, but flies from one violet, say, to another violet, and never meddles with another flower until it has got back to the hive . . .' (*History of Animals*, IX, 40, trans. D'Arcy Wentworth Thompson).

he published between 1761 and 1766. His writings record a great deal of careful and critical observation, and some remarkable advances in floral biology. Kölreuter found that insect visits were necessary for the successful pollination of cucumbers and their relatives, Irises, and many *Malvaceae*, and he says 'In flowers in which pollination is not produced by immediate contact in the ordinary way, insects are as a rule the agents employed to effect it, and consequently to bring about fertilisation also; and it is probable that they render this important service if not to the majority of plants at least to a very large part of them, for all of the flowers of which we are speaking have something in them which is agreeable to insects, and it is not easy to find one such flower, which has not a number of insects busy about it.' He examined the nectar in many flowers, and concluded correctly that it was the source of the bees' honey, and that its significance to the flower lay in the attraction of insects. With remarkable patience he counted the numbers of pollen grains produced by various flowers, and found by experiment how many were needed to fertilise all the ovules in the flower; and he described the structure of the pollen grain with surprising accuracy considering the crude microscopes of the time. Among other observations he described the sensitive stamens and stigmas which occur in a number of plants, and he noticed that the stamens of the willow-herb and other plants ripen before the stigma – a fact whose importance in floral biology was soon to be realised.

The founder of the systematic study of the relations between flowers and insects was Christian Konrad Sprengel. Sprengel, the son of a clergyman, was born in Brandenburg in 1750. He was not a botanist by training. He studied theology and philology, and for much of his working life was a teacher, first at the school of the Friederichs-hospital in Berlin, and then from 1780 to 1794 as Rector of the great Lutheran school at Spandau. According to his own account, Sprengel was drawn to the study of insect pollination of flowers in 1787 by his observation of hairs on the bases of the petals of *Geranium sylvaticum*. 'Convinced that the wise Creator of nature has brought forth not even a single hair without some particular design, I considered what purpose these hairs might serve.' Sprengel came to the conclusion that, as the nectar was provided for the nourishment of insects, the hairs served to protect the nectar from being spoilt by rain. In the following year he examined the flowers of a forget-me-not, and recognised in the yellow ring surrounding the centre of the flower a honey-guide, leading insects to the nectar in the short tube at the centre of the sky-blue flower. From these and many other observations in the next few years, Sprengel was led to distinguish four parts of the flower concerned with the secretion of nectar: the nectary itself, which prepares and secretes the nectar; the nectar reservoir, which receives and contains the nectar secreted by the nectary; the nectar cover, protecting the nectar from rain; and the parts that enable insects readily to find the nectar –

corolla, odour and honey-guides. In 1793 he published his classic book *Das entdeckte Geheimniss der Natur im Bau und in der Befruchtung der Blumen* – The secret of Nature revealed in the structure and fertilisation of flowers – in which he described the floral adaptations of some 500 species of flowers, often with admirable lucidity, accuracy and detail. Sprengel pointed out the very wide occurrence of protandry (ripening of the anthers before the stigmas), and he was the first to describe the opposite condition of protogyny, which he found in the Cypress Spurge (*Euphorbia cyparissias*) in 1791.

Sprengel was an excellent observer, and he left little to add to his descriptions of the structural adaptations of many flowers to insect pollination. He also discussed wind-pollinated flowers, pointing out the vastly greater quantity of pollen produced by them than by insect-pollinated flowers,* and the significance of their exposed anthers and large, often feathery, stigmas. From his observations he came to the conclusion that 'Nature seems unwilling that any flower should be fertilised by its own pollen.' His near contemporary, Thomas Knight (1758–1838), for many years president of the Horticultural Society of London, also concluded from his experiments on peas that cross-fertilisation was beneficial; among the progeny of his hybridisations he found '. . . a numerous variety of new kinds produced, many of which were, in size, and in every other respect, much superior to the original white kind, and grew with excessive luxuriance . . .'†

Sprengel's work remained little noticed for over half-a-century although his ideas seem to have been quite widely known and discussed; for instance, they are mentioned quite explicitly and at some length in all seven editions of Kirby and Spence's *Introduction to Entomology* between 1815 and 1867.‡ The next important contributions to the study of flower

---

* But see p. 256.

† Knight's experiments on peas tantalisingly foreshadow those of Mendel. Knight observed that purple flower-colour was dominant to white and tallness to dwarf-ness, and that reciprocal crosses produced the same results. He noticed segregation for flower- and seed-colour in back-crosses between hybrid and white-flowered plants. It does not seem to have occurred to him to investigate further the inheritance of these striking characters, probably because that seemed irrelevant to the interest in plant-breeding which led him to carry out his experiments. The episode is an intriguing illustration of the role of chance in scientific discovery.

‡ Darwin wrote of Sprengel's book 'This author's curious work, with its quaint title of "Das Entdeckte Geheimniss der Natur", until lately was often spoken lightly of. No doubt he was an enthusiast, and perhaps carried some of his ideas to an extreme length. But I feel sure, from my own observations, that his work contains an immense body of truth. Many years ago Robert Brown, to whose judgment all botanists defer, spoke highly of it to me, and remarked that only those who knew little of the subject would laugh at him.' (*Fertilisation of Orchids*, p. 275, footnote.)

FIG. 1. Sprengel's title-page illustrating some of the floral mechanisms he observed. Note: the drawings of the ichneumon on a Twayblade flower (II), the bee on a *Salvia* (XV) and the wasp visiting a Figwort (XXV). Compare Plates 29, 32, and 41.

pollination came from Charles Darwin (1809–82), who published many observations on the subject from 1857 onwards. In 1858, the year before the *Origin of Species*, Darwin showed that various Papilionaceous flowers set seed less vigorously if they are covered with a net to prevent the visits of insects. In 1862 he published an account of the pollination mechanism of the Primrose and other species of *Primula* that have flowers of two kinds, the first of several contributions on heteromorphic flowers (see p.

224); and his classic book on the fertilisation of Orchids appeared in the same year. Like Sprengel and Knight, Darwin was drawn to the conclusion that 'Nature tells us in the most emphatic manner that she abhors perpetual self-fertilisation.' The results of his experiments and observations on *The effects of cross- and self-fertilisation in the vegetable kingdom* appeared in 1876.

The *Origin of Species* and Darwin's work on pollination stimulated a rebirth of interest in the biology of pollination, and in the relations between plants and insects; and the succeeding few decades are the classical period of floral biology, during which the bulk of our knowledge of pollination mechanisms and insect visitors was gathered. Flower pollination was a topic of lively current interest, and those who contributed observations included some of the best-known botanists of the period, and many lesser-known men besides. A few names stand out above the others, as those who gave shape and direction to the study.

Asa Gray in North America and Fritz Müller in South America both followed closely on Darwin, and each published many papers on pollination from the 1860s onwards. The mass of scattered information was rapidly growing, and Friedrich Hildebrand, Professor of Botany in Freiburg, published the first comprehensive textbook on floral biology in 1867. Hildebrand classified all the floral arrangements known to him, and a few years later Federico Delpino in Italy produced a very much more elaborate classification. Perhaps the greatest of observers of the relationships between insects and flowers was Hermann Müller (1829–83), brother of Fritz Müller mentioned above, who taught for most of his working life at the Realschule in Lippstadt. Hermann Müller was 37 when he became acquainted with Darwin's *Origin of Species* and *Fertilisation of Orchids*, and from then on he devoted his energies to the study of pollination. He observed and recorded a vast number of individual visits of insects to flowers, and published his results in three important books between 1873 and 1882. His observations enabled him to describe the pollination mechanisms and insect visitors of many central European plants; more than this, he was able to state the general principle that flowers to which insect visits are constant and sufficient are adapted exclusively for cross-pollination by insects and that, on the other hand, in proportion as insect visits are uncertain the floral arrangements permit or favour spontaneous self-pollination. Müller seems to have stimulated other botanists to follow his example to an even greater extent than Darwin twenty years earlier, and the literature of the remaining years of the nineteenth century abounds in investigations of pollination in particular districts, or in particular groups of plants. Some of those dealing with particular plants are mentioned later in this book; general observations in particular districts of the British Isles were made by Willis and Burkill (1895–1908, south and east coasts of Scotland, Cambridge, mid-Wales), Scott-Elliot (1896, Dumfriesshire), Gibson (1893, St Kilda), Buchanan-White (1898, Perthshire) and various others. No doubt many became interested in the pollination of flowers through the account in Anton Kerner von Marilaun's *Natural History of Plants*, still a mine of readable information. Ernst Loew (1894) summarised the work on floral biology carried out in central and northern Europe in the decade following the death of Hermann Müller, and the whole period fittingly culminates in the publication of the very valuable *Handbook of Flower Pollination* by Paul Knuth, Professor in the Ober-Realschule at Kiel, and himself the author of many papers on floral biology in the north German islands and elsewhere.

After the turn of the century, interest in this classical floral biology waned. There were probably several reasons for this. It was a period of ascendancy for experimental and laboratory botany: palaeobotany and morphology, plant physiology, and the new sciences of genetics and cytology. Plant ecology was developing as a vigorous branch of botany, demanding the attention of those interested in plants in their natural habitats. It is probably also true to say that floral biology had reached the limit of its development in Europe in the state of biology at the time; the main lines of the subject had been sketched out, and although much detail remained to be filled in, further major advances in the understanding of flowers had to wait for the development of cytology and genetics, ecology and the study of animal behaviour. For the time being floral biology passed largely from the field of active research to the textbooks; a state of affairs exemplified by A. H. Church's magnificent but uncompleted *Types of Floral Mechanism* (1908), in which it is easy to feel that the beautiful drawings of the details of the flowers completely overshadow observation of their functions. A professional biologist who bridged the gap between the post-Darwinian and the modern periods is the Austrian, Fritz Knoll, whose outstanding contribution was a thorough elucidation of the insect-trapping mechanism of *Arum*. This was published in the same year as an account of remarkably similar arrangements for trapping insects for pollination in an unrelated plant, *Ceropegia woodii*, by Leopoldine Müller (1926; see Chapters 6 and 10). But for the most part, the tradition of Darwin and Hermann Müller lingered on in the hands of amateur naturalists, and produced a notable twentieth-century addition to floral biology in the discovery of 'pseudocopulation' in the pollination of the Orchids that mimic insects; and in so doing, solved a problem which had greatly puzzled Darwin and all his successors (see p. 244).

Many of the most significant of the newer observations in 'classical' floral biology have come from outside Europe. Pollination by both birds and bats is recorded in Knuth's *Handbook*, but the importance of birds as pollinators in America and the warmer regions of the Old World was only slowly recognised, and was not fully appreciated until well into the present century. Recognition of the significance of bats as pollinators has come even more recently, though thanks to the observations of Porsch, van der Pijl, H. G. Baker, Vogel and others we now know that bat-pollination is widespread and important in tropical countries.

If the centre of interest passed from floral biology as such, it moved to subjects which illuminated many aspects of the functions of flowers and their relationships with their pollinating agents. In the chapters that follow we have tried to give an account of flower pollination in the context of the biological knowledge that has grown up in the last sixty years or so, and only some of the main lines of development of particular relevance to floral biology need be sketched here. When Darwin was writing about floral biology, little was known about the details of the way in which fertilisation was brought about once the pollen had reached the stigma. The fusion of a nucleus from the pollen-tube with one from the embryo-sac of the ovule was discovered by Strasburger in 1884, some four years after he and Fleming had independently elucidated the main features of the usual mode of division of cells and their nuclei. The reduction division, or *meiosis*, in which the number of chromosomes is halved in the formation of the nuclei which afterwards take part in sexual fusion, was first observed in plants by Overton in 1893. The main cytological details of the formation of the sexual parts and the process of fertilisation in flowers had been worked out by 1900, and nuclear cytology had become an established science. The breeding experiments on peas carried out by Gregor Mendel a few years after the publication of Darwin's *Origin of Species* were retrieved from obscurity in 1903, and provided the foundation on which a new science of genetics was built. The analogy between the behaviour of the various characters in Mendel's peas, and the behaviour of chromosomes at meiosis, suggested that the chromosomes of the cell nucleus might be the bearers of the hereditary units, or genes as they were later called. Evidence soon accumulated to confirm this conclusion, which bound cytology and genetics inseparably together. Cytology and genetics have impinged in many ways on our understanding of floral biology. Genetics can account for the inheritance of characters significant to flower pollination, such as the pin- and thrum-eyed conditions in Primroses, or the incompatibility factors which render individuals of many plants sterile to their own pollen; and conversely, the floral biology of a species does much to determine the pattern of distribution of genes in a population. Genetics has also provided rational theories of the way in which natural selection might work on genetical characters, on the old problem of hybrid vigour, and on the reasons for the prevalence of mechanisms favouring cross-pollination; theories which could be tested, and modified in the light of new findings from experiments.

Pollination impinges on both genetics and ecology in the study of the genetical composition and micro-evolution of natural plant populations, which was pioneered by the Danish botanist Göte Turesson from 1922 onwards, and is often called 'experimental taxonomy'. Pollination is frequently the main means by which genes move through a population of flowering plants, and gene-flow depends on an interaction between the floral adaptations of the species concerned, and the ecological factors of the habitat in which it lives. Not uncommonly, species of similar floral biology are isolated and prevented from interbreeding by differences in their ecology; a state of affairs most obvious when the isolation breaks down, as has happened in the case of the red and white campions which were studied by H. G. Baker. Sometimes the reverse state of affairs is found, when species growing in rather similar habitats are kept distinct by their different adaptations for pollination, a situation nicely illustrated by the two British Butterfly Orchids (see p. 242).

Experimental methods also made their impact on the study of pollination itself. Darwin and his successors leaned heavily on observation in their study of the visits of insects to flowers, and made little use of experiment; indeed they could hardly have done otherwise. However, by the time Knuth's *Handbook* was published, Plateau had already performed an extensive (if sometimes rather crude) series of experiments on the attractiveness to insects of the colours and odours of flowers, and Lubbock had carried out experiments on the colour sense of bees which anticipated the more detailed and critical investigations made thirty or forty years later. From this time on, many experiments were made on the responses of insects – mainly bees – to the various elements of colour, shape, pattern and odour in flowers, of which the best known are probably those of von Frisch (1913 onwards) and Clements and Long (1923). These experimental investigations of the senses and behaviour of bees and their reactions to flowers lead naturally to much more general questions of insect behaviour. This again is a subject about which there was little satisfactory knowledge during the classic period of floral biology; but animal behaviour had also become the subject of experiment, and of increasingly critical study, and by the 1930s had developed into a coherent science with an extensive literature of its own.

After following different courses for half a century, plant and animal ecology have tended to converge in the last decade or two on several fundamental points of common interest. The pollination of a flower by an insect (or a bird or bat) takes place not in isolation, but in a habitat often of bewildering complexity, populated by numerous species of plants and animals, competing and interdependent – in short, an ecosystem. Ultimately, the food of all animals – and the energy they expend – comes from the energy of sunlight, absorbed by the green plants and used to build carbon dioxide and water into sugars and other elaborate organic substances. Nectar and pollen are both important food sources for substantial groups of animals. Is the expenditure of material by the plant related to the efficiency of pollination? How is the efficiency of pollination affected by differences in the distribution or abundance of a plant? To what extent do plants and their pollinators influence one another's distribution and abundance? Many questions of this kind cannot be answered until we have far more factual information than we have now; but there are some which will emerge – and a few which we have discussed – in the later chapters of this book.

CHAPTER 2

THE FLOWER IN THE LIFE OF THE PLANT

THIS chapter is of the nature of a discursive glossary. The reader may prefer to leave it aside, and read parts of it only as he needs to do so to clarify matters considered in later chapters.

THE STRUCTURE OF FLOWERS

It is generally much easier to recognise a flower as such than to give a definition of what a flower is. Typically, a flower is made up of four kinds of members: *sepals*, making up the *calyx*; *petals*, forming the *corolla*; *stamens*, sometimes collectively referred to as the *androecium*; and the ovary or *gynoecium*, made up of one or more *carpels*. These are borne on the *receptacle* – the conical or thickened tip of the flower-stalk or *pedicel*. Thus in the flower of a Buttercup (Fig. 2) there are five green sepals, which enclose and protect the developing bud, and five glossy yellow petals, each with a minute flap-like *nectary* at the base, which form the most conspicuous

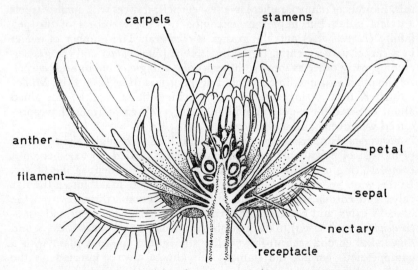

FIG. 2. Half-section of a Buttercup (*Ranunculus repens*).

part of the flower. Next, there are a large number of stamens, each consisting of a *filament* bearing an *anther* which will open to release the powdery, yellow *pollen*. Finally, in the centre of the flower there is a cluster of carpels, each one with a receptive *stigma* at the tip, and each containing an *ovule* which, after fertilisation, can develop into a *seed*.

It is possible to recognise a similar basic structure in most flowers, but there is enormous variation in detail. Firstly, there is variation in the number of parts. The Buttercups usually have five sepals and five petals, a characteristic they share with a large proportion of dicotyledons, but other numbers are found, for instance, among the Poppy family (*Papaveraceae*) and Crucifers (*Cruciferae*) which regularly have their parts in twos and fours, and amongst monocotyledonous plants which typically have their parts in threes. Larger numbers are found, though less commonly; the Lesser Celandine, unlike the related Buttercups, has about 8–12 petals (but only three sepals), and in Water-Lilies and Cacti the petals may be very numerous. A Buttercup has a large (and indefinite) number of stamens and carpels, but many plants have quite small and regular numbers of both. The stamens often occur in rings (or 'whorls'), the members of each whorl usually equalling in number the sepals and petals, and the parts in successive whorls alternating in position. It is quite common also to find as many carpels as corolla lobes. Thus in the Geranium family (*Geraniaceae*) the flower-parts are in fives throughout (with two whorls of stamens); in the Lily and Iris families (*Liliaceae*, *Iridaceae*) they are in threes. Often, however, there are fewer carpels than this. Flowers of many families have two carpels; flowers with single carpels are less usual, but important examples are the members of the pea family (*Leguminosae*) and the grasses (*Gramineae*). The number of ovules in a carpel varies greatly. In the Buttercups (*Ranunculus*) and the grasses, among many other examples, there is only a single ovule in each carpel. At the other extreme, in an Orchid of the genus *Maxillaria*, Fritz Müller estimated that a single capsule (comprising three carpels) contained about one-and-three-quarter million seeds; each carpel must have produced well over half-a-million ovules.

Flower-parts often appear to be joined. This is brought about by the primordia of neighbouring flower-parts being carried up together on a complete ring of growing tissue early in the development of the flower. In this way, 'sepals' and 'petals' often appear to be 'fused' into a tubular calyx and corolla, dividing into separate lobes some distance from the base. A calyx and corolla of this kind are called *gamosepalous* and *gamopetalous* (or *sympetalous*) to distinguish them from the *polysepalous* calyx and *polypetalous* corolla seen in flowers like the Buttercup. In a plant with a gamopetalous corolla the stamens are almost always inserted on the inside of the corolla-tube; they have become 'fused' to it during development by exactly the same process. Carpels are often fused into a *syncarpous* ovary, contrasting with the *apocarpous* ovary of the Buttercup. The stigma may be *sessile* on the carpel, as in the Buttercup, or it may be borne at the tip of a more-or-less elongated *style*. In a syncarpous ovary there may be a single style and stigma, or a single style branched at its tip bearing several stigmas, or there may be several styles. In the last two cases the number of styles or stigmas usually indicates the number of carpels.

The form of the ovary is closely bound up with the growth of the receptacle. Many flowers have a narrow, more-or-less conical receptacle like that of the Buttercup. In others, the receptacle develops into a disc, with the sepals, petals and stamens inserted around the edge, and the carpels in the centre. This is easily appreciated by comparing a Buttercup flower with a Strawberry (*Fragaria*, Fig. 3) or a Saxifrage. In the flower

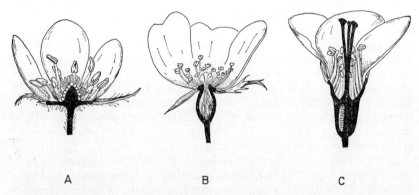

A   B   C

FIG. 3. Diagrammatic sections of flowers of (A) Strawberry (*Fragaria × ananassa*), (B) Dog Rose (*Rosa canina*) and (C) 'Japonica' (*Chaenomeles speciosa*). Solid black indicates the extent of the receptacle and, in C, the carpel tissues fused to it.

of a Plum or Cherry (Fig. 36, p. 186) the receptacle forms a shallow cup, secreting nectar, with the single carpel in the centre. If a Rose is cut in half, it will be seen that the receptacle has formed a deep flask enclosing the carpels, with a narrow opening at the top through which the styles project in the centre of the flower. This suggests the way in which it may have come about that in many flowers the carpels are completely embedded in the tissues of the receptacle, with the sepals, petals and stamens apparently borne on top of the ovary. A flower like the Buttercup is said to be *hypogynous* and to have a *superior* ovary. The Saxifrage, Strawberry or Rose is said to be *perigynous*. The flower of an Apple or a Daffodil (Fig. 49, p. 196) is said to be *epigynous*, and to have an *inferior* ovary.

So far we have assumed that a flower will contain all four kinds of floral members. This need not necessarily be so. There may be only one whorl of members surrounding the stamens, or if there is more than one whorl they may be similar in colour and texture; in such cases we speak of a *perianth* (made up of *tepals*) rather than of a calyx or corolla. There may be clear evidence that particular flower-parts are missing; thus there may be an obvious calyx, but no corolla, even though a normal corolla is found in related plants. In some plants the perianth may be missing altogether. Stamens are sometimes reduced to sterile *staminodes*, or may be lost without trace. The loss of a whorl of stamens may account for cases where two adjacent whorls of floral members appear opposite one another (like the stamens and petals of the Primrose) instead of alternating in the normal way.

The Buttercup has both stamens and carpels in the same flower; the flowers are *hermaphrodite* or *bisexual*, and this is the usual condition in flowering plants. However, there are many plants which have stamens and carpels borne in separate *staminate* and *pistillate* (or *ovulate*) flowers – often loosely referred to as 'male' and 'female' flowers. In some families unisexual flowers are the rule, and seem to be a characteristic feature of long standing, as in many of our catkin-bearing trees. In other cases (for example the Red Campion (*Silene dioica*, Fig. 39, p. 187) and the Shrubby Cinquefoil (*Potentilla fruticosa*)) the flowers resemble those of related hermaphrodite species, the pistillate flowers often have vestigial stamens and vice versa,, and it is clear that they must be of relatively recent origin from normal hermaphrodite ancestors.

If male and female flowers are borne on the same individual, as in Hazel (Fig. 86, p. 266) or Marrow (*Cucurbita pepo*), the plant is said to be *monoecious*. If they are borne on separate individuals, as in Red Campion or the Willows, the plant is *dioecious*. A species which bears hermaphrodite and female flowers on the same individual is *gynomonoecious* – and on different individuals, *gynodioecious*. The corresponding terms *andromonoecious* and *androdioecious* are applied to plants that produce hermaphrodite and male flowers on the same individual and on different individuals respectively. In some species the distribution of sex in the flowers is even more diverse, and there may be male, female and hermaphrodite flowers either on the same or on different individuals (for example Salad Burnet (*Poterium sanguisorba*)). Such species are described as *polygamous*.

Individual flowers are often grouped together in *inflorescences* (Fig. 4). These are of two main kinds. In *racemose* inflorescences the oldest flowers are at the base, and the youngest at the apex. If the individual flowers are stalked the inflorescence is a *raceme*, if they are not it is a *spike*. A branched raceme is called a *panicle*, and a raceme in which the pedicels of the lower flowers elongate to produce a flat-topped inflorescence with all the flowers at the same level is called a *corymb*.

In *cymose* inflorescences the stem apex terminates in a flower, and subsequent growth is produced by side-branches lower down the stem.

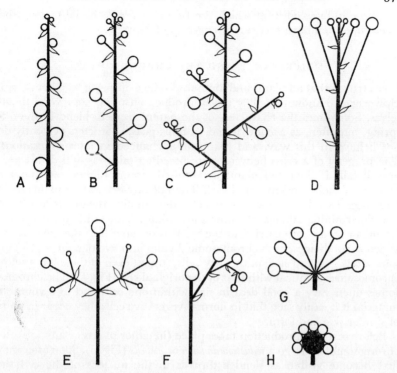

FIG. 4. Types of inflorescences. A-D, racemose inflorescences (A, spike; B, raceme; C, panicle; D, corymb). E-F, cymose inflorescences (E, dichasial cyme; F, scorpioid cyme). G, umbel. H, capitulum or head.

Two common types of cymose inflorescence are the *dichasial cyme* (or *dichasium*), in which two branches are produced below each flower (many *Caryophyllaceae* and *Gentianaceae*), and the *scorpioid cyme* in which the single branch below each successive flower is always produced on the same side, so that the upper part of the inflorescence curls like a scorpion's tail (for example Forget-me-not (*Myosotis*) and other *Boraginaceae*).

There are two types of inflorescence, the *umbel* and the *head* (or *capitulum*), in which it is often not obvious whether the inflorescence is fundamentally racemose or cymose. In an umbel the individual flower stalks radiate from a point like the ribs of an umbrella; an umbel may be *compound*, consisting of an umbel of smaller umbels. In a head or capitulum the flowers are sessile (or nearly so), clustered tightly together at the tip of the stem (for example, Clovers, *Compositae*).

Inflorescences may be mixed in character. Thus many *Labiatae* have racemose inflorescences of which the branches are dichasial cymes, and *Compositae* often have racemes or corymbs of capitula.

INHERITANCE AND THE CHROMOSOMES*

It is a truism that all plants and animals produce offspring like themselves. However, we know that they rarely produce offspring *exactly* like themselves. Sometimes the characters of the parent seem to blend in the offspring, but often, as Mendel found with his peas, characters evidently do not behave in this way, and reappear unchanged in later generations. The progeny of a cross between pure-breeding tall peas and dwarf peas are all tall. If these tall plants are crossed among themselves, about a quarter of their progeny are dwarf. The tall first-generation plants must have been carrying – intact – the factor determining dwarfness as well as that determining tallness. We now know what Mendel did not; that these factors or 'genes' are carried in the nucleus of the cell in the *chromosomes*. When a cell divides, each chromosome divides into two daughter chromosomes, one going to each of the new cells, so that each of these has a set of chromosomes identical with that of the original cell (Fig. 5). The chromosomes often vary a good deal in size within one cell, and in favourable material it is easily seen that in normal vegetative cells they occur in pairs of similar size and form.

Before sexual reproduction takes place (in either plants or animals) the chromosomes undergo a *reduction division* or *meiosis* (Fig. 6). Soon after they first become visible as slender threads in the nucleus of the cell the corresponding chromosomes come together in pairs; there is evidently a point-to-point attraction so that the paired chromosomes come to lie closely together along their whole length. While this is taking place, the chromosomes shorten and thicken, so that by the end of this stage they are readily stained and seen with the microscope. Each chromosome then divides longitudinally into two threads or *chromatids*, which remain attached at the unstained *centromere*, and the two chromosomes become twisted round one another like a piece of electric flex. When splitting is complete, the chromosomes begin to move apart. They do not immediately separate, however. At one or more points along the length of the paired chromosomes, breaks occur in one chromatid of each chromosome. Each broken end reunites with the remainder of the chromatid from the other chromosome, so that there are once again four complete chromatids. However, there has been an exchange of genetic material between the chromosomes, and the pairs of chromosomes are now linked at these points of crossing-over or *chiasmata*. As the paired chromosomes move

* A useful elementary account of the matters dealt with in the following paragraphs (and of various topics touched on elsewhere in this book) is given by Briggs and Walters (1969).

apart, the chiasmata appear to move towards the ends of the arms of the chromosomes, so that the chromosomes come to lie end to end or, if there are chiasmata in both arms of the chromosomes, the pair form a ring. The chromosomes continue to shorten and thicken, the membrane of the nucleus breaks down, and a structure called the *spindle* appears in the cell. Composed of what appear to be delicate transparent longitudinal fibrils,

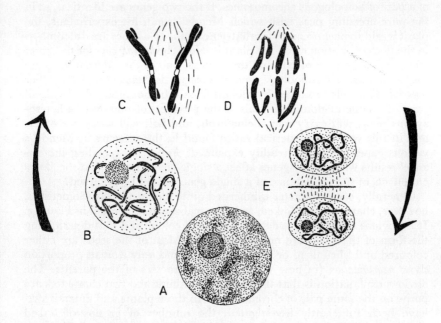

FIG. 5. Diagram illustrating the behaviour of two chromosomes during mitosis. A, interphase. B, prophase. C, metaphase. D, anaphase. E, telophase. Compare Plate 1a-e.

its name describes its shape. The chromosomes take their place round the equator of the spindle, and one chromosome of each pair moves to each pole. At this point there may be a short resting stage, or meiosis may go on with scarcely a break to its final stages. A new spindle forms in each of the two nuclei produced by the first division, and one chromatid from each chromosome goes to each pole.

There are now four nuclei, each with half the number of chromosomes present in the cell from which the process started. The number of chromosomes found in the normal body cells of most plants and animals is called the *diploid* number; that found in the cells following meiosis is called the *haploid* number. Sexual fusion restores the diploid number, and reconstitutes the pairs of homologous chromosomes of the normal vegetative cells.

The behaviour of the chromosomes at meiosis is exactly what would be required of the carriers of the factors or genes to explain the results of Mendel's experiments. In the normal vegetative cells of a plant, a simple 'Mendelian' character is controlled by a pair of genes, one borne by each of a pair of homologous chromosomes. If the two genes are identical, as in the pure-breeding peas with which Mendel began his experiments, the plant is said to be *homozygous* for that gene. If the two genes are different, as in the first-generation cross, the plant is said to be *heterozygous* for the gene; often, as in the case of tall and dwarf peas, one of the alternative genes (or, in the usual terminology, one *allelomorph*) is *dominant* and the other *recessive*. The cells produced by meiosis from a homozygous plant will all carry the same allelomorph. Half of the products of meiosis of a heterozygous plant will carry one allelomorph, and half will carry the other, and in this way the numerical ratios found in the progeny in Mendel's various experiments are readily explained. Apparent 'blending inheritance' results when many genes affect a single character, so that the clearcut effects of the inheritance of a single gene cannot be observed.

Generally, each character is inherited quite independently. Sometimes, however, characters are linked, as in the maize cob illustrated in Plate 1g. It is obvious that the gene determining seed colour and that determining the form of the grain are not independent. Most of the seeds are either coloured and 'shrunken' or white and 'full', and only a small proportion show *recombination* (in new ways) of the characters of the parents. The obvious explanation is that the genes determining the two characters are borne on the same pair of chromosomes. In those plants and animals that have been sufficiently investigated, the number of groups of linked characters is equal to the number of pairs of chromosomes, and this is one of the best pieces of evidence that the chromosomes are in fact the bearers of the hereditary characters. Chiasma formation at meiosis is important for two reasons. It plays an essential part in the mechanics of the process, holding together the pairs of chromosomes so that one member of each pair regularly goes to each pole at the first division; it also allows recombination between characters carried on the same pair of chromosomes.

In animals or plants which have the sexes separate, sexual reproduction will tend to bring about free interchange and recombination of genes within the population; individuals will tend to be heterozygous for many genes, and there will be a good deal of diversity among the offspring of any pair of individuals – a situation well exemplified in our own society. On the other hand, in a plant which is regularly self-fertilised the individuals of successive generations will become progressively more homozygous; gene-interchange cannot take place, and ultimately there

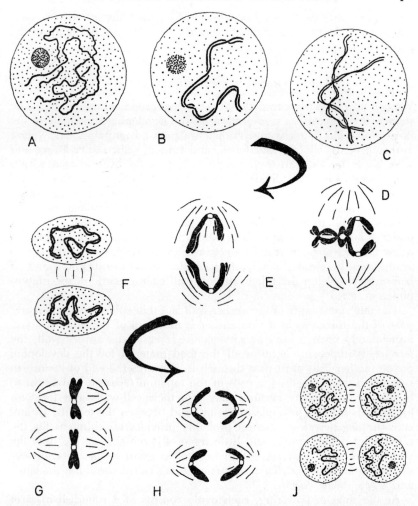

FIG. 6. Diagram illustrating the behaviour of a pair of homologous chromosomes during meiosis. A-C, prophase I (A, leptotene; B, pachytene; C, diplotene). D, metaphase I. E, anaphase I. F, telophase I. G, metaphase II. H, anaphase II. J, telophase II. Compare Plate 1f. The nucleolus would normally be associated with a pair of chromosomes, not shown here.

will be no gene-recombination either. Habitual inbreeding – breeding of closely related individuals – has the same effects in a less extreme form. Most plants (and indeed most animals) have breeding systems somewhere between the two extremes. The way in which breeding systems are controlled by pollination mechanisms, and their significance in the life of the plant, are considered in later chapters.

POLLINATION AND FERTILISATION

We can now look in more detail at the actual process of pollination. The stamen first appears as a projection on the developing receptacle of the flower, and the filament and anther are soon recognisable. The young anther becomes slightly four-lobed, and rows of cells, rather larger and with larger nuclei than their neighbours, become differentiated within each lobe. These cells divide by walls parallel with the surface of the anther; the outer cell-layers form the inner parts of the anther wall, while the inner layers divide a number of times to produce the *pollen mother-cells* (often abbreviated 'PMCs'). The pollen mother-cells then undergo meiotic division; each produces a *tetrad* of four haploid cells. In a few families of plants the four cells remain together so that the pollen grains occur in tetrads; usually, the individual cells separate and round off before they develop the thick resistant wall characteristic of the mature pollen grain.

The outermost layer of the anther wall is a thin epidermis; the greater part of the thickness in a ripe anther is made up of the 'fibrous layer' immediately underneath. The innermost layer of the anther wall, the *tapetum*, is important because all the food material for the developing pollen mother-cells must pass through it. Towards the end of meiosis in the pollen mother-cells, the cells of the tapetum separate and begin to break down; in some families they lose their cell-walls and form an amorphous mass of protoplasm around and between the tetrads. In any case the degenerating tapetum probably provides nourishment for the developing pollen grains, and little trace of it remains by the time the pollen grains are mature. Before the pollen grain is shed, its nucleus divides to form a *vegetative nucleus* (sometimes called the 'tube nucleus') and a *generative nucleus*.

At the time of flowering, each ovule consists of a roundish mass of tissue, the *nucellus*, closely surrounded by one or two *integuments* attached at the base of the ovule and leaving only a narrow open channel, the *micropyle*, at the apex. The opposite end of the ovule, the *chalaza*, is attached to the carpel by a short stalk, the *funicle*. Some time before the flower opens, the nucleus of a cell near the centre of the nucellus undergoes meiosis. The lowest of the resulting four cells enlarges, crushing the others out of shape, to form the *embryo-sac*.* Its nucleus divides into two, and

* The type of embryo-sac described here is found in over 70 per cent. of the flowering plants that have been investigated. Other flowering plants differ considerably in details (but generally not in the essentials) of embryo-sac develop-

each of the products divides twice to give two groups of four nuclei. Two nuclei, one from each quartet, come together and fuse to form a large single nucleus (the 'diploid fusion-nucleus') in the centre of the embryo-sac. Of the three nuclei remaining at the micropylar end of the embryo-sac, one enlarges to form the egg. The remaining five nuclei in the embryo-sac (two 'synergids' at the micropylar end and three 'antipodals' at the chalazal end) take no direct part in fertilisation, and need not concern us further. The embryo-sac enlarges at the expense of the neighbouring cells by a process of digestion, so that by the time it is ready for fertilisation it forms a large cavity in the nucellus, bounded by a very thin cell wall, and lined with dense, granular cytoplasm (Plate 2c).

Subsequent events depend on the transfer of pollen grains to the stigma. This process of *pollination*, and the means by which it is brought about, are the subject of this book. Pollination is not an end in itself; it is merely a necessary prelude to *fertilisation* of the embryo-sac, and the development of the ovule into a *seed*.

On reaching the stigma, the pollen grain germinates, producing a *pollen-tube*. Germination is easily observed by putting pollen grains in a sugar solution of suitable strength for a short time. Many pollen grains have obvious germination pores. In dicotyledonous plants there are often three pores or longitudinal furrows arranged round the equator of the grain. In some plants, such as the Milkworts (*Polygala*) the germ-spores may be very numerous; in others there may be only a single pore (as in grasses) or none (as in sedges). But with a few exceptions (such as the Mallow, Bellflower and Cucumber families (*Malvaceae, Campanulaceae, Cucurbitaceae*), where a number of short pollen tubes help to attach the grain to the stigma), only a single pollen tube develops whatever the number of germ-pores. The vegetative nucleus of the pollen grain passes into the growing pollen tube; it is followed by two *gametes* formed by division of the generative cell either before or after germination of the grain. Experiments show that pollen tubes will grow away from a source of free oxygen, towards a source of moisture, and towards the chemical secretions of the style and stigma, and under natural conditions the pollen-tube grows between the papillae of the stigma into the tissues of the style, and so down through the style to the ovules.

The pollen-tube usually enters the ovule through the micropyle (more rarely through the chalaza), and then pierces the nucellus and the wall of the embryo-sac. The remaining living contents of the pollen-tube are discharged into the embryo-sac, where one of the two male gametes fuses with the egg, and the other with the diploid fusion nucleus.

---

ment and fertilisation. The gymnosperms (which include the conifers and cycads) differ in important respects from the true flowering plants (angiosperms) in the development of their pollen grains and embryo-sacs (see p. 264).

The fertilised egg can now develop to form the embryo of the future seed. The nucleus formed by fusion of the second male gamete with the diploid fusion-nucleus divides repeatedly in the course of formation of the *endosperm*, which provides for the nutrition of the young embryo, and often (as in the cereals) constitutes the food reserve of the ripe seed.

*   *   *

The reproduction of flowering plants is thus an exceedingly complicated process. To the question of how far it may be compared with sexual reproduction in animals two answers may be given. In its place in the life of the organism as it exists to-day, and in its overall cytological and genetical consequences, it is precisely equivalent. But in detail, the processes of sexual reproduction in flowering plants and in animals are clearly very different, and reflect entirely different origins, of which something will be said in Chapter 12.

CHAPTER 3

INSECT-POLLINATED FLOWERS – I

In the great majority of plants there is provision not only for the transfer of pollen from anthers to stigma but also for the transport of the pollen from one flower to another, which means that there is at least a possibility of transfer from one *plant* to another (cross-pollination). All flowers owe their construction to the need for cross-pollination, for those that are habitually self-pollinated are merely sporadically evolved derivatives of cross-pollinated types (see Chapter 9).

The means used for pollen-transport by plants are either the physical elements (Chapter 8), or the movements of animals (Chapters 3–7). It is usual for animal-pollinated plants to provide a convenient alighting place, food, and conspicuous colouring and scent. No single floral type provides for all types of potential animal pollinator, and in fact it is by variations in the above attributes, and in the size, shape and arrangement of the flowers, that different plants adapt themselves to pollination by more or less restricted groups of animals. It is also usual in animal-pollinated plants for both stamens and pistils to be present in every flower.

One of the foods provided is, in fact, pollen, which was also present, of course, in the wind-pollinated ancestors of animal-pollinated flowers. In most wind-pollinated plants the flowers are unisexual, and insects feeding on their pollen would not normally visit the completely different female flowers. In becoming adapted to insect-pollination, therefore, a plant of this kind must either have changed over to producing organs of both sexes in a single floral structure, or it must have evolved a means of producing insect food in the female flowers (but see also p. 362). In fact most insect-pollinated flowers are hermaphrodite, so this appears to be the favoured condition. At the same time these flowers have evolved the production of an alternative food to pollen; this is usually nectar, but solid food is provided in some beetle-pollinated flowers (p. 364), where it takes the form of special nutritive tissues, usually attached to the stamens.

NECTAR

Nectar is an aqueous solution consisting almost exclusively of sugars, and only three of these occur in quantity, namely sucrose (cane sugar), fructose and glucose. Sucrose is a di-saccharide, which can be converted into equal parts of the two mono-saccharides, glucose and fructose, by the action of the enzyme invertase. Nectars may contain sucrose only, or mixtures, in various proportions, of all three or any two of the sugars. The nectar of nearly 900 species was analysed by a rough chromatographic method by Percival (1961), and classified into three prevailing types: those in which sucrose is predominant, those with large proportions of all three main sugars, and those in which glucose and fructose together predominate. The results suggest some degree of specialisation in nectar type according to the type of insect visitor to which the flower is adapted, but exceptions occur (Table 1).

The constitution of nectar may change somewhat with age, possibly owing to the presence of invertase (Jaeger, 1957; Percival, 1961), and sometimes variation in nectar-type from one individual of a species to another, or from one population to another, is found. In two species of Sallow (*Salix*) the male plants have sucrose-dominated nectar and the females glucose-plus-fructose-dominated nectar.

Those flowers which are highest on a plant, or in an inflorescence, produce the least nectar, while the rate of nectar-production and the sugar concentration sometimes vary with the age of the flower and independently of each other. Most plants also show daily peaks of nectar secretion and often of sugar concentration as well. Stress on plants caused by water shortage is greatest in the middle of the day, so that maximum secretion is easier at other times; often, however, it is in the middle of the day when maximum secretion takes place, which is good evidence that the peak is an adaptation to the time of pollinator activity. In addition to these variations, the concentration of nectar may be increased by evaporation, or decreased by dilution with rain (Jaeger, 1957).

Carbohydrate, the source of energy, is the principal food requirement of adult winged insects, and is the main constituent of nectar. It seems likely that, before the bees evolved, nectar flowers must have had a considerable advantage over pollen flowers. Bees, however, differ from other insects in that their larvae feed on flower foods collected by the adults. Since the larvae, in addition to nectar, require much protein for growth, this creates a great demand for the protein-rich pollen. In order to feed their young as well as themselves the bees make far more visits to flowers than do other insects, and bees are therefore extremely effective pollinators. Their evolution must have brought about a widespread increase in pollen production by insect-pollinated flowers, in spite of the fact that nectar is more economical to produce.

COLOUR

The 'colour' of white petals is sometimes produced without the aid of any pigment, the effect of whiteness being a result of reflection and refraction

Table 1. Types of nectar in various flowers

---

s=sucrose predominating
sgf=sucrose, glucose and fructose about equal
fg=fructose and glucose predominating
( )=spp. closely allied to those mentioned in the text
*=spp. adapted to pollination by short-tongued insects

Plants of Chapter 3	S	SGF	FG
DICOTYLEDONES			
Anemone pulsatilla	x		
Ranunculus acris	x*		
Helleborus foetidus	x		
Stellaria holostea			x*
Geranium pyrenaicum			x*
Geranium pratense		x	
Rubus fruticosus		x*	
Rubus idaeus			x
Euphorbia paralias			x*
(Barbarea verna)			x*
Sorbus aucuparia			x*
Heracleum sphondylium			x*
Myrrhis odorata		x*	

Plants of Chapters 5, 6 and 7	S	SGF	FG
DICOTYLEDONES			
Cheiranthus cheiri			x
Viola odorata			x
(Viola tricolor)	x		
Corydalis lutea	x		
Aquilegia vulgaris	x		
Prunus avium		x	
Ribes nigrum	x		
Ribes uva-crispa	x		
Lythrum salicaria	x		
Silene dioica			x
(Vicia sativa)	x		
Primula vulgaris	x	x	
Vinca minor	x		
Euphrasia officinalis			x
(Lamium maculatum)	x		
Campanula rotundifolia	x		
(Taraxacum officinale)			x*
Tussilago farfara			x*
Senecio jacobaea			x*
Eupatorium cannabinum			x
Cirsium acaulon	x		
Centaurea nigra	x		
(Carduus nutans)	x		
MONOCOTYLEDONES			
Endymion nonscriptus			x
Fritillaria meleagris			x*
Polygonatum multiflorum	x		
(Narcissus sp. – Trumpet Daffodil)			x
Iris pseudacorus	x		
Epipactis palustris	x		
Listera ovata			x*

at numerous cell surfaces and, in particular, the surfaces between the cells and adjoining air spaces within the tissue. (The whiteness of snow has a similar cause.) Pigments occur in the floral parts of plants either dissolved in the cell sap, or in bodies called plastids in the cytoplasm. The cytoplasm lines the inner wall of the cell, and the rest of the volume is taken up by the cell sap.

The pigments found in the cell sap are known as flavonoids, and form two main classes – anthocyanins and anthoxanthins. All purple and blue shades and most red colours are due to anthocyanins, which occur commonly in leaves and stems as well as flowers. The core of an anthocyanin molecule is called an anthocyanidin; there are three common anthocyanidins – pelargonidin giving scarlet colours, cyanidin giving red and magenta, and delphinidin giving mauve, purple and blue (Fig. 7). In nature the anthocyanidin molecule is always compounded with one or more sugar molecules, so forming an anthocyanin, which is usually somewhat bluer than the corresponding anthocyanidin. The colour of anthocyanins can also be affected by complexing with metal ions (usually iron or aluminium), as, for example, in the blue flowers of Cornflower (*Centaurea cyanus*) and *Hydrangea macrophylla*. Anthoxanthins are similar to anthocyanins, but the various molecules, with a series of slightly different core structures (Fig. 7) produce a range of colours from pale ivory to deep yellow. The anthoxanthins which appear ivory white to us absorb strongly in the ultra-violet region of the spectrum, so that such pigments apear blue-green to ultra-violet-sensitive insects (see p. 170). Anthoxanthins and anthocyanins may occur together in the same petal and jointly contribute to the colour, giving reddish and brownish shades, as in some garden Snapdragons (*Antirrhinum majus*); sometimes the two types of pigment in such flowers occur in different cells, but they can be found mixed in the same cell. The anthoxanthins that produce ivory shades can, in fact, form a loose chemical combination with anthocyanins, and this causes a marked increase in blueness. This phenomenon, called copigmentation, is very frequently involved in the production of blue shades.

The pigments found in plastids are not soluble in the cell sap but dissolve in oils and in fat-solvents. The best known plastid pigment is chlorophyll, which is the leaf-pigment responsible for absorbing light energy in the process of photosynthesis. It also occurs regularly in green sepals and may sometimes influence the petal colour or even, in the case of some green flowers, constitute the main floral pigment. The plastids which contain chlorophyll are known as chloroplasts, and they normally contain four pigments: chlorophyll-a and chlorophyll-b, which are green, carotene, which is orange, and xanthophyll, which is yellow. The last two also occur independently of chlorophyll as flower pigments; both belong to a class of pigments known as carotenoids (Fig. 7), which include red and brown colours as well as orange and yellow. Familiar examples of

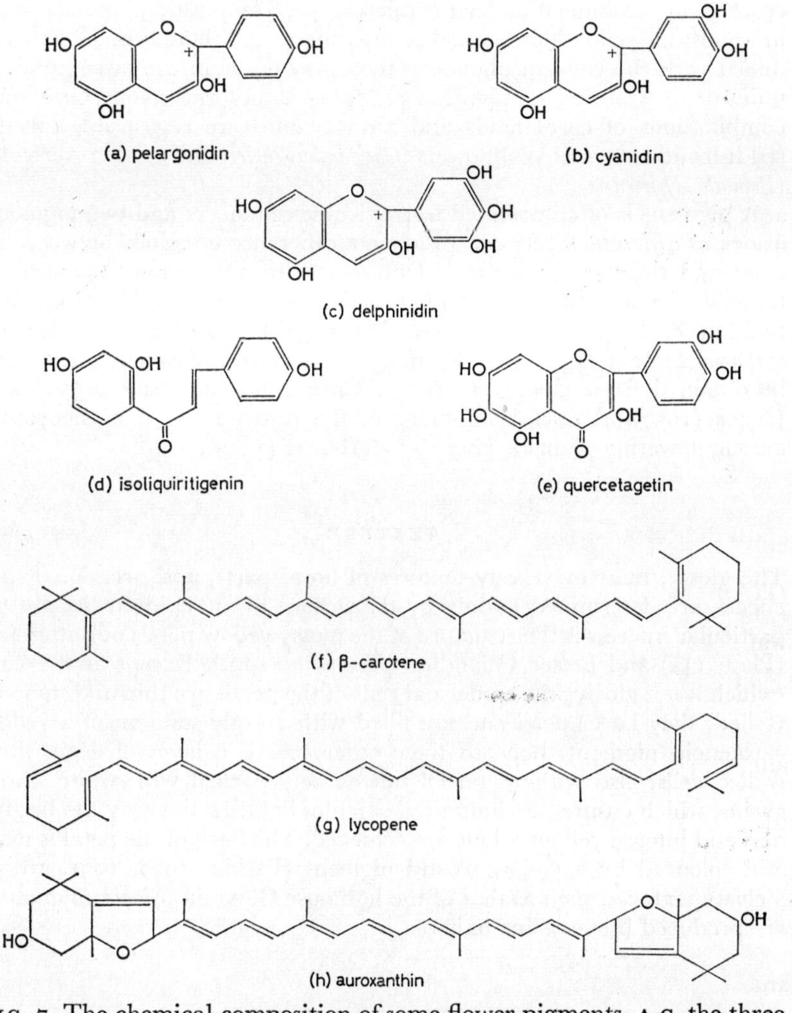

(a) pelargonidin

(b) cyanidin

(c) delphinidin

(d) isoliquiritigenin

(e) quercetagetin

(f) β-carotene

(g) lycopene

(h) auroxanthin

FIG. 7. The chemical composition of some flower pigments. A-C, the three main anthocyanidins. D, E, examples of anthoxanthins. Isoliquiritigenin is one of the pigments of Common Gorse (*Ulex europaeus*), and quercetagetin is the principal pigment of the Primrose (*Primula vulgaris*). F-H, examples of carotenoids: two carotenes and a xanthophyll.

intense carotenoid pigmentation are the carrot and tomato, and the pigmented plastids of the tomato are easily seen if a little of the flesh is crushed and examined under the microscope. The plastid pigments occur in almost all yellow flowers and many others, and though not physically mixed with the cell-sap pigments, they may occur in the same petal. A uniform distribution of pigment produces a uniform colour, and such combinations of carotenoids and anthocyanins are responsible for the reddish and brown of Wallflowers (*Cheiranthus cheiri*) and garden Auriculas (*Primula × hortensis*).

A pigment is often confined to particular cell layers and two pigmentations in different layers can produce a difference of colour between the front and the back of a petal. Different distributions over the surface, moreover, can produce patterning, as in the Heartsease (*Viola tricolor*), London Pride (*Saxifraga × urbium* (Plate 13c)) and Foxglove (*Digitalis purpurea* (Plate 31a, p. 238)). Further information on flower pigments can be obtained from Goodwin (1965), Crane and Lawrence (1952) and Jaeger (1957). A useful summary of the distribution of anthocyanins among flowering plants is given by Harborne (1963).

TEXTURE

The glossy, matt or velvety textures of floral parts, and occasional shot effects, are determined mainly by the surface of the epidermal cells. Of particular interest is the structure of the glossy yellow petals of Buttercups (Plate 14a) and Lesser Celandine (*Ranunculus* spp.). Except at the base (which is not glossy) the epidermal cells of the petals are thin and smooth-walled; they lack nuclei and are filled with an oily solution of a yellow carotenoid pigment. Beneath these outer cells is a layer of deep, thin-walled cells, also without nuclei but densely packed with white starch grains, which ensure maximum reflection of light. In this way the brightness and intense yellow colour are achieved. The back of the petal is matt and coloured by a yellow plastid pigment (Parkin, 1928, 1931, 1935). Velvety textures, such as that of the hothouse Gloxinia (*Sinningia speciosa*) are produced by very fine hairs.

THE ALIGHTING PLACE

The alighting platform for insects is usually provided by the perianth. Unlike many wind-pollinated flowers, which are normally in pendulous and flexible catkins or have pendulous filaments or pivoted anthers, a greater rigidity of floral parts is found in insect-pollinated flowers. This gives the insects something firm to grip, while rigid stamens and styles will also ensure forcible contact with the visitors' bodies.

SOLITARY AND LOOSELY CLUSTERED FLOWERS

The structure of some of the simpler types of insect-pollinated flowers can be seen in the accompanying diagrams (Fig. 8). These flowers are all radially symmetrical and, from the point of view of pollination, most of them are essentially alike in construction. However, in some the perianth consists of petals and sepals, and in others it consists of tepals which are in most cases more or less petaloid. Four of these examples have been chosen from the family *Ranunculaceae*, in which there are many genera with this simple type of flower. The examples also show, however, that simple flowers of this kind occur in a variety of families and that they are not confined to those more primitive families in which the petals are separate. A union of petals is particularly associated with the long-tubed type of flower to be described in a later chapter. The occurrence of cup- or saucer-shaped flowers in families with united petals is an example of a phenomenon which we shall show to be very general; it is that the basic structure of the flower, which is usually characteristic for families or genera, does not govern very closely the type of pollination mechanism; any structural arrangement can by various distortions be adapted to perform in a variety of different ways.

A glance at the diagrams will immediately show that some of the flowers have nectaries and others have not; those with nectaries are likely to be visited for their nectar and for their pollen, while those without will only be visited for their pollen, with the proviso that most flowers are occasionally visited by insects seeking shelter, or investigating unfamiliar flowers, or making mistaken visits. The insects that visit these flowers are chiefly the shorter-tongued species of *Diptera* (flies) and *Hymenoptera* (bees, wasps, sawflies and ichneumons). *Coleoptera* (beetles) and some *Lepidoptera* (butterflies and moths) and *Hemiptera* (bugs) may also be expected.

To study these diagrams is only to begin to have an understanding of the pollination mechanisms, and the behaviour of all these examples during anthesis (that is, the period during which the flower is open) will now be described. First, however, we must restate the ends to which the flower is adapted. These are, firstly, to effect pollination, which leads to fertilisation, and secondly (in most cases) to ensure that some of this is cross-pollination between flowers on different individual plants. In fact, in most of the examples cross-pollination between different flowers on the same individual (geitonogamy) is not prevented, but an insect will sooner or later pass from one plant to another so that at least a proportion of cross-pollinations must be between flowers on different plants.

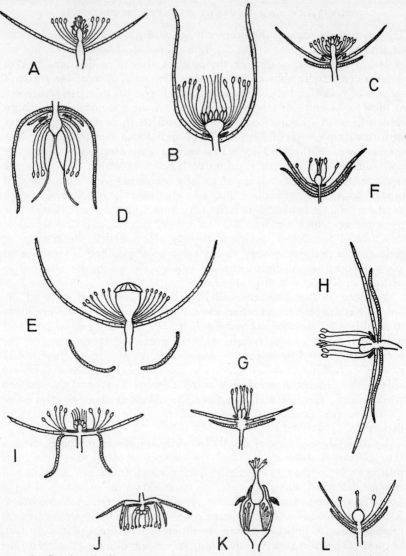

FIG. 8. Sectional diagrams of flowers, in their approximate natural orientation. *Nectaries* indicated by thick lines or black areas; *sepals* darkly stippled; *petals* lightly stippled. A, *Anemone nemorosa*. B, *A. pulsatilla*. C, *Ranunculus acris*. D, *Helleborus foetidus*. E, *Papaver rhoeas*. F, *Stellaria holostea*. G, *Geranium pyrenaicum*. H, *G. pratense*. I, *Rubus fruticosus*. J, *R. idaeus*. K, *Euphorbia paralias*. L, *Lysimachia vulgaris*.

Wood Anemone (Anemone nemorosa) *family* Ranunculaceae

When the flower opens the stamens are crowded over the stigmas and prevent them from being pollinated, though pollen is being shed from the outer stamens at this stage. Later the stamens diverge and the stigmas can be pollinated. No nectar is produced, but the flowers are visited for pollen by beetles, by flies of several families (including *Muscidae, Syrphidae* (hoverflies), *Conopidae* and *Bibionidae*) and by bees, chiefly honey-bees, bumble-bees, and solitary bees of the genus *Andrena* (Edwards, 1956; Hamm, 1934; Knight, 1963; Knuth; Willis and Burkill). Thrips (*Thysanoptera*) and bugs of the genus *Anthocoris* have also been recorded. Self-fertilisation in the Wood Anemone is prevented by self-incompatibility; that is, a flower can only be fertilised by pollen coming from a different plant (Knight, 1963).

Pasque Flower (Anemone pulsatilla) *family* Ranunculaceae

The long stigmas are receptive slightly before the first pollen is shed. The outer anthers produce no pollen and are modified to act as nectaries. Insects visit the flowers for both pollen and nectar, and their activities may cause either self- or cross-pollination. The chief visitors are honey-bees, bumble-bees and the solitary bees *Andrena* and *Halictus*, but beetles, bugs and thrips visit the flowers, and ants (order *Hymenoptera*, family *Formicidae*) steal nectar without being likely to cause pollination.

Meadow Buttercup (Ranunculus acris) *family* Ranunculaceae

The stigmas are receptive first and can be pollinated by insect visitors; later the anthers shed pollen, starting with the outermost, and by the time the innermost have dehisced the stigmas are shrivelled and the carpels have started their development (Plate 14, p. 115; Fig. 2). Forms with varying degrees of reduction of the anthers are found, the extreme being a completely female form. This species too is self-incompatible (Coles, 1971). Nectar is secreted beneath a small flap-like scale at the base of each petal. Insects visit the flowers for pollen and nectar, and thrips, bugs, *Lepidoptera*, beetles, *Hymenoptera* and flies are recorded. A small moth (see p. 104) and two small beetles (family *Chrysomelidae*, p. 69) are the most characteristic associates and these feed on pollen but bees and flies are also important visitors (Harper, 1957).

Stinking Hellebore (Helleborus foetidus) *family* Ranunculaceae

The flower is in the form of a pendulous green bell made up of green sepals with red edges. The petals do not have the display function that they have in the Buttercup but their function as nectaries is enhanced. The stigmas are receptive before the anthers mature and curve outwards in the mouth of the calyx; later the filaments elongate, beginning with the outer, and the dehiscent anthers come to fill the now somewhat expanded mouth of the flower. The stigmas still appear receptive at the beginning of the male stage, but Knuth states that self-pollination at this stage is probably ineffective. This species flowers from January to March and is visited by the few longer-tongued insects that are about at that season, that is by honey-bees (collecting pollen and nectar), bumble-bees, the bee *Anthophora*, and the hoverfly *Tubifera* (*Eristalis*). The flower scent is disagreeable but a similar scent is found in some other bee-pollinated flowers (some *Crocus* and *Cytisus* spp. – P.F.Y.).

Corn Poppy (Papaver rhoeas) *family* Papaveraceae

The flowers of this plant last only a day, opening in the early morning, and the anthers ripen and the stigmas are receptive simultaneously. Insect visitors usually alight in the centre of the flower and so readily effect pollination. There is no nectar, and the insects come for pollen; beetles, flies (hoverflies and *Muscidae* (house-fly family)) and *Hymenoptera* (especially bumble-bees and honey-bees) are the chief visitors. The bees have a characteristic way of dealing with flowers of this kind with a large number of stamens: they lie on their sides and scramble round the flower, drawing the anthers between their legs and working the pollen into their pollen-baskets (Plate 5a). Bumble-bees and, especially, honey-bees have the power of distinguishing between the Corn Poppy and other species of poppy found in Britain, and on any one flight most individuals visiting an experimental mixed planting were observed to be constant to one species, or almost so. It is not known whether any scent differences perceptible to bees are present, but the Corn Poppy differs from the other species of poppy in its flower shape, and in the pigments (cyanidins and pelargonidins) of its petals (McNaughton and Harper, 1960). The flower colours of the Corn Poppy seen by man are not those seen by the bees, as these insects can see ultra-violet light but not red. The red petals, in fact, are 'bee-ultra-violet', but the stamens are 'bee-black'; sometimes there is a dark mark at the base of the petals, and this is seen by bees as black with a white edging (Daumer, 1958; see Chapter 5).

Like the Wood Anemone, the Corn Poppy is self-incompatible.

Greater Stitchwort (Stellaria holostea) *family* Caryophyllaceae

Anthesis in this flower (Plate 5b) may be divided into three stages: (1) the stigmas are curled inwards over the ovary and are not yet receptive; the five outer stamens bend towards the middle of the flower and shed their pollen; (2) the outer stamens bend outwards again and their place is taken by the inner stamens which now shed their pollen; the styles uncurl and the stigmas become receptive, though they still face the centre of the flower; (3) the stamens wither and the styles bend outwards, the stigmas becoming recurved. During the first stage insects can only take pollen away and may thus pollinate other flowers that have reached a later stage; in the later stages they may come into contact with the stigmas and effect cross- or self-pollination; in the absence of insects, self-pollination is usually effected in the third stage because the stigmas come into contact with the withering anthers. Sometimes purely female plants are found as well as hermaphrodite ones; in addition, plants have been found in which the stigmas and anthers mature in each flower at the same time, and these regularly fertilise themselves. The normal form, on the other hand, has a greater chance of cross-pollination and only pollinates itself at the end of anthesis. The flowers of the Greater Stitchwort are visited for pollen and nectar by beetles, *Lepidoptera*, flies (for a British list, see Parmenter, 1952a) and *Hymenoptera* (chiefly bees).

Mountain Cranesbill (Geranium pyrenaicum) *family* Geraniaceae

The flower of this well-established alien to the British flora behaves in much the same way as that of the Greater Stitchwort: first the five inner stamens are bent outwards while the five outer stand erect and shed their pollen; next day the inner stamens do the same; after one or two further days the stigmas curl outwards, having previously been erect with their stigmatic surfaces pressed face to face. If there is still some pollen in the flower, insects can cause self- or cross-pollination, but if insect visits have already been numerous there may be no pollen left and then only cross-pollination is possible. Plants with only female flowers occur, as well as hermaphrodite plants. Pollen and nectar attract beetles, flies (*Syrphidae* and *Muscidae*) and *Hymenoptera* (bees and some wasps) to the purplish pink flowers.

Meadow Cranesbill (Geranium pratense) *family* Geraniaceae

This flower is structurally the same as that of *G. pyrenaicum*, but it differs in having the axis of the flower nearly horizontal, in its larger size, and in the violet-blue petals with darker radiating veins (Plates 6a and b). It behaves in much the same way as *G. pyrenaicum*, five stamens at a time becoming erect, shedding their pollen and then returning again to lie close to the petals. However, the stigmas become receptive so late in relation to the pollen-shedding of the inner anthers that self-pollination does not take place. All observers have found bees to be the chief visitors to the flowers; these comprise many genera, but the honey-bee has been noted as particularly abundant by Müller, and bumble-bees have also been recorded. Nearly all recorded visits by bees have been for nectar. Bees alight on the central organs, so that the stamens need to be erect in order to be touched by them.

The family *Malvaceae* (Mallows) is not shown in the illustrations; it is related to the *Geraniaceae* and has a very similar flower-structure, but instead of the filaments being flattened they are united to form a tube.

Bramble (Rubus fruticosus) *family* Rosaceae

Pollen is shed and the stigmas are receptive at the same time; the stamens spread outwards and the outermost anthers ripen first; when the innermost ripen, their filaments become erect and are thus enabled to effect self-pollination. The flowers (Plate 7a) are visited abundantly by insects; beetles eat nectar and pollen, but some of them also eat parts of the flower, so that they are rather destructive. The dipteran (fly) visitors represent several families but the most important are the *Syrphidae*; the honey-bee and the bumble-bees are abundant, and many other genera of bees and wasps and some *Lepidoptera* also visit the flowers. (British records of visitors to Bramble flowers are given by E. and H. Drabble (1927) and Willis and Burkill.)

Raspberry (Rubus idaeus) *family* Rosaceae

The flowers are constructed in the same way as those of the Bramble but they hang down and are often somewhat hidden amongst the foliage (Plate 7b). The flower is in any case inconspicuous because the white petals are small and erect. The stiff erect stamens tend to keep the insects away from the abundant nectar and to allow them only to insert their tongues. Bees are the chief visitors, seeking pollen and nectar, and honey-bees and bumble-bees are often very abundant. Wasps, sawflies, flies (hoverflies and *Empididae*), ants and *Lepidoptera* have also been recorded, as well as beetles feeding destructively. Willis and Burkill recorded visitors to Raspberry flowers in Scotland and found moths to be frequent among them. The relative positions of stamens and stigmas in the inverted flowers make self-pollination easy.

Grass of Parnassus (Parnassia palustris) *family* Parnassiaceae

The white, bowl-shaped flowers possess five normal stamens and five sterile ones (staminodes) of remarkable form, each being divided into several filaments tipped with a glistening knob (Plate 7c). The nectar is fully exposed, being produced on the flat basal part of each staminode by two green patches on the surface. Pollen is shed before the stigmas are receptive. The glistening knobs on the staminodes attract flies which at first lick these knobs but later transfer their attention to the nectaries (see Chapter 4, part 2). The flowers have a honey-like scent, and the scent source has been found to be in the basal parts of the staminodes. This was discovered by dividing many flowers into their constituent parts and placing them in separate jars; after a time the jars were opened and it was found that only the staminodes had produced scent. Then a series of staminodes was divided up into the basal portions, the filaments and the knobs, and tested in a similar way (Daumann, 1932). The development of the flower is protandrous; the filaments are at first very short, but one soon elongates and becomes bent over the top of the ovary so that its anther, which then dehisces, is directly in the centre of the flower; after about a day the first stamen bends outwards and another takes its place, and so on. After all the filaments have completed this process the stigma, which is sessile, becomes receptive. Only the larger visitors cover the middle of the flower and effect pollination: these are chiefly hoverflies and other *Diptera* (especially families *Muscidae* and *Calliphoridae*) but a wide variety of short-tongued *Hymenoptera* has been recorded (Daumann, 1932; Knuth; Kullenberg, 1953).

Sea Spurge (Euphorbia paralias) *family* Euphorbiaceae

What appears to be the flower in this species is really a compound structure built up of one reduced female flower and several reduced male flowers surrounded by a cup, called the cyathium, resembling a perianth but derived from bracts. This 'flower' is green or yellowish green and the rim of the cup carries large horizontal, yellowish green glands, which secrete nectar over the whole of their upper surface. The styles of the stalked female flower emerge first; after pollination the stalk elongates and the ovary comes right out of the cup and hangs down. After this the stamens, each one representing a flower, emerge and shed their pollen. The main insect visitors are small flies of various families, though beetles, wasps and bees are occasional visitors. Sawflies, ichneumons and ants also visit the flowers of this and other species of *Euphorbia* (Plates 8a and b). At the beginning of the flowering period some purely male 'flowers' appear, providing pollen for the first stigmas, and at the end some purely female 'flowers' appear which are able to receive pollen from the stamens of the last hermaphrodite ones.

Yellow Loosestrife (Lysimachia vulgaris) *family* Primulaceae

This is a plant in which the corolla is sympetalous; the stamens are also united into a tube at the base and two of them are slightly longer than the other three. The flowers are visited for pollen by flies, bees and certain wasps (*Odynerus*). The bees concerned are the solitary bees *Halictus*, *Andrena*, and *Macropis labiata*, the last being particularly associated with *Lysimachia vulgaris*, although it does visit other flowers (see pp. 149 and 151).

*   *   *

The descriptions of floral behaviour just given show that, in addition to morphology, developmental behaviour and physiology play a part in the pollination mechanism of insect-pollinated flowers. In commenting on this, use of the formidable battery of cumbersome technical terms which describe the different types of floral behaviour will be reduced to a minimum.

In many of the examples it was mentioned that the maturity of the stigmas does not entirely coincide with the shedding of the pollen; this is called dichogamy. The result is that during part of the receptive period of the stigmas self-pollination cannot occur, and so they have an improved chance of being cross-pollinated by the insects that are attracted to the flower. In most of the examples given so far there is some overlap between the maturity of the stigmas and the shedding of the pollen, so that there is a period when cross-pollination is favoured and a period when self-pollination is favoured. In this way the two aims of floral behaviour stated earlier (p. 51) are effectively achieved. Where the stigmas are receptive first (i.e. flowers protogynous), there is a period during which only cross-pollination can take place, followed by a period when selfing is possible. In Greater Stitchwort and Mountain Cranesbill pollen is shed before the stigmas are receptive, and the flowers are said to be protandrous. In this case, if there is no pollen left in the flower when the stigmas become receptive, there will be no final chance of self-pollination, and if there is some left there is evidently no initial period in which self-pollination of the stigmas is impossible. However, if there is no pollen left it is because insects are visiting the flowers and there will be a good chance of crossing, whereas if there *is* pollen left in the flower because insects are inactive, self-pollination will be desirable and possible.

In dichogamous species the very first and last flowers to open may have their anthers or stigmas wasted, according to which organs mature first. The wastage can be minimised by having unisexual flowers as well as hermaphrodite ones (p. 36). This is well exemplified by the unisexual 'flowers' of the strongly protogynous Sea Spurge. In Greater Stitchwort purely female plants sometimes occur; if the first flowers of these open at the same time as the first flowers on hermaphrodite plants, they are able to act as receptors for the pollen of the latter which initially have no mature stigmas. It was Hildebrand's view that unisexual forms of normally hermaphrodite species had the function of eliminating this wastage, but the advantage given in this way is so slight that one may doubt whether these unisexual forms are maintained in response to this need. The behaviour of *Euphorbia* is more effective in this respect than that of *Stellaria* as it involves the production merely of unisexual 'flowers', and not of unisexual plants.

Dichogamy cannot be effective in pollen flowers such as the Poppy, because there is little inducement for insects to visit the flowers outside the period when pollen is being shed. The abundant pollen of the Poppy, scattered by the bees, would certainly lead to a high proportion of self-pollination were it not that the plants are self-incompatible. Self-incompatibility is achieved by a physiological mechanism preventing pollen from growing on the stigmas of the same flower or any other flower on the same plant; it is discussed in more detail in Chapter 6. Many other species, besides those mentioned in this chapter, also prevent self-fertilisation in this way.

In many species of plants there is what may be regarded as partial self-incompatibility; the pollen tubes of an individual's own pollen grow less rapidly down its styles than those from other plants of the species. This has the same effect as dichogamy in that it favours cross-fertilisation while not preventing eventual self-fertilisation. This 'prepotency of pollen from a distinct individual over that of the same individual' was discovered by Darwin (1876) but appears to have received little study since then.

DENSELY CLUSTERED FLOWERS

In some of the plants so far described the flowers are solitary on long stalks, though there are normally several flowers open simultaneously on any one plant, while in others the flowers are grouped, as in the Bramble and Greater Stitchwort. In many plants of the latter type the flowers open in succession, at such intervals that only one or two in each inflorescence are open together. Successional flowering requires the least amount of adaptation on the part of the plant, as flower initials are normally laid down in succession and their successive maturation spreads the demand on food reserves over a period. If a plant can adapt itself to open all its flowers at approximately the same time it will be capable of making a greater floral display which will be more attractive to insects and so increase the proportion of insect-pollination. This, however, brings the disadvantage that insects in their search for flowers are less often forced to fly from plant to plant, so that the proportion of cross-pollination is lessened.

Another way in which plants become more attractive is by increase in flower-size, but this is not usually carried very far, probably because the floral mechanism requires that the flower-size should be fairly closely related to the size of insects. Very large flowers, therefore, are less frequently found than aggregations of small ones.

There are in fact many British plants in which numerous flowers in an inflorescence open simultaneously. In most the flowers are small and the inflorescence functions more or less as a unit for the purposes of pollination. Some examples will now be described (Fig. 9). Though in all of them the flowers are small, in some the inflorescence is larger than the largest individual flowers found in British plants. In all cases most of the visitors are much larger than the size of the flower and walk at random over the surface of the inflorescence.

Common Meadow Rue (Thalictrum flavum) family Ranunculaceae

This is an example of a pollen-flower. There are no petals and only small sepals, the visual attraction being provided by the yellow anthers. There is no nectar, no scent, and the perianth provides no alighting platform. The insect visitors – flies (Syrphidae and Muscidae) and bees (bumble-bees and honey-bees) – are supported by the mass of the anthers, which open at the same time as the stigmas are receptive, or a little after. The clustering of the flowers probably improves the effectiveness of the rather unstable support for insects. In this essentially entomophilous species there may also be some wind-pollination, and the genus Thalictrum is interesting because it contains both entomophilous and anemophilous species, in addition to species with both entomophilous and anemophilous forms (see Chapter 8).

Baneberry (Actaea spicata) family Ranunculaceae

The stigmas are mature before the anthers. The filaments are club-shaped but the reason for this is not known. Small insects visit the flowers for pollen, but self-pollination in their absence is no doubt easy.

Common Winter-cress (Barbarea vulgaris) family Cruciferae

The flower contains two short stamens, which are equal in length to the pistil, and four longer ones. In sunny weather the flowers open and insects probing for nectar or collecting pollen cause both self- and cross-pollination. In dull weather the short stamens pollinate the stigmas by bending towards them. The flowers are visited by beetles, flies (Muscidae, Empididae and Syrphidae), short-tongued bees, honey-bees, bumble-bees, and Microlepidoptera, some insects taking nectar or pollen only, others taking both (Plate 8c).

Wild Candytuft (Iberis amara) family Cruciferae

The corymbose flower heads are nearly flat but the outer flowers open before the inner. However, the outer flowers do not fade until after the

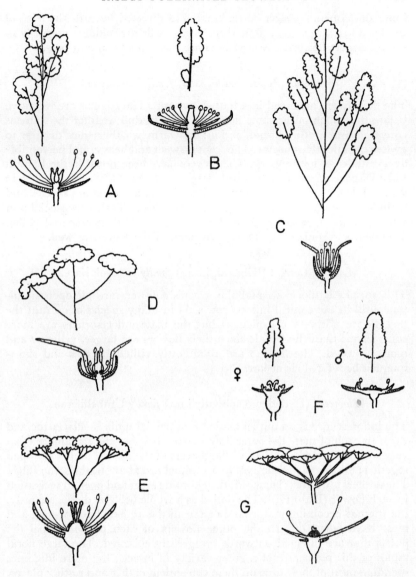

FIG 9. Outlines of inflorescences and sectional diagrams of flowers. *Nectaries* indicated by thick lines or black areas; *sepals* darkly stippled; *petals* lightly stippled. A, *Thalictrum flavum*. B, *Actaea spicata*. C, *Barbarea vulgaris*. D, *Iberis amara*. E, *Sorbus aucuparia*. F, *Ribes alpinum*. G, *Heracleum sphondylium*.

inner ones open. In each flower the petals directed towards the edge of the head are larger than those directed towards the middle, especially in the outermost flowers. Insect visitors are flies (*Muscidae*) and bees.

Rowan (Sorbus aucuparia) *family* Rosaceae

The flowers are arranged in a flattish corymb. The stigmas are receptive before the pollen of the same flower is shed. In dull weather the stamens converge and cause self-pollination; in warm weather they diverge to expose the abundant nectar. The scent is sweet and heavy, and perceptible at a distance of many yards. The flowers have been noted in the English Lake District attracting exclusively large numbers of blow-flies (*Calliphora* spp. – P.F.Y.). However, the recorded visitors are of many species and include beetles (several families), flies (several families), bees (social and solitary) and other *Hymenoptera*, *Lepidoptera*, etc. Both nectar and pollen are taken and some of the beetles eat parts of the flowers as well.

Mountain Currant (Ribes alpinum) *family* Grossulariaceae

This species is dioecious and the greenish flowers are all open simultaneously in the conical inflorescence. As in many species of currant the flowers are not very conspicuous but the male inflorescences are more conspicuous than the female since their flowers are larger, yellower and more crowded. The flowers are abundantly visited by flies and short-tongued bees for their nectar.

Hogweed (Heracleum sphondylium) *family* Umbelliferae

The inflorescence is an umbel made up of partial umbels (Plates 10c and 11c). In each of these the outer flowers open first but the remainder open very soon afterwards, so that all the flowers of the partial umbel are open together; all the partial umbels in an umbel are at about the same stage. The anthers are at first incurved; they become erect and open in succession in each flower (Plate 12b). The umbel as a whole behaves dichogamously, the stigmas becoming mature only after all the anthers have opened and most have fallen off. In the outer flowers of each partial umbel the petals that are directed outwards are greatly enlarged. The large floral tables of this plant attract a great variety of insects, the more idle ones spending most of their time on these convenient eating and resting places. These visitors are thrips, bugs, beetles (many families), very many flies of various families (British lists are given by E. and H. Drabble, 1917, 1927, and Grensted, 1946), and the following *Hymenoptera*: sawflies, ichneumons, solitary wasps (including *Pompilidae* (spiderhunting-wasps)), social wasps, and solitary and social bees. Most of the insects take food (pollen or nectar) but some visit the flowers in the course of other activities. Thus some wasps visit to feed from the flowers, while others are seeking prey amongst the other insects and may alight momentarily to rest or for longer periods if the sun goes in. The visits of these wasps, especially the larger common wasps (*Vespula*), stir up the throng of insects on the heads and no doubt increase the movement of insects from one plant to another. The common yellow dung fly (*Scopeuma stercorarium*) visits the umbels frequently for nectar and insect prey (Corbet, 1970).

In a florally similar plant, Sweet Cicely (*Myrrhis odorata*), purely male flowers are developed towards the end of the flowering period; they provide pollen for the last of the hermaphrodite inflorescences, which would otherwise not be pollinated.

*   *   *

In this group of plants we again see developmental behaviour, as well as structure, acting to promote cross-pollination. Thus the Common Wintercress and the Rowan are protogynous, while the entire umbel of the Hogweed is protandrous. In the dioecism of the Mountain Currant we see another method of absolutely preventing self-pollination, the evolution of which presupposes a very dependable method of cross-pollination. Cross-pollination is no doubt assisted by the larger size of the male flowers, as it is desirable that insects should visit these first, and Müller found that, when insects were presented with plants of the same species differing in flower-size, they tended to fly first to the larger flowers.

The examples in which the flowers in an inflorescence open simultaneously show it behaving as a unit. The corymb of the Wild Candytuft and the partial umbel of the Hogweed are organised as a unit in another way, for they show some differentiation of structure and function, with their enlarged marginal flowers and petals providing more than their fair share of the display area. This differentiation is carried to its extreme in the Guelder Rose (*Viburnum opulus*), in which the marginal flowers of the corymb are sterile and several times larger than the central fertile flowers. The use of the term 'blossom' has been proposed for the functional unit of pollination by Faegri and Van der Pijl (1966); the 'blossom' usually consists of one morphological flower, but in the examples given here it consists of many, while on the other hand a single flower may comprise more than one 'blossom' (see pp. 197 and 220).

RELATION BETWEEN FLOWER CHARACTER AND POLLINATOR

Much attention has been paid by workers on pollination to the relationship between the properties of flowers and the types of insect that visit them. The properties of the flower that concern us, in dealing with the small group of plants so far considered, are the type of food offered, accessibility and visibility of food, colour and scent. The type and visibility of food, flower colour and types of insect visitors of the flowers described are summarised below.

First group – flowers not closely aggregated

Plant	Nectar	Flower colour	Insect visitors
Anemone nemorosa	None	White to pink	Various
Papaver rhoeas	None	Red, black centre	Various, chiefly bees and flies
Lysimachia vulgaris	None	Yellow	Hoverflies, bees and wasps
Euphorbia paralias	Exposed	Yellow-green	Various, chiefly small flies
Ranunculus acris	Partly concealed	Yellow	Various
Stellaria holostea	Partly concealed	White	Various, bees perhaps most important
Anemone pulsatilla	Concealed	Purple, yellow centre	Chiefly bees
Geranium pyrenaicum	Concealed	Purple	Chiefly bees and flies
Geranium pratense	Concealed	Violet-blue	Chiefly bees
Rubus fruticosus	Concealed	White or pink	Various, chiefly bees
Rubus idaeus	Concealed	White	Chiefly bees
Helleborus foetidus	Concealed	Green	Chiefly bees

Second group – flowers closely aggregated

Plant	Nectar	Flower colour	Insect visitors
Thalictrum flavum	None	Yellow	Bees and flies
Actaea spicata	None	White	Small insects?
Ribes alpinum	Exposed	Yellow-green	Bees and flies
Heracleum sphondylium	Exposed	White	Various
Barbarea vulgaris	Partly concealed	Yellow	Various, chiefly bees and flies
Iberis amara	Partly concealed	White or mauve	Bees and flies
Sorbus aucuparia	Partly concealed	White	Various

These examples are too few to be a basis for generalisations, but we can see how they fit in with the conclusions of Knuth, based on the very large amount of information available to him. All the examples of pollen-flowers here have the pollen visible and fully accessible. Of the nectar-producing flowers, Meadow Buttercup (*Ranunculus acris*) and Bramble

(*Rubus fruticosus*) supply a large amount of pollen and the rest only small quantities. An increasing concealment of nectar and its increasing inaccessibility are among the main features of floral evolution, going hand in hand with a corresponding evolution in the length of insect mouthparts and in insect behaviour. Shorter-tongued insects still abound to feed on the simpler types of flowers dealt with in this chapter. As accessibility of nectar gradually decreases, so the balance in insect visitors changes, with the less well-adapted forms becoming less frequent visitors though not disappearing altogether. The first stages of this evolution are seen in the present examples. Where the nectar is exposed, all comers can feed on it, as in the Hogweed (*Heracleum sphondylium*), which is visited by a variety of insects, and the Sea Spurge (*Euphorbia paralias*), which is particularly favoured by small flies and other short-tongued insects like ichneumons. The Mountain Currant (*Ribes alpinum*) is perhaps primarily adapted to such small flies also, but not much is known about its pollinators. Where the nectar is partly concealed, there is still a wide range of visitors, but in some cases bees predominate, and when it is quite concealed bees are the most important visitors. The Pasque Flower (*Anemone pulsatilla*), Stinking Hellebore (*Helleborus foetidus*) and Raspberry (*Rubus idaeus*) have even less accessible nectar than the other examples with fully concealed nectar and are more exclusively visited by bees.

Regarding colour, insects with short and medium-length tongues, including hoverflies and short-tongued bees, visit the white and yellow flowers, although the Bramble (*Rubus fruticosus*) (often white but sometimes pink) is also visited by the longer-tongued bees. The longer-tongued bees are the main visitors of the purple Pasque Flower (*Anemone pulsatilla*), the bluish Meadow Cranesbill (*Geranium pratense*), the red (with ultra-violet) pollen-flower of the Corn Poppy (*Papaver rhoeas*) and also the green Stinking Hellebore (*Helleborus foetidus*).

Knuth states that pollen-flowers are usually white, yellow or red, and are visited by beetles, short-tongued flies and short-tongued bees, but that long-tongued bees and flies visit especially red, violet and blue pollen-flowers. Flowers with exposed nectar he says may be white, greenish-yellow, yellow, pink or sometimes red, and are visited by beetles (which, however, avoid greenish flowers), flies and short-tongued bees; flowers with partly concealed nectar he says are white or yellow, and short-tongued insects are less common on these than insects with medium-length tongues, such as hoverflies and bees; Knuth sums up flowers with concealed nectar as being rarely white or yellow, and generally shades of red, blue or violet, their visitors being the longer-tongued insects, such as hoverflies, bee-flies (order *Diptera*, family *Bombyliidae*) and bees (including the honey-bee and the bumble-bees). Our examples thus illustrate Knuth's generalisations quite well.

These first stages in the evolution of increased concealment and inaccessibility represent the beginnings of specialisation in pollination, in which the number of types of insect which can visit or pollinate a flower becomes restricted. The significance of specialisation is dealt with in Chapter 12, after some of the many specialised flower-types have been described in Chapter 6. The degree of structural adaptation involved in specialisation varies greatly, and among the examples described so far specialisation is achieved by quite simple modifications of the flowers. Thus the convergence of the tepals in the Pasque Flower and the sepals in the Stinking Hellebore to form a bell-like flower reduces the visibility and accessibility of nectar, while the nodding flowers of the Stinking Hellebore, Meadow Cranesbill and Raspberry are attractive to the industrious bees but discourage the flies, which appear to prefer a horizontal disc on which they can loiter in the sun as well as feed. In fact, Willis and Burkill (1903b, pp. 566–7) say that from their statistics, 'it is easy to show that a greater exclusion of undesirable or little desirable insects is effected by the simple inversion of the flower than by lengthening the tube.' These authors' criterion of desirability of visitors appears to have been largely a matter of length of proboscis (Willis and Burkill, 1903a, pp. 315–16). Müller, discussing the drooping flowers of *Borago*, implies that many insects are actually unable to hang as bees can. Kugler (1940) found that the bee *Halictus fulvicornis* was able to climb up the side of a specimen tube which could not be climbed by the hoverfly (*Scaeva pyrastri*); this is not a very fair test, however, as the fly is much larger than the bee, and the subject appears to require further study.

Many of the genera described in this chapter, and to be described in Chapter 6, include other species in which self-pollination is the rule. They are often very similar to their insect-pollinated relatives and differ from them only slightly in structure or behaviour; they are described in Chapter 9.

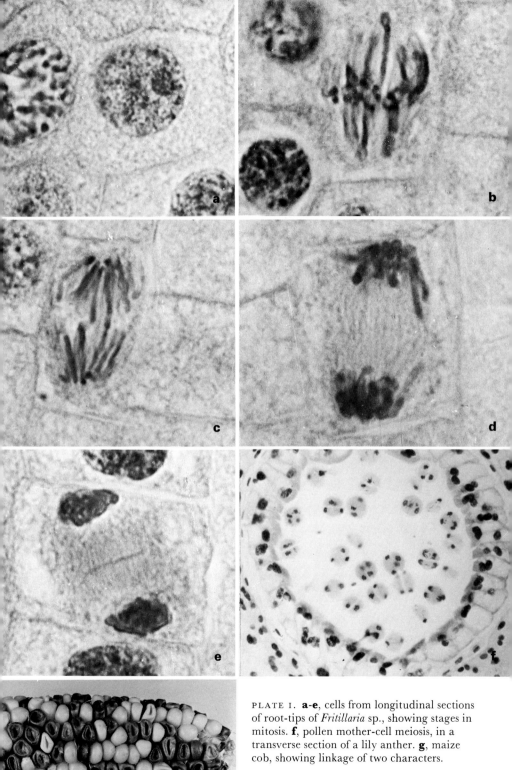

PLATE I. **a-e**, cells from longitudinal sections of root-tips of *Fritillaria* sp., showing stages in mitosis. **f**, pollen mother-cell meiosis, in a transverse section of a lily anther. **g**, maize cob, showing linkage of two characters.

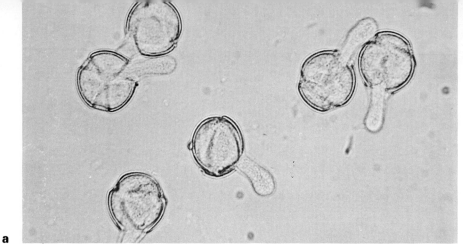

a

b

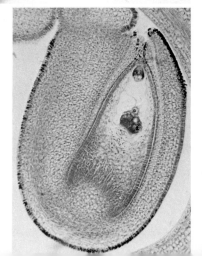

PLATE 2. **a**, pollen grains of Stinking Hellebore *(Helleborus foetidus)* germinating in 10 per cent sugar solution, showing pollen-tubes.
b, germinating pollen grains on stigma of Chickweed *(Stellaria media)*. **c**, longitudinal section of an ovule of a species of Columbine *(Aquilegia sp.)*, showing the embryo-sac.

c

THE INSECT VISITORS – I

Coleoptera (beetles)

THIS is the largest order of insects, though not the largest in the British Isles. The beetles are not, however, of great importance in flower pollination and little seems to have been added to our knowledge of flower-visiting beetles in Europe in the present century.

The development of the beetle proceeds through the stages of egg, larva, pupa and adult. The larva has a well-developed head and, usually, six legs. The larval habits of the different species are diverse and include feeding on the internal parts of plants. The larvae do not, however, become involved in pollination. The adults are winged, and while many species spend most of their lives out of sight others live in the open and may readily take flight. Amongst those that live in the open are the flower visitors.

The mouth-parts of beetles are of the biting and chewing type which is regarded as primitive. Two examples of primitive insect mouth-parts are shown in Fig. 10. The mandibles are the biting and chewing parts while the maxillae and labium taste and manipulate the food; the labium also helps, in conjunction with the labrum, to form a chamber in which the food can be confined while the chewing is taking place. Thus a beetle bites and chews the food while it is still outside the mouth, and its mouth-parts perform roughly the functions which in the mammal are performed by mouth, jaws, tongue and, in some cases, fore-limbs.

The beetles are classified on structural and anatomical characters into two main sub-orders: *Adephaga* and *Polyphaga*. The first of these is mainly predatory and its members do not visit flowers. The second is much the larger group and includes the flower-devourers and the pollinators.

Examples of beetles which feed destructively on flowers are the click-beetles (family *Elateridae*), and the cockchafer (*Melolontha melolontha*) and rose chafer (*Cetonia aurata*), both belonging to the family *Scarabaeidae* (Plate 42). These species feed on the soft parts of plants and thus sometimes destroy the flowers. The value to the plant of any pollination which they may perform is presumably offset by the damage they do. However, *Cetonia aurata* has been recorded feeding on pollen, apparently in a non-destructive manner, at the flowers of Dogwood (*Thelycrania sanguinea*, family *Cornaceae*) (Parmenter, 1956), and *Athous haemorrhoidalis* (family *Elateridae*) visits the Orchid *Listera ovata* (Plate 42) (Hobby, 1933).

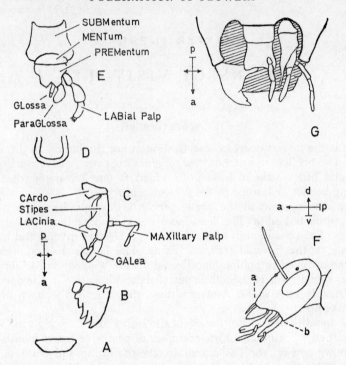

FIG. 10. A-E. Diagrams of separated mouth-parts of cockroach (*Blatta*). Capital letters indicate abbreviations used in other figs. (see Fig. 11). (A, LABRum-epipharynx [single]; B, Mandible [paired]; C, MAXilla [paired]; D, HYPopharynx [single]; E, LABium [single; appendages paired].) F, diagram of head of beetle; the organs shown in A-E are attached in alphabetical order from point a to point b; distance a-b exaggerated. G, head of beetle from below, paired parts shown singly; the organs A, B, C and E are recognisable; D is concealed. A-E, partly after Imms, *Outlines of Entomology*; G, after Crowson (1956). 'Compass points': a = anterior, p = posterior, d = dorsal, v = ventral.

According to Müller a gradual transition towards a floral diet has been shown to exist in many families of beetles. In some instances the nearest relatives of anthophilous (flower-frequenting) species are not known to visit flowers, while in others whole genera or even families feed only on flower food in the adult state. The inference drawn from this by Müller is that the habit of eating flower foods has arisen more recently in some evolutionary lines of beetles than in others. Müller took a series of beetles from the family *Cerambycidae* (longicorns) (Plate 11c) to demonstrate increasing degrees of adaptation to floral diet. The species which he illustrated are listed below, with page references to illustrations in Imms's (1947) volume in this series:

Leiopus nebulosus	never on flowers	(p. 79, no. 33)
Clytus arietis	occasionally on flowers	(p. 223)
Leptura livida	feeding at flowers only	(p. 254, no. 10)
Strangalia attenuata	feeding at flowers only	(p. 79, no. 21, *S. maculata*)

The modifications towards flower feeding consist of a tilting upwards of the head which brings the mouth-parts forward, the development of the head behind the eyes to form an additional, neck-like region, transformation of the usually short and broad first segment of the thorax into a long and narrow segment, and a lengthening of the hairs on the lobes of the maxillae. These modifications increase the insects' ability to reach sunken nectaries (*Strangalia* being further developed in this direction than *Leptura*) and to lick up the nectar efficiently.

Three families of which the adults are exclusively anthophilous are (according to Müller) *Mordellidae*, *Oedemeridae* (Plate 11a) and *Malachiidae* (now included in *Melyridae*). Two beetles especially attracted to the flowers of *Ranunculus* (see p. 53) are *Hydrothassa marginella* and *Cryptocephalus aureolus*, which feed on pollen. They belong to the *Chrysomelidae* (leaf-beetles), in which flower-visiting species are numerous. Flower visitors are also numerous among the *Cantharidae*; examples are the soldier beetle (*Rhagonycha fulva*) (Plate 11b), which is frequent at *Umbelliferae*, and *Cantharis*, which pollinates the Frog Orchid in Finland (see Chapter 7). Very few species of weevils (family *Curculionidae*) visit flowers, their long snouts being an adaptation not to the winning of nectar but to the boring of holes in which the eggs are laid. The tiny beetles *Meligethes* (family *Nitidulidae*) (Plate 11e), and *Anthobium* and *Omalium* (family *Staphylinidae*) are found in many flowers, often in abundance, and creep about eating pollen and nectar and probably sometimes causing pollination. When studying pollination in the Lesser Celandine (*Ranunculus ficaria*), Marsden-Jones (1935) proved that the movements of *Meligethes*, though not seen by him, were quite frequent; he did this by removing the beetles from all the flowers of certain plants and inspecting the same plants a few hours later, when their flowers were found to be again occupied by *Meligethes*. Some beetles also eat the nutritive tissues which, apparently, are provided especially for them in certain non-British beetle-pollinated flowers (see Chapter 10).

Though the flower-visiting beetles are more active than many beetles with different habits they are usually less active than insects such as flies and *Hymenoptera*, and therefore less useful to the plants they visit. Beetles tend to protect themselves by their horny exterior and their repellent secretions rather than by flight, and so may linger in the same flower or inflorescence for hours.

The greatest depth to which European beetles can reach with their fore-parts for nectar is about 6 mm., which is achieved by *Strangalia*. In South America, however, the genus *Nemognatha* is adapted to reach deep-seated nectar, having its maxillae modified to form a slender tube 12 mm. long (Müller).

Verne Grant (1950b) has remarked that flowers specially adapted to beetle-pollination would be expected to have the ovules well protected against the jaws of the beetles. This, he suggests, might be provided by a form of perigyny in which the ovules are more or less sunken into the receptacle, by epigyny (inferior ovaries) or by the close massing of the flowers together so that the ovaries cannot be reached, as in the *Compositae* (see p. 220). In a list of beetle-pollinated flowers which Grant compiled, such characters were much more common than in a comparable list of flowers pollinated by long-tongued insects. European plants pollinated – not necessarily exclusively – by beetles, and which possess such protection, are *Nymphaea*, *Rosa*, *Spiraea*, *Thelycrania*, several *Umbelliferae*, *Sambucus*, *Knautia*, *Achillea*, *Adenostyles*, *Chrysanthemum*; examples without such protection are *Fragaria*, *Rubus* and *Pyrola*. Grant has drawn attention to the fact that beetles are more important pollinators in semi-desert areas such as Persia, South Africa and California, than in moister regions.

Scents specifically associated with beetle-pollination (in preference to pollination by short-tongued insects in general) are not readily found in the British flora; examples from elsewhere are the spicy scent of certain crab-apples (*Malus* spp.), of *Paeonia delavayi* (a Tree Peony) and Winter-sweet (*Chimonanthus praecox*), and the smell of fermenting fruit produced by *Calycanthus* spp. (Grant, 1950a).

Diptera (flies)

This is another major order of insects, and in the British Isles it substantially outnumbers the *Coleoptera* in species, with about 5,200 as against 3,700. Structurally the true flies are easily distinguished by their possession of only one pair of wings, the second pair being reduced to stalked knobs, the halteres or balancers.

The stages of the life-cycle are the same as in the beetles. The habits of the larvae are very varied, but the larvae are not normally found in flowers (except within the tissues) and do not take any part in pollination. In the adult state the vast majority of species have well-developed wings and fly readily. Very many flies are flower-visitors and the order is therefore an important one in flower-pollination.

Nearly all flies have suctorial mouth-parts, and there are two main types of these; one type is adapted for sucking the internal fluids of animals and is capable of penetrating the tissues of the food-organism, while the other is adapted for taking up exposed fluids and lacks penetrating organs. The second is the commoner and the piercing type has probably arisen repeatedly from the sucking type. Many of the flies that feed on exposed fluids can also eat small solid particles, including pollen grains. The ability of flies to take in solid particles through their tubular proboscides is dependent on the suspension of the particles in a liquid, and this is always available because saliva is conveyed nearly to the apex of the proboscis, usually by a long hypopharynx. In some groups of flies the suctorial proboscis has become elongated in adaptation to feeding at the longer-tubed flowers.

The *Diptera* are divided into three sub-orders, of which the *Nematocera* are considered to be the most primitive. Both the piercing-and-sucking and the sucking types of mouth-parts are found in this sub-order. The main organ of the sucking mouth-parts of *Diptera* is the labium, with its labella which may correspond to the paraglossae of other orders (see Fig. 10). The labella collect the food and the labium conveys it to the mouth. In most of the families of the *Nematocera* the proboscis is quite short; it varies considerably in the details of its appearance, but in the sucking insects its basic construction and functioning are much as in *Tipula* (crane-fly or daddylonglegs, family *Tipulidae*), the mouth-parts of which were well described as long ago as 1874 by Hammond. In this fly the mouth-parts (Fig. 11) are borne on a tubular prolongation of the head; the mandibles are missing, and there is little of the maxillae apart from their very long, five-jointed palps; the remaining structures are the labium, with no palps, and the labrum. The labrum is attached above the mouth and is roughly triangular in outline, though it is bi-lobed at the apex; it is strongly convex above and concave beneath. The labium is attached below the mouth and is a more complex structure than the labrum. It is partly covered with soft skin (cuticle), but its short and stout basal portion is supported at the back by a sclerite (a hard plate) corresponding to the mentum of beetles. On the front of this basal region of the labium is a furrow which leads out from the mouth, beneath the labrum, to the gap between the labella, where the salivary duct opens into the furrow. The labella are two large bladders which contain in their surface a number of sclerites between which they can be folded up; Hammond did not know by what means they were distended but it is presumably by blood-pressure as in the blow-fly (*Calliphora*) which will be described later. The lower and hindersurfaces of the labella are covered with a system of microscopic furrows called pseudotracheae (because of their resemblance to the tracheae, or breathing tubes, of insects) (Fig. 11E). The furrows are kept in shape by transverse, incomplete rings of hard chitin; there are four major pseudotracheae in *Tipula* and numerous minor ones lead into

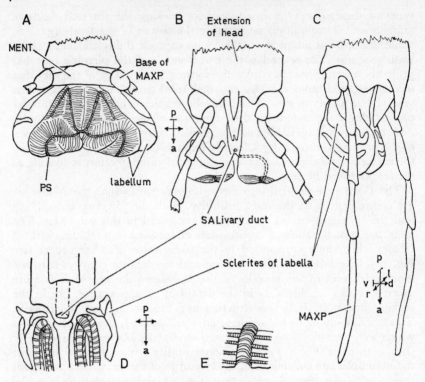

FIG. 11. Mouth-parts of crane-fly (*Tipula*). Capital letters indicate abbreviations used in other figs. (see Fig. 10). A, from behind (i.e. the underside, because mouth-parts point downwards, not forwards). B, from in front. C, from right side and in front. D, meeting of main PSeudotracheae at the end of labial furrow. E, portion of a main pseudotrachea and minor ones seen from within the labellum. A, C-E, after Hammond (1874); B, based on A and C.

them. The labella act as a filter by which liquids can be taken up from a wet surface and solid particles left behind. The fluid is presumably drawn into the pseudotracheae by capillary action; the main pseudotracheae lead into the groove on the front of the labium (Fig. 11D), and the liquid is drawn into the mouth by the sucking action of the pharynx exerted through a tube formed by the labrum and the groove on the front of the labium. The mouth-parts of *Tipula* are thus adapted to mopping up exposed fluids, and exemplify a system which is very common and widespread among the *Diptera*. Whereas, however, most *Diptera* have the hypopharynx developed as a free tongue-like or lance-like structure with the salivary duct inside it and opening at its tip, *Tipula* does not.

The flower-visitors of the sub-order *Nematocera* comprise about eleven families. These are set out in the following list, together with a selection of the British flowers they visit.

TIPULIDAE (Crane Flies)
Tipula visit:

Parnassia palustris	Grass of Parnassus	Nectar well exposed
Valeriana dioica	Marsh Valerian	} Small, tubular flowers
Valeriana officinalis	Common Valerian	
Hedera helix	Ivy	} Nectar well exposed
Umbelliferae		

Nephrotoma visit:

Filipendula ulmaria	Meadowsweet	Nectarless flowers
Euphorbia	Spurge	} Nectar well exposed
Umbelliferae		

Limonia visit:

Saxifraga hypnoides	Mossy Saxifrage	Nectar well exposed

Limnophila visit:

Potentilla palustris	Marsh Cinquefoil	Nectar partly concealed
Listera cordata	Lesser Twayblade	Pollinia seen on head of insect; nectar well exposed

(Records given by Knuth, Müller, Parmenter – 1952b, and Willis and Burkill)

CULICIDAE (Mosquitoes and Gnats)
Anopheles visit:

Mentha aquatica	Water Mint	Small, tubular flowers

Culex pipiens visits:

Frangula alnus	Alder Buckthorn	} Nectar well exposed
Heracleum sphondylium	Hogweed	

(Records given by Knuth, Corbet – 1970)

BIBIONIDAE (St Mark's Flies and Fever Flies)
Bibio hortulanus visits:

Acer pseudoplatanus	Sycamore	
Umbelliferae		
Euphorbia	Spurge	} Nectar well exposed
Euonymus europaea	Spindle-tree	
Polygonum bistorta	Bistort	
Cruciferae		} Small, tubular flowers

Bibio johannis visits:

Salix cinerea	Grey Sallow	Catkins
Ranunculus ficaria	Lesser Celandine	Nectar partly concealed

Bibio pomonae visits:

Polygonum viviparum	Alpine Bistort	} Small, tubular flowers
Senecio jacobaea	Common Ragwort	
Umbelliferae		Nectar well exposed
Trifolium pratense	Red Clover	Seeking nectar at long, tubular flowers
Trifolium repens	White Clover	Probably feeding on pollen at long, tubular flowers

Dilophus febrilis visits:

Malus sylvestris	Apple	} Flies sometimes visiting in large numbers. Nectar partly concealed
Sorbus aucuparia	Rowan	
Crataegus	Hawthorn	
Ranunculus flammula	Lesser Spearwort	Nectar partly concealed
Saxifraga hypnoides	Mossy Saxifrage	} Nectar well exposed
Hedera helix	Ivy	
Euphorbiaceae	Spurge and Mercury	

(Records given by Drabble – 1927, Knuth, Parmenter – 1952b, and Willis and Burkill)

MYCETOPHILIDAE (Fungus-gnats) visit:

Chrysosplenium	Golden Saxifrage	} Nectar well exposed
Adoxa moschatellina	Moschatel	
Hedera helix	Ivy	
Umbelliferae		
Compositae		Small, tubular flowers

CECIDOMYIIDAE (Gall-midges) visit:

Parnassia palustris	Grass of Parnassus	} Nectar well exposed
Chrysosplenium	Golden Saxifrage	
Adoxa moschatellina	Moschatel	
Armeria maritima	Thrift	} Small, tubular flowers
Achillea millefolium	Yarrow	

SIMULIIDAE (Black-flies) visit:

Chrysosplenium	Golden Saxifrage	} Nectar well exposed
Adoxa moschatellina	Moschatel	
Hedera helix	Ivy	
Malus sylvestris	Apple	Nectar partly concealed
Salix viminalis	Osier	Catkins

CHIRONOMIDAE (Non-biting Midges) visit:

Saxifraga hypnoides	Mossy Saxifrage	} Nectar well exposed
Chrysosplenium	Golden Saxifrage	
Adoxa moschatellina	Moschatel	
Filipendula ulmaria	Meadowsweet	Nectarless flowers
Certain *Salix* spp.	Sallows	Catkins
Orchidaceae	Orchids	Flies sometimes stick to stigmata and die

CERATOPOGONIDAE (Biting-midges) visit:
Potentilla palustris   Marsh Cinquefoil   Nectar partly concealed
Senecio jacobaea   Common Ragwort   Small, tubular flowers
Salix herbacea   Least Willow   Catkins
(Records for the last five families given by Hagerup – 1951, Kimmins – 1939, Knuth, Smart – 1943, and Willis and Burkill)

SCATOPSIDAE
Scatopse visit:
Loiseleuria procumbens Wild Azalea   } Small, tubular flowers
Cochlearia   Scurvy Grass   }
Rosaceae   }
Saxifraga oppositifolia Purple Saxifrage   } Nectar partly concealed
Sedum rosea   Rose-root   }
Hedera helix   Ivy   Nectar well exposed
(Records given by Parmenter – 1952b, and Willis and Burkill)

ANISOPODIDAE
Anisopus visit:
Heracleum sphondylium   Hogweed   Nectar well exposed
(Record given by Grensted – 1946)

PSYCHODIDAE (Owl-midges or Moth-flies) visit:
Arum maculatum   Lords and Ladies   (see p. 227)

Most of these flies are very small; all except *Culicidae* have very short mouth-parts, even the relatively large *Bibio* (Plate 12a) and *Tipulidae* having a proboscis only a millimetre or two in length. Most of the flowers visited by them are either flat or bowl-shaped and have well-exposed or partly concealed nectar, or are tubular but so small that the nectar is easily reached. (These small tubular flowers are frequently gathered into dense clusters to form 'brush-blossoms' (see p. 218), or into a well-defined head or 'capitulum' (see pp. 220–4).) Nectarless flowers are presumably visited for their pollen. *Nematocera* appear, from the records of Willis and Burkill, to visit flowers chiefly for nectar, but these authors record that *Bibionidae, Mycetophilidae* and *Scatopsidae* eat pollen as well. From the *Mycetophilidae, Sciara* may be singled out as the chief flower-visiting genus.

The piercing-and-sucking *Nematocera* are mosquitoes and gnats, black-flies, and biting-midges. The largest of these are the mosquitoes, *Culex* and *Anopheles*. They have a long narrow labium with small, equally narrow labella; the labrum and hypopharynx are long and sword-like and are placed together to form a feeding tube which lies in the groove on the front of the labium. There are also two needle-like maxillae and, in the female, two mandibles of the same form. All these long slender organs are used by the female to pierce the skin of the food-animal. The male mouth-parts are not used for blood-sucking, the mandibles being absent. Thus the mosquitoes (and, similarly, the black-flies and biting-midges) lack the mopping-up type of labella of the crane-fly, but the proboscis can be used for sucking up drops of nectar. *Anopheles claviger* has been recorded taking nectar – from unnamed flowers – at 1 a.m. (Parmenter, 1958) and *Culex pipiens* and *Theobaldia annulata* (mostly males but some females) taking nectar from Hogweed from 7.30 p.m. until midnight (Corbet, 1970).

Nematocera are the chief visitors to the flowers of Moschatel and Golden Saxifrage, plants which inhabit damp and shady places, have tiny flowers with freely exposed nectar and very short stamens, and appear to be specially adapted to pollination by very small flying insects. The owl-midges, and sometimes other insects, are responsible for pollination in *Arum* species (family *Araceae*) (see p. 227). The females, in particular, are attracted to the plant by its evil smell, a response probably connected with egg-laying, since they breed in excrement and other decaying matter. Outside the British Isles *Nematocera* are the special pollinators of various plants. For example, mosquitoes pollinate a small North American Orchid, *Habenaria obtusata* (Dexter, 1913; Thien, 1969a and b), and these and other *Nematocera* are the pollinators of the Orchid genus *Pterostylis* (p. 310) and several of the other exotic insect-trapping plants mentioned in Chapter 10, as well as of the cocoa plant (p. 312).

The next sub-order of the *Diptera* to be considered is the *Brachycera*. Several families of this sub-order visit flowers to feed, and are given in this next list.

STRATIOMYIDAE (Soldier-flies) visit:

Umbelliferae		Nectar well exposed
Chrysanthemum leucanthemum	Ox-eye Daisy	⎫
Matricaria recutita	Scented Mayweed	⎬ Small, tubular flowers
Senecio jacobaea	Common Ragwort	
Cirsium arvense	Creeping Thistle	
Mentha	Mint	⎭

(Records given by Harper and Wood – 1957, and Knuth)

THEREVIDAE (Stiletto-flies)

Thereva visit:

Potentilla anserina	Silverweed	⎫ Nectar partly concealed
Crataegus	Hawthorn	⎭
Galium mollugo	Hedge Bedstraw	⎫
Umbelliferae		⎬ Nectar well exposed
Euphorbia cyparissias	Cypress Spurge	⎭

(Records given by Knuth)

RHAGIONIDAE (or LEPTIDAE – Snipe-flies) visit:

Cerastium arvense	Field Mouse-ear	Nectar partly concealed
Oenanthe	Water Dropwort	Nectar well exposed
Hypochaeris radicata	Common Catsear	Small, tubular flowers

(Records given by Knuth, and Willis and Burkill)

DOLICHOPODIDAE (Longheaded-flies) visit:

Ranunculus acris	Meadow Buttercup	Nectar partly concealed
Potentilla erecta	Tormentil	⎫
Parnassia palustris	Grass of Parnassus	⎬ Nectar well exposed
Galium spp.	Bedstraws	⎭
Umbelliferae		
Compositae		⎫ Small, tubular flowers
Myosotis	Forget-me-not	⎭

(Records given by Knuth, Parmenter – 1942, 1949, and Willis and Burkill)

EMPIDIDAE visit:

Thelycrania sanguinea	Dogwood	⎫
Acer campestre	Maple	
Umbelliferae		⎬ Nectar well exposed
Euphorbia amygdaloides	Wood Spurge	
Listera ovata	Common Twayblade	⎭
Certain *Cruciferae*		⎫
Compositae		⎬ Small, tubular flowers
Origanum vulgare	Marjoram	
Veronica officinalis	Heath Speedwell	⎭
Dactylorhiza spp.	Spotted and Marsh Orchids	Flies sometimes stick to stigmata and die
Caltha palustris	Marsh Marigold	⎫
Stellaria spp.	Stitchworts	⎬ Nectar partly concealed
Prunus avium	Gean	⎭
Convolvulus arvensis	Field Bindweed	⎫ Large, tubular flowers
Menyanthes trifoliata	Bog Bean	⎭

(Records given by Kimmins – 1939, Knuth, Parmenter – 1951, and Willis and Burkill)

ASILIDAE (Robber-flies) visit:

Ranunculus spp.	Buttercups	⎫ Nectar partly concealed
Rubus fruticosus	Bramble	⎭
Umbelliferae		Nectar well exposed
Knautia arvensis	Field Scabious	Small, tubular flowers
Vaccinium myrtillus	Bilberry	Nectar concealed

(Records given by Knuth)

BOMBYLIIDAE (Bee-flies)
Bombylius visit:

Cardamine pratensis	Lady's Smock	⎫
Viola riviniana	Dog Violet	
Oxalis rubra	Pink Oxalis	
Pulmonaria	Lungwort	
Ajuga reptans	Bugle	⎬ Large, tubular flowers
Vinca minor	Lesser Periwinkle	
Primula elatior	Oxlip	
Primula vulgaris	Primrose	
Primula veris	Cowslip	⎭

Stellaria holostea   Greater Stitchwort   ⎫ Nectar partly concealed
Malus sylvestris   Apple   ⎬
Limonium   Sea Lavender   ⎭
Tussilago farfara   Coltsfoot   ⎱ Small, tubular flowers
Myosotis   Forget-me-not   ⎰
Certain *Salix* spp.   Sallows   Catkins
Hypericum perforatum   Common St John's Wort Nectarless flowers
(Records given by Christy – 1922, Knight – 1968, Knuth, Müller, Parmenter – 1952a, Scott – 1953, and Verrall – 1909)

Phthiria pulicaria visits:
  Hieracium pilosella   Mouse-ear Hawkweed ⎱ Small, tubular flowers
  Hypochaeris radicata   Common Catsear   ⎰
  (Records given by Verrall – 1909)

Anthrax flava visits:
  Umbelliferae   ⎫
  Galium mollugo   Hedge Bedstraw   ⎬ Nectar well exposed
  Galium verum   Lady's Bedstraw   ⎭
  Cirsium arvense   Creeping Thistle   ⎱ Small, tubular flowers
  Thymus   Thyme   ⎰
  Hypericum perforatum   Common St John's Wort Nectarless flowers
  (Records given by Knuth)

LONCHOPTERIDAE (Pointedwing-flies)
Lonchoptera visit:
  Ranunculus acris   Meadow Buttercup   Nectar partly concealed
  Chrysosplenium   Golden Saxifrage   Nectar well exposed
  Valerianella   Cornsalad   ⎱ Small, tubular flowers
  Mentha aquatica   Water Mint   ⎰
  (Records given by Harper – 1957, and Knuth)

PHORIDAE
Phora visit:
  Ranunculus ficaria   Lesser Celandine   ⎱ Nectar partly concealed
  Cerastium holosteoides Common Mouse-ear   ⎰
  Saxifraga hypnoides   Mossy Saxifrage   ⎫
  Parnassia palustris   Grass of Parnassus ⎬ Nectar well exposed
  Umbelliferae   ⎭
  Compositae   Small, tubular flowers
  (Records given by Knuth, and Willis and Burkill)

The flowers in the above list have been classified in the same way as those visited by *Nematocera*, but there is an additional category of large, tubular flowers with deeply concealed nectar. The brachyceran families containing the largest numbers of species that regularly visit flowers are *Stratiomyidae*, *Empididae* and *Bombyliidae*.

Stratiomyidae are medium-sized flies, often brightly coloured, rather slow in flight, frequenting damp places and waterside habitats (Plate 12b).

The proboscis is well-developed but rather short; the maxillae are scarcely developed; the labrum is long and narrow and lies above the labium, which has large labella with numerous pseudotracheae. Thus these flies are equipped with the mopping-up type of labella of the crane-fly. Knuth records that they have been seen at the flowers of twenty-two different families; British records, however, are rather scarce.

The next two families in the list, stiletto-flies and snipe-flies, visit flowers to some extent, but the main source of nourishment in the adult is obscure. *Rhagio*, a genus of snipe-flies, has large labella with numerous pseudotracheae.

The *Dolichopodidae* are a large family of small flies with specialised feeding habits; the labella are well-developed and used for crushing small insects such as aphides. These flies appear to visit flowers rather infrequently (Plate 12d).

The *Empididae* are a still larger family of small flies, of sizes up to about 10 mm. long (Plates 9a and 17c, p. 102 and 150). They are predatory on other insects, chiefly *Diptera*. They catch their prey in their legs, then pierce it with their mouth-parts and suck its juices. The proboscis (Fig. 12A and B) is rigid and, in some species, relatively long; for example, it is 4 mm. long in the large species *Empis ciliata*; in others it is short or extremely short. Many species visit flowers, and the proboscis is well-adapted for taking nectar, especially from small, tubular flowers, by sucking up the drops in the manner of the mosquitoes. Empids can also pierce soft plant tissues and suck the juices. They visit all types of *Compositae*, including the longer-tubed types, *Centaurea* (knapweed) and *Cirsium* spp. (Thistles). *Empididae* are no doubt effective pollinators of small, tubular flowers, but the smaller species probably frequently fail to cause pollination when visiting widely open flowers with exposed or partly concealed nectar, because there is room for them to reach the nectar without touching the stigmata. The chief flower-visiting genera of the family are *Empis*, *Hilara* and *Rhamphomyia*; the records of Willis and Burkill indicate *Empis* as being much the most important of these in Britain, having been seen at the flowers of about fifty species of plants, while Hobby and Smith (1961) list 20 species of flowers visited by the one species of fly *Empis tessellata* (Plate 9a).

The robber-flies, which have a long, rigid proboscis, are similar in their method of catching prey to the *Empididae*, but they are mostly much larger insects. They appear to be less frequent flower-visitors than the empids, however.

The *Bombyliidae* include some of the most highly specialised flower-feeders in the *Diptera*. The genera *Bombylius* and *Phthiria* both have a long, slender, rigid proboscis normally held pointing forwards horizontally but having some flexibility in its joint with the head. *Bombylius* species (Plate 12c) are rather large and visit chiefly large, tubular flowers, such as the Primrose, which has a long, narrow tube at the base. Some flowers of this type in the list, although found in gardens, either do not occur in the wild in the British Isles or are very rare. The list also shows that *Bombylius* visit flowers with more exposed nectar. It is interesting to note that one of them, Greater Stitchwort, has similar flowers to *Cerastium litigiosum* which Knoll (1921) found to be one of the most favoured flowers of *Bombylius fuliginosus* in Dalmatia (see p. 113). *Phthiria pulicaria* has a relatively long proboscis, but as this insect is only 3 mm. long it cannot reach deep-seated nectar. *Bombylius* and *Phthiria* each have a proboscis similar to that of the *Empididae* but used exclusively for feeding at flowers. In *Bombylius major* the proboscis is 10 mm. long, and in *B. discolor* it is 11 to 12 mm. long; the tube is formed by the labrum and hypopharynx (Fig. 12C–F), supported by the labium, the labella of the latter being long and narrow and forming an extension of the tube. According to Müller there are no pseudotracheae, but Becher (1882) observed two pseudotracheae, joined at the base, on each labellum, while Gouin (1949) reported a few on each labellum. The margins of the pseudotracheae in this genus are raised above the general surface of the labella (Peterson, 1916). The maxillae are fine bristles and the hypopharynx is stouter; Müller considered that the hypopharynx and labella were used for boring into soft tissues, as he had often seen these flies putting their tongues into the nectarless flowers of Common St John's Wort. *Bombylius* usually hover when feeding, keeping their wings in motion all the time; while hovering, the fly may sometimes hold the flower with its legs, using them to tilt its body so that the proboscis can enter the tube, but this may not be the case with flowers which have horizontal tubes (Knoll, 1921; Simes, 1946). *Bombylius* move rapidly from flower to flower and are clearly highly developed nectar-feeders, though some are known to eat pollen (Knoll, 1921; Beattie, 1972). These bee-flies resemble small bumble-bees, and are on the wing early in the year. They may be quite important as pollinating agents for some spring flowers, although they are more or less ineffective as pollinators of some wide-throated flowers which they visit, such as Ground Ivy (*Glechoma hederacea*).

In the genus *Anthrax* of the *Bombyliidae*, the proboscis is short, with large, considerably flattened labella. Thus these flies have the mopping-up type of mouth-parts and do not visit large, tubular flowers.

Lonchopteridae and *Phoridae*, the remaining two families in the list of *Brachycera*, are transitional to the next sub-order from the point of view of classification. *Lonchoptera*, which has the mopping-up type of labella and is the only British genus of its family, and species of *Phora* visit various flowers with easily accessible nectar and pollen.

The third and last sub-order of the *Diptera* is called *Cyclorrhapha*. It is by far the largest and is itself subdivided into series, the first of which to be dealt with is the *Aschiza*. This series comprises three families in Britain;

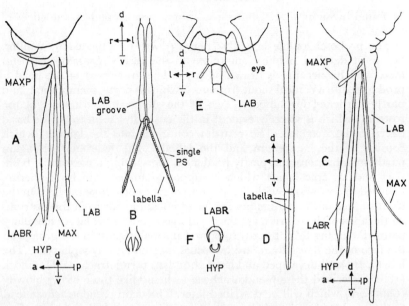

FIG. 12. Mouth-parts of flies. A-B, *Empis* (A, complete; B, labium and 3-toothed tip of labrum). C-F, *Bombylius* (C, complete; D, labium from beneath; E, the same, showing junction with the head; labium incomplete; F, transverse section of food tube, parts slightly displaced). The folds at the base (C, E) allow some extension and retraction of the labium. A-C, after Peterson (1916); F, D, based on Kugler; E, after Gouin (1949).

of these, one has apparently not been recorded visiting flowers; another, *Dorilaidae* (*Pipunculidae*), consists of small species, some of which have been recorded at the flowers of white *Umbelliferae*; while the third, *Syrphidae*, is the most important family of flower-visitors among the *Diptera* (Plates 7a and c, 12e and f, 13, 17).

The *Syrphidae*, or hoverflies, comprise nearly 250 species in the British Isles. They get their popular name from their habit of remaining stationary in the air, a habit which they share with *Bombyliidae* and *Pipunculidae*. Many of them are brightly coloured, while the darker species are often highly polished. The bright colouring often consists of a pattern of yellow and black which gives them a resemblance to wasps, or more rarely it consists of red and black, which again makes them resemble certain kinds of wasp, bee or ichneumon. Some species are stout and furry, and in their colouring closely mimic various species of bumble-bees; while others are brown with a furry thorax and resemble honey-bees. Many species will be found in sunny places, but some are frequent in and on the edges of woods.

The proboscis of the hoverfly (Fig. 13) was investigated by Müller. His most detailed publications relate to the species *Eristalis* (*Tubifera*) *tenax*, and his account is drawn on here. The lower part of the head is produced into a conical snout (rostrum), which is partly membranous and partly hardened for support. Two of the supporting structures in the rostrum (which is scarcely evident in the crane-fly) seem to be the basal parts of the maxillae. The rostrum continues into the labium, which consists of the haustellum and the labella. The haustellum is again partly membranous and partly hardened, there being a contractile basal part and an apical part which is supported underneath by a sclerite, believed to correspond with the mentum of more primitive insects. On the front surface of the haustellum is a groove leading out from the rostrum. At the apex of the rostrum are the small maxillae and the larger maxillary palps. Emerging from the rostrum, and normally lying in the groove on the face of the haustellum, are the usual labrum and hypopharynx. The labella are well-developed and have abundant pseudotracheae. Dimmock (1881) described these pseudotracheae as being like those of the blowfly (*Calliphora*) which will be described later. The main pseudotracheae lead to the gap between the labella and so into the groove on the upper surface of the haustellum.

Müller described firstly the method of feeding on pollen. A quantity of pollen is seized by the labella; movements of the labella rub the pollen down into single grains and pass them back into the groove on the front of the haustellum. The labrum and hypopharynx, which together form a tube, seize the pollen in the labial groove and pass it backwards into the mouth. A few seconds elapse while the pollen is swallowed, and the action is then repeated. Müller was apparently not aware that the pollen is conveyed along the tube by means of saliva. The salivary duct passes down the hollow inside of the hypopharynx and discharges at its tip (Becher, 1882); the insect then sucks the saliva back into its mouth and the pollen with it. *Eristalis* were seen by Müller eating the pollen of Evening Primrose (*Oenothera*), which is held together by elastic threads, and using its feet to break the threads. In watching hoverflies of the genus *Syrphus* (*Syrphidis*) feeding on pollen, I have seen how supple the labella are; they appeared to be in continuous motion, but I could not see that they were being rubbed together and there was no pause, such as Müller saw, after a portion of pollen had been taken in.

When hoverflies feed on nectar occurring on a flat or convex surface (as in Hogweed), the pseudotracheae are used to mop it up, and this is clearly their main function, unbeknown to Müller. He described a method of sucking nectar from tubular flowers by which the labella were pressed together and the labium retracted by the contractile basal part,

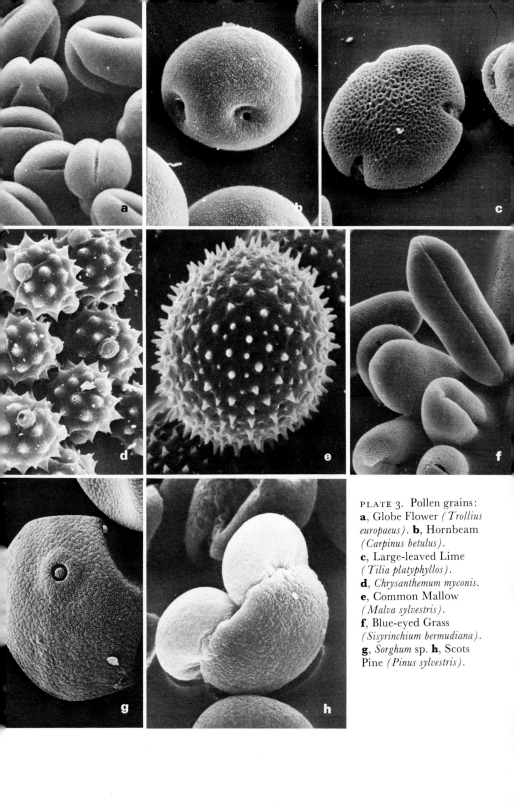

PLATE 3. Pollen grains:
a, Globe Flower (*Trollius europaeus*). b, Hornbeam (*Carpinus betulus*).
c, Large-leaved Lime (*Tilia platyphyllos*).
d, *Chrysanthemum myconis*.
e, Common Mallow (*Malva sylvestris*).
f, Blue-eyed Grass (*Sisyrinchium bermudiana*).
g, *Sorghum* sp. h, Scots Pine (*Pinus sylvestris*).

a

PLATE 4. **a**, honey-bee collecting pollen from flower of *Helleborus orientalis*. **b**, drone-fly *(Eristalis intricarius)* on flower of Marsh Marigold *(Caltha palustris)*.

b

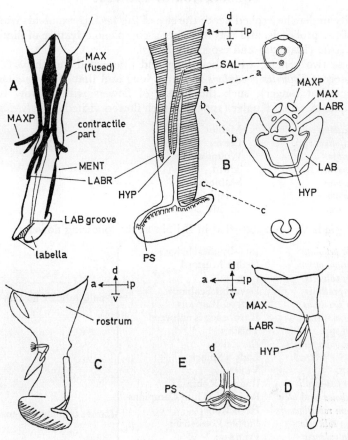

FIG. 13. A, proboscis of a hover-fly (*Tubifera tenax*), ¾-front view, sclerites in black. B, the same in longitudinal and transverse sections (a, b, c). C, *Sepsis*, proboscis. D-E, *Conops* (D, complete; E, labella, from the front; d, structure believed to represent glossae, fused). A, B, based on Müller, Peterson (1916) and Gouin (1949) (a, c, inferential); C, D, E, after Peterson.

so that the labrum and hypopharynx protruded from between the labella and sucked up the nectar directly; he found that this method was always used by the genus *Rhingia*, which has a long proboscis and narrow labella (Plates 12e and 17a) and is thus similar to the bee-flies. In *Rhingia*, however, the proboscis attains its great length chiefly by elongation of the rostrum, not the labium. All the *Syrphidae* can retract their proboscis into a cavity under the beaked head; the size of this beak corresponds with the size of the proboscis, and it is therefore a conspicuous feature of flies such as *Volucella* (Plate 7a) and especially *Rhingia*.

These two genera, with their long and attenuated mouth-parts (the proboscis of *Volucella bombylans* is 7 mm. long and that of *Rhingia rostrata* is somewhat longer), suck the nectar of flowers with narrow tubes. Examples given by Kugler (1955a) of such flowers visited by *Volucella* are:

Melilotus	Melilot
Trifolium spp.	Clover
Stachys	Woundwort
Armeria	Thrift
Knautia	Scabious
Centaurea	Knapweed
Cirsium	Thistle

Rhingia have been recorded in Britain at the following flowers:

Alliaria petiolata	Jack-by-the-Hedge	
Sisymbrium officinale	Hedge Mustard	
Calluna vulgaris	Ling	
Succisa pratensis	Devilsbit Scabious	
Senecio squalidus	Oxford Ragwort	Small, tubular flowers
Centaurea spp.	Hardhead, Knapweed	
Cirsium spp.	Thistles	
Taraxacum officinale	Dandelion	
Cardamine pratensis	Lady's Smock	
Viola sp.	Violet	
Lychnis flos-cuculi	Ragged Robin	
Silene dioica and *alba*	Red and White Campions	
Geranium robertianum	Herb Robert	Large, tubular flowers
Lythrum salicaria	Purple Loosestrife	
Primula vulgaris	Primrose	
Pedicularis palustris	Red-rattle	
Labiatae	Deadnettles, etc.	
Endymion nonscriptus	Bluebell	
Campanula rotundifolia	Harebell	Large, bell-shaped flowers
Convolvulaceae	Bindweeds	
Ranunculus spp.	Buttercup, Celandine, Spearwort	Flat or saucer-shaped flowers; nectar partly concealed
Potentilla anserina	Silverweed	
Prunus spinosa	Blackthorn	
Crataegus spp.	Hawthorns	
Heracleum sphondylium	Hogweed	Nectar well exposed

(Records given by Baker – 1957, Colyer and Hammond – 1951, E. and H. Drabble – 1927, Hamm – 1934, Parmenter – 1948, and Willis and Burkill.)

The numerous large tubular flowers in this list have deep-seated nectar which is doubtless accessible to *Rhingia* and inaccessible to most other hoverflies. Of the other flowers in the list, most are favoured by hoverflies in general, and they represent all the families chiefly visited by them, namely *Ranunculaceae, Cruciferae, Caryophyllaceae, Rosaceae, Umbelliferae* and *Compositae.*

Most hoverflies have a proboscis length of from 2 to 4 mm. However, *Eristalis tenax* has a proboscis 7 mm. long, though it nevertheless visits chiefly flowers with exposed nectar; according to Kugler (1955) this is a consequence of the fly's having broad and clumsy labella (compare Plate 4).

E. and H. Drabble (1917 and 1927) recorded visits by 124 species of flies, nearly all hoverflies, to flowers of 35 species of plants: some of their results are referred to in connection with flowers described elsewhere in this book. They concluded that *Compositae* were especially favoured by hoverflies, but found inexplicable differences in attractiveness; thus Autumn Hawkbit (*Leontodon autumnalis*) was very attractive, and Dandelion (*Taraxacum officinale*) and a hawkweed (*Hieracium boreale*) were well visited, but the similar flowers of Nipplewort (*Lapsana communis*), Common Catsear (*Hypochaeris radicata*) and Smooth Hawksbeard (*Crepis capillaris*) were visited by only one or two species each. Only Nipplewort, however, showed this lack of visits in the observations of Willis and Burkill.

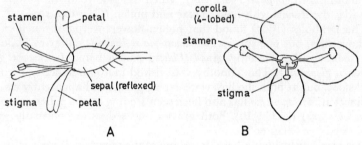

FIG. 14. A, flower of Enchanter's Nightshade (*Circaea lutetiana*), from the side. B, flower of Germander Speedwell (*Veronica chamaedrys*), from the front. After Kugler (1939).

Two plants much visited by hoverflies, Birdseye Speedwell (*Veronica chamaedrys*) and Enchanter's Nightshade (*Circaea lutetiana*), were investigated by Kugler (1938). These are members of different families but have a similar arrangement of stamens and stigma (Fig. 14). In the Speedwell the corolla is bright blue and the stamens are weak near the base. When an insect clings to one or both stamens they droop so that the underside of the insect's body comes into contact with the stigma. The anthers also touch the underside of the insect so that repeated visits to

different flowers of this species are likely to lead to cross-pollination
(Plate 13a). Although hoverflies frequently alight on the stamens, they
sometimes cling to the surface of the corolla, in which case they may
avoid coming into contact with anthers and stigma. Small bees, however,
which frequently visit the flowers, almost always alight on the stamens and
are therefore more reliable pollinators than the hoverflies. The stamens
of the white-petalled Enchanter's Nightshade droop in the same way
when an insect alights upon them, and pollination also takes place in the
same way. In a natural woodland habitat Kugler found that *Diptera* were
the only visitors to this flower, but in park-land where there was open
ground near at hand small bees also visited it. The hoverflies best suited
to these two flowers are small species which inhabit shady situations (for
example, species of *Baccha*, *Neoascia* and *Syritta* (Plates 7c, 13a and b)).

In England, the Drabbles found that Birdseye Speedwell and Brooklime
(*Veronica beccabunga*) received few visits from hoverflies, but that these
insects operated the pollination mechanism.

The common British Bindweeds (family *Convolvulaceae*) (Plate 21b)
have been repeatedly noted as favourites of hoverflies, and especially of
Rhingia, which has been recorded feeding on the pollen (Bennett, 1883;
E. and H. Drabble, 1927; Verdcourt, 1948; Parmenter, 1948; and Baker,
1957). A plant which is visited almost exclusively by the smaller hoverflies
is Small Balsam (*Impatiens parviflora*), which has pale yellowish flowers and
provides the insects with both nectar and pollen (Coombe, 1956).

The Drabbles (1927) found that pollen-flowers were not much visited
by hoverflies, but one species (*Melanostoma mellinum*) (compare Plate 5b)
was seen visiting Timothy grass (*Phleum pratense*) and Cocksfoot grass
(*Dactylis glomerata*). The Timothy was visited more abundantly than the
Cocksfoot, but at both the flies were eating pollen and apparently causing
pollination. *Melanostoma* has also been seen feeding on the pollen of a sedge
(*Carex binervis*) (M.C.F.P.). Both grasses and sedges are normally wind-
pollinated (see Chapter 8).

It is worth mentioning here that some of the views on *Diptera* expressed
by the early workers on pollination are not acceptable to-day. Thus,
Loew considered that the hairs on the head and antennae of some
Syrphidae were of importance in conveying pollen, which they no doubt are,
but he apparently regarded them as adaptations to pollen-transference.
The correlated evolution of flowers and the insects that pollinate them is
a very interesting subject, but it does not seem advisable to assume that
insects have evolved a feature which is of no help to them but only to the
flowers they visit, for it is difficult to see how selection could operate in
such a way. Müller wrote anthropomorphically of hoverflies 'delighting'
in the colours of the flowers that attracted them. We can accept Müller's
statement that 'the power to detect hidden honey advances parallel to the
structural adaptations for securing it' but it seems to be a mistake to

PLATE 5. **a**, bee collecting pollen from flower of Corn Poppy *(Papaver rhoeas)*.
b, hover-fly *(Melanostoma* sp.*)* feeding on pollen of Greater Stitchwort *(Stellaria holostea)*.

PLATE 6. **a**, honey-bee collecting nectar from a recently-opened flower of Meadow Cranesbill *(Geranium pratense)*. **b**, honey-bee on an older flower of Meadow Cranesbill; five of the anthers are withered, the remaining five are dehiscing, and the stigma-lobes have begun to diverge.

a

b

regard it, as Müller did, as a matter of 'intellectual acuteness'. The behaviour of *Syrphidae* at the zygomorphic flowers of *Scrophulariaceae* and *Labiatae* is more appropriate to the structure of the flower than the behaviour of *Muscidae* and *Calliphoridae*, flies which tend to walk all over the flower at random, but this is not evidence that hoverflies are more intelligent; it is safer to assume that they have instinctive reactions which tend to promote appropriate behaviour, and that other families, not being such specialised flower-feeders, lack these reactions.

The only remaining series of the sub-order *Cyclorrhapha* containing flower-visitors is called *Schizophora*. This series is itself divided into two sections, as shown in the diagram.

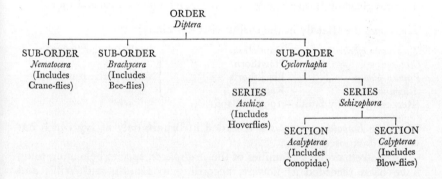

In the *Acalypterae* there are numerous families, one of which will be dealt with on its own and the rest as a group. The family to be singled out is the *Conopidae*, a small family, some of whose members have a long proboscis. In the genus *Sicus*, for example, the basal part of the slender proboscis, the labium, is directed forwards and is about 2.5 mm. long; the terminal part is about 3 mm. long and consists of the greatly elongated labella, which are fused at the base. This part can be directed downwards or, when out of use, folded right back underneath the labium. The total length of the proboscis, including the rostrum, is about 6 mm. The slender proboscis of the species *Conops quadrifasciatus* is about 4 mm. long, the labella being short (Fig. 13D and E). These flies (*Conops* and *Sicus*) feed on nectar, but are said not to eat pollen; the nectar flows between the labella, into the groove on the labium and then up a tube formed by the labrum and hypopharynx. Sometimes these genera visit tubular flowers of the type favoured by bee-flies, the hoverfly *Rhingia* and bees; examples are:

Succisa pratensis	Devilsbit Scabious	} Small, tubular flowers
Scabiosa	Scabious	
Trifolium pratense	Red Clover	} Large, tubular flowers
Echium	Viper's Bugloss	

Most *Conopidae*, however, show a considerable partiality for *Compositae* and other flowers with exposed or easily reached nectar such as *Umbelliferae* and *Rosaceae*. Examples of flowers visited by species which fly from June and July onwards are:

Senecio jacobaea	Ragwort	} Small, tubular flowers
Chrysanthemum leucanthemum	Ox-eye Daisy	
Rubus fruticosus	Bramble	Nectar partly concealed
Heracleum sphondylium	Hogweed	} Nectar well exposed
Angelica sylvestris	Wild Angelica	

(Records given by Harper and Wood – 1957, and Smith – 1959 and 1961)

The *Conopidae* that fly in the spring visit especially:

Taraxacum officinale	Dandelion	Small, tubular flowers
Crataegus sp.	Hawthorn	
Prunus spinosa	Blackthorn	} Nectar partly concealed
Allium ursinum	Ramsons	

(Records given by Smith – 1959 and 1961)

The rare *Leopoldius signatus* is recorded in Britain only at Ivy which has very well exposed nectar.

Of the remaining 39 families of the *Acalypterae*, species belonging to 23 have been recorded at flowers, according to Knuth, and Willis and Burkill. Several of these families consist of rather small flies which have a habit of waving their wings, often alternately, as they walk about. In some of these families the wings are banded or marbled, and flies of this kind have been recorded at flowers, including those that attract by imitating carrion or decaying material. One family of this type, the *Trypetidae*, have frequently been recorded at *Compositae*, but the flies concerned lay their eggs in the flower-heads (capitula), so that although they may transfer pollen the capitula are later damaged by the larvae. *Trypetidae* have, however, been recorded also at Water Mint (*Mentha aquatica*) and Great Water Parsnip (*Sium latifolium*); and three genera are mentioned by Knuth as specialised flower-visitors. Six families which have the wing-waving habit, but in which the wings are clear, visit flowers, and *Sepsidae* is the most important of these. The genus *Sepsis* (Fig. 13c) consists of small dark brown flies with the first abdominal segment narrow and thus they resemble ants. They are frequently seen on the heads of *Umbelliferae* and have been recorded at a number of flowers, as have other genera of the *Sepsidae*. Among the rest of the acalypterate families the *Lauxaniidae* and *Chloropidae* appear to be the most important flower-visitors. *Lauxaniidae* have been recorded at several rather diverse flowers by Willis and Burkill, while *Chloropidae* visit especially *Umbelliferae* and *Compositae*, though visits to *Myosotis* are recorded both in Britain and on the Continent. Chloropid flower-visitors include *Oscinella frit*, the frit fly, the larvae of which are a pest of cereals, and *Chlorops*, a small yellowish fly with longitudinal black and yellow stripes on the thorax. Members of several acalypterate families are attracted to *Arum* (see p. 227) and, on the Continent, to the florally similar *Aristolochia*. This seems to be especially the case with *Drosophilidae* (*Drosophila* spp. on the Continent), which breed in decaying fruit, carrion and excrement (among other things), and with *Sphaeroceridae*, which are often dung-frequenting, the genus *Limosina* being recorded at Lords and Ladies (*Arum maculatum*) and Large Cuckoo Pint (*A. neglectum*) in Britain (Grensted, 1947; Prime, 1954) and *A. italicum* on the Continent (Knuth).

The flies covered by the previous paragraph are medium-sized or, more often, small or very small, and have a proboscis of the mopping-up type which is usually quite short, though the genus *Siphonella* (*Chloropidae*), which has been recorded at flowers, has a long proboscis which is much the same in structure as that of the hoverfly genus *Eristalis*.

There are about 279 genera in the *Acalypterae*; the other section of the series *Schizophora*, the *Calypterae*, includes about 288 genera which are grouped into four families. However, the *Acalypterae* recorded at flowers by Willis and Burkill represent 13 genera, whereas the *Calypterae* represent about 40. The *Calypterae* are larger in size on the average and perhaps more active in the open than the *Acalypterae*; they are undoubtedly much more important as flower-visitors and pollinators. They are nearly all unspecialised in feeding-habits, and feed on many kinds of liquid and semi-liquid foods. In many genera the proboscis is provided with 'prestomal teeth', which can be used for scraping and are situated between the labella; these teeth are not, however, an adaptation to feeding at flowers. In all four families, feeding on both nectar and pollen is recorded.

Of the first of these families, *Cordiluridae*, only one genus appears to have been recorded at flowers in Britain, although the family contains 22 genera. This exceptional genus is *Scopeuma* (=*Scatophaga*), the larvae of which live in dung and the adults of which are predatory on insects, piercing them with their prestomal teeth. *S. stercorarium* (Plate 43c, p. 302) is the common yellow dung fly, the males of which are golden-yellow furry insects. *Scopeuma* visit flowers to seek prey and to feed on pollen and nectar. Colyer and Hammond (1951) record it at Blackthorn (*Prunus spinosa*), Hawthorn and Bramble, while Knuth gives a long list of flowers visited, including many *Ranunculaceae*, *Cruciferae*, *Umbelliferae* and *Compositae*.

Larvaevoridae (=*Tachinidae*) is a family of rather stiffly bristly flies, often of a greyish colour, whose larvae are internal parasites of insects and other arthropods. Much the commonest member of this family in Willis and Burkill's lists is *Crocuta geniculata* (*Siphona geniculata*), which is only about 6 mm. long. Its slender proboscis (Fig. 15A and B) is long for the size of the insect, being about twice as long as the head, with long rostrum and labium and relatively very long labella: it is said to feed on pollen as well as nectar. It visits rayed and purely ligulate *Compositae* (p. 220), and many

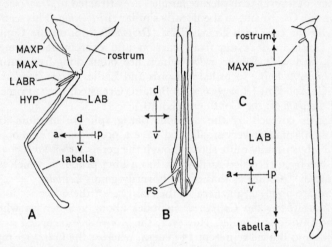

FIG. 15. Mouth-parts of flies. A, *Crocuta*, proboscis. B, *Crocuta*, labella, fused together, from the front. C, *Calirrhoe*, proboscis. A, B, after Peterson (1916); C, after Kugler.

other plants with small tubular florets, such as Mint (*Mentha*). Although it also visits flowers with partly concealed or well-exposed nectar, Willis and Burkill found that it avoided certain flowers with freely exposed nectar which were especially favoured by other *Diptera*. *Crocuta geniculata* is, therefore, evidently rather specialised. Three other species of the same family recorded at flowers are *Calirrhoe siberita*, *Eriothrix rufomaculatus* and *Larvaevora* (=*Echinomya*) *fera*. The last two are common at flowers, and occur at *Compositae*, *Umbelliferae*, *Mentha*, etc. *Eriothrix* is a smallish fly with the fore part of the abdomen black in the centre and red at the sides; *Larvaevora fera* is a large robust fly with the centre of the thorax and abdomen black and the sides rusty red; according to Kugler, it has a proboscis 5–6 mm. long and is found at tubular flowers more often than the shorter-tongued members of the family. *Calirrhoe siberita*, a light brownish fly about a centimetre in length, is of interest in that it is a specialised flower-feeder with a long slender proboscis (Fig. 15c). It has a short rostrum, long labium, and short labella (Kugler); these proportions are different from those of *Crocuta geniculata*. Colyer and Hammond (1951) record it as often visiting Field Scabious (*Knautia arvensis*); Knuth's list consists of *Clematis recta*, Zigzag Clover (*Trifolium medium*), Purple Loose-strife (*Lythrum salicaria*), Field Scabious and Marjoram (*Origanum vulgare*). These flowers, with the exception of *Clematis*, have their nectar concealed in narrow tubes. Records of flower-visits by members of fourteen other genera of this family can be found in Colyer and Hammond 1951, and in papers by Andrews 1953, Parmenter 1941, 1952a and b, Grensted 1946, Harper and Wood 1957, and Willis and Burkill.

Closely related to the *Larvaevoridae* are the *Calliphoridae*; Willis and Burkill recorded species belonging to 8 of the 42 genera of this family visiting flowers. Among them were *Sarcophaga carnaria*, the common flesh fly, a familiar large fly, patterned in black and grey and having red eyes and large feet, and *Lucilia caesar*, a metallic green fly with red eyes. The proboscis in *Calliphoridae* is short (2–4 mm.) and examples of the flowers visited are:

Saxifraga aizoides	Yellow Mountain Saxifrage	} Nectar well exposed
Umbelliferae		
Ranunculus	Buttercups	} Nectar partly concealed
Sorbus	Rowan, etc.	
Compositae		
Mentha aquatica	Water Mint	} Small, tubular flowers
Polygonum viviparum	Alpine Bistort	

(Records given by Willis and Burkill)

Among the thistles (family *Compositae*) only *Cirsium arvense*, in which the nectar is most accessible, is visited by the *Calliphoridae*. These flies have also been seen feeding on the pollen of certain large tubular flowers, the nectar of which is inaccessible to them. Kugler mentions *Pollenia rudis*, *P. atramentaria*, *Lucilia* (green-bottles), *Onesia* and *Sarcophaga* as being predominantly flower-feeders, and *Calliphora* (blue-bottle or blow-fly) and *Cynomya* as feeding chiefly on carrion and excrement, although they also visit flowers; *Pollenia vespillo*, he says, feeds at flowers and sap only. Most of these *Calliphoridae* also appear in the lists of Willis and Burkill.

It is on *Calliphora* (blow-fly) that the most detailed studies of the proboscis of *Cyclorrhapha* have been made. The most recent and thorough of these is the 58-page account by Graham-Smith (1930). The basic structure of the mouth-parts (Fig. 16) is the same as in the hoverflies and *Sepsis*, but the maxillae are further reduced and only the palps remain.

The labrum and hypopharynx are helped to form a closed tube by the fact that the edges of the labrum are grooved to receive the edges of the hypopharynx (Fig. 16Cb); this tube ends just short of the end of the labium, but here the labial groove is roofed over by pairs of interlocking

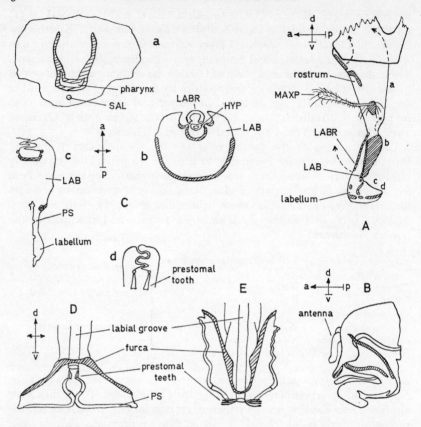

FIG. 16. Mouth-parts of the blow-fly (*Calliphora*); sclerites shaded. A, proboscis, extended; arrows show movements involved in retraction. B, vertical section through head, with proboscis retracted. C, sections through A, at levels indicated by letters a-d. D, section through labella in filtering position. E, the same, in direct feeding position. After Graham-Smith (1930).

folds which arise from either side of the groove and continue the tube (Fig. 16Cc).

The pseudotracheae are completely closed over when the labella are used for filtering, and the liquid passes into each pseudotrachea through a series of sloping passages at right angles to its length (Fig. 17). The diameter of the pseudotracheae in *Calliphora erythrocephala* ranges from .01 to .02 mm. and the sloping lateral passages are .004 to .006 mm. in

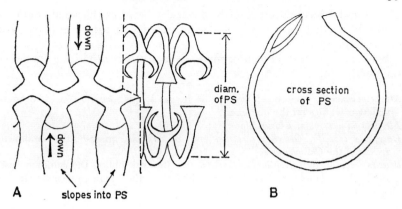

A   slopes into PS   B

FIG. 17. Mouth-parts of the blow-fly, continued. A, portion of a slightly gaping pseudotrachea as seen on surface of labellum and, on right, with surface removed to show sclerites. B, one of the sclerites of the pseudotracheae. After Graham-Smith (1930).

diameter. Between the labella there are four rows of prestomal teeth (Fig. 16D and E).

Several different feeding positions of the labella are described by Graham-Smith. He fed films of drying milk and drying solutions of sugar containing finely ground Indian ink to the flies, and it was from the impressions left by their proboscides in these and other foods that he made his discoveries about these feeding positions. In the filtering position the labella are inflated and spread horizontally on the food surface; they are also pressed together so that the labial tube has no connection with the food other than through the pseudotracheae (Fig. 16D). Other feeding positions allow the prestomal teeth to reach the food and are presumably not used for feeding at flowers. Filtering is entirely abandoned when the direct feeding position is adopted, as it is for viscid materials; here the labella are thrown back still further, causing the teeth to diverge horizontally and allowing the food to pass directly into the labial tube. The movements of the labella are produced by muscles acting on the furca (see Fig. 16D and E) but the labella are inflated by blood pressure. When not in use, the proboscis is folded up and retracted into the head (Fig. 16B). The pseudotracheae and the tubes of the haustellum can all be opened along one side; this may be of importance in allowing for cleaning if the passages become blocked.

The very large family *Muscidae*, with 84 genera and about 450 species in Britain, is the most important flower-visiting family among the *Diptera* apart from the hoverflies. Willis and Burkill's records cover about 28 genera and I have found records of a few others but this total could no doubt be further increased by a search of the literature and by field observation.

Some of the better known or more distinctive flower-visiting species of the family are:

Musca autumnalis, a close relative of the house fly (*M. domestica*) with orange-yellow markings on the abdomen;
Mesembrina meridiana, a large black fly;
Dasyphora cyanella, *Orthellia caesarion* and *O. cornicina*, green-bottles similar to *Lucilia* of the *Calliphoridae*;
Polietes lardaria and *Graphomya maculata*, two flies resembling the common flesh Fly in their colouring, though this applies only to the female in *Graphomya*; and *Fannia canicularis*, the lesser house fly.

The *Muscidae* do not appear to be specialised flower-visitors, but many of them are occasional or regular ones. Willis and Burkill found that they visited species of thistle besides *Cirsium arvense*, which has the most accessible nectar, and recorded them feeding on pollen at flowers with inaccessible nectar. The Drabbles recorded one species feeding on the pollen of grasses (cf. *Syrphidae*, p. 86).

These four families of the *Calypterae* comprise the great majority of the short-tongued *Diptera* that visit flowers. Willis and Burkill emphasise that in Britain, as in Norway, the short-tongued *Diptera* form a greater proportion of the flower-visitors than they do in either North or South Germany, where bees and beetles preponderate. This difference is evidently due to a relatively greater abundance of flies in the British fauna. In summer, at Scarborough, 61 per cent. of 1800 individual visits to six species of plant were made by short-tongued *Diptera*; in spring the percentage was 48, 27 species of plant being observed. Willis and Burkill's results show that the less specialised flowers most visited by short-tongued *Diptera* are also those most visited by insects in general. The *Compositae* most visited by short-tongued *Diptera* in Scotland were some of the purely ligulate yellow ones, followed by two rayed composites, Daisy (*Bellis perennis*) and Ragwort (see p. 220). Of the *Umbelliferae*, Hogweed and Spignel (*Meum athamanticum*) were most visited, while Wild Angelica (*Angelica sylvestris*) was hardly visited at all.

In their second paper Willis and Burkill classify insects according to their 'desirability', which is an unsatisfactory expression because it does not tell us the type of flower to which a given insect is desirable, and in fact the classification seems to be based mainly on proboscis-length; it results in the families of the *Calypterae* being classed as indifferent for pollination. The authors seem to have formed a low opinion of these flies as pollinators, judging by this classification and by their view that short-tongued flies are unimportant for pollinating Autumn Hawkbit (*Leontodon*

b

PLATE 7. **a**, hover-fly *(Volucella bombylans)* visiting flower of Bramble *(Rubus fruticosus)*. **b**, honey-bee collecting nectar from flower of Raspberry *(Rubus idaeus)*. **c**, hover-fly *(Syritta pipiens)* probing one of the false nectaries of Grass of Parnassus *(Parnassia palustris)*.

c

PLATE 8. **a**, ichneumon feeding on nectar from one of the 'glands' of the inflorescence of Wood Spurge *(Euphorbia amygdaloides)*. **b**, ant feeding at an inflorescence of Portland Spurge *(Euphorbia portlandica)*. **c**, solitary bee *(Halictus* sp.*)* feeding at flower of Common Wintercress *(Barbarea vulgaris)*.

a

b

c

autumnalis), although numerous as visitors to it. However, the Drabbles (1927) say: 'We have satisfied ourselves that the *Diptera* do play a very large part in the pollination of this plant and other Composites.' Clearly many plants have unspecialised flowers adapted to pollination by short-tongued insects and, in their first paper, even Willis and Burkill infer that short-tongued flies are of considerable importance in pollination.

Many of the flowers visited by short-tongued insects such as these *Diptera* are scented. The scents are of various different kinds: for example, Meadowsweet (*Filipendula ulmaria*) and Rowan (*Sorbus aucuparia*) have a heavy sweet scent, some have a lighter sweet scent, such as the Crab Apple (*Malus sylvestris*), while a different type of sweet scent is produced by Sallows (*Salix* spp.). A sallow-like scent is also produced by Lady's Bedstraw (*Galium verum*), Thrift (*Armeria maritima*) and perhaps White Mignonette (*Reseda alba*). Some *Umbelliferae* are sweet-scented but have a tang of stale (human?) dung, as in Hedge Parsley (*Anthriscus sylvestris*), while in Alexanders (*Smyrnium olusatrum*) (Plate 12a) the balance of these scents is reversed. In two cruciferous genera, represented by some European yellow-flowered species of *Alyssum* and the swede (*Brassica napus*), a smell of horse dung is apparent. The yellowish *Umbelliferae*, some other *Cruciferae*, such as Treacle Mustard (*Erysimum cheiranthoides*) and Sweet Alison (*Lobularia maritima*), and some *Compositae* (for example, certain forms of Goldilocks (*Aster linosyris*, Plate 12f)) have a more or less sugary sweet smell with a tang of urine or ammonia; some people perceive a similar tang in the heavily scented flowers of Hawthorn (*Crataegus monogyna*) (Plate 9a). These scents, even to human noses, therefore show a fairly evident correlation with the prevailing habits of these unspecialised flower visitors.

Diptera with long tongues are scattered through various families of the *Brachycera* and *Cyclorrhapha*; they can reach nectar not available to shorter-tongued insects and show some diminution of attention to flowers with well exposed nectar. This is particularly true of *Bombylius*, *Rhingia* and *Crocuta* and is a sign of their specialisation.

Lepidoptera (butterflies and moths)

This large order is represented by about 2000 species in the British Isles. The life-cycle consists of the same stages as that of *Diptera* and *Coleoptera*; nearly always the larvae feed on angiospermous or gymnospermous plants (see Chapter 12), usually living externally but sometimes in wood, in flower-heads or as leaf-miners. With few exceptions the adults feed only on liquids, and the range of these is much the same as in *Diptera*, namely nectar, fruit juice, exudates from plants and fluids occurring on excrement and carrion. Nectar, however, is the main food of the *Lepidoptera*, but some species occasionally feed on the other foods mentioned, for example *Nymphalis io* (the peacock butterfly), and yet others feed on them predominantly or exclusively, for example *Nymphalis antiopa* (Camberwell beauty butterfly) and *Acherontia atropos* (the death's head hawk-moth).

In the general classification of the order in the ninth edition of Imms's *General Textbook of Entomology* (1957), the butterflies constitute two superfamilies in the midst of a long series of superfamilies of moths. Three sub-orders are recognised, and one (the *Ditrysia*) contains all the butter-flies and nearly all the moths. Any generalisations which follow apply to this sub-order rather than to the other two, which are small and will be dealt with separately. The so-called *Microlepidoptera* are distributed among all three sub-orders in this classification.

The *Lepidoptera* are perhaps the order of insects best known to the British public and they are certainly the most popular among naturalists for collection and study (Plates 14 and 15). They have been honoured by the devotion of two volumes of the 'New Naturalist' series to them (Ford, 1945 and 1955). Also they are very uniform in mouth-parts and mode of feeding. For all these reasons they can be dealt with more briefly here than the *Diptera*.

The lepidopteran proboscis is very differently constructed from that of the *Diptera*; its form commits its owners to an exclusively suctorial method of feeding. Whereas the sucking tube of the *Diptera* is formed by the great development of the labrum, hypopharynx and labium, that of the *Lepidoptera* is formed by the terminal lobes, or galeae, of the two maxillae (Fig. 18). The slender galeae are enormously elongated and lie touching one another, their contiguous surfaces being hollowed out so that a tube, circular in cross-section, is formed between them. This is the food canal (Fig. 18D and E). The base of this tube connects with the mouth and the apex is open for taking in the food. The galeae are themselves hollow internally, and their cavities are continuous with the general body cavity which contains the blood. The proboscis is rolled up under the head when not in use (Fig. 18c) and is kept in this condition by a longitudinal elastic bar in each galea which is coiled when at rest. The outer surface of the galea is constructed to facilitate this rolling, since it consists of alternate transverse bands of thin membrane and thickened rings (Fig. 18F); the wall of the food tube is comparatively rigid, but it has a laminated structure which permits coiling in the vertical plane while preventing movement from side to side. During uncoiling small oblique muscles inside each galea contract and cause an alteration in the cross-sectional shape of the proboscis, such that the upper surface changes from flat to convex. Convexity of the upper surface is incompatible with coiling, and the proboscis is therefore forced to unroll (see Fig. 18D and E). The effect is analogous to the effect of the convexity of a coiled steel ruler when released from its casing. The operation of the mechanism is also dependent

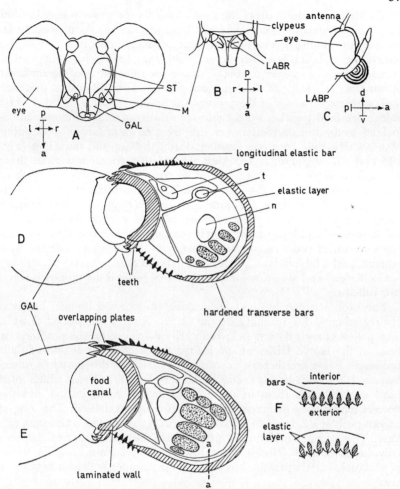

FIG. 18. Mouth-parts of the large white butterfly (*Pieris brassicae*). A, underside of head, with base of proboscis. B, upper side of base of proboscis. C, head with coiled proboscis. D, transverse section of proboscis taken near the base when in coiled state with upper surface flat. E, the same, taken near the middle when in extended state with upper surface curved. g, gland cell; n, nerve; t, trachea. Muscle is stippled. F, vertical longitudinal sections of wall of underside of proboscis at a . . . a in E; upper: proboscis extended; lower: proboscis coiled. After Eastham & Eassa (1955).

on the rigidity of the walls of the food canal, the presence of longitudinal partitions inside the galeae, and the maintenance of blood pressure in them by the closure of a valve at the base of each. Some distance from the base of the proboscis is a 'knee-bend' (Plate 15a), kept in being by a set of muscles which counteract at this point the uncoiling effect of the main set of muscles. The angle of the bend can be varied at will and its presence makes for easy transfer of the proboscis from one flower to another. The galeae are held together on their lower side by a series of closely inter-locking teeth, and on their upper side by a series of larger overlapping plates which slide over one another during coiling and uncoiling (Fig. 18D and E). A glandular secretion is produced which lubricates these plates in their movements and helps to seal the joint which they form along the upper edge of the food canal. This description of the proboscis is based on the work of Eastham and Eassa (1955) on *Pieris brassicae* (large white butterfly); these authors rejected earlier theories of the mechanism of lepidopteran proboscis movements. The rest of the mouth-parts consist of two structures which may be rudimentary mandibles, a labrum, and a labium which bears two palps of moderate size which may act as 'feelers' and are often furry so that they conceal the proboscis when it is rolled up.

The proboscis can be used for sucking up exposed liquids, but it is clearly adapted for reaching nectar at the base of long-tubed flowers. The pollen of such flowers is involuntarily conveyed on the proboscis or head of the insect. Different species vary greatly in the length of their proboscis. Some are therefore confined to shallower flowers while others can reach the deep-seated nectar of the very long-tubed ones which must have evolved with them in mutual adaptation. Such insects and the flowers which they pollinate are clearly highly specialised. The longest known proboscis belongs to a South American species and is about 25 cm. long. The longest lepidopteran proboscis in Europe is that of the con-volvulus hawk-moth (*Herse convolvuli*), which is 65–80 mm. long. Examples of proboscis-lengths provided by some of the commoner British *Lepidoptera* are:

Pyrausta (a genus of moths); 4–9 mm.
Plusia gamma (Silver-Y moth); 15–16 mm.
Coenonympha pamphilus (Small Heath butterfly); 7 mm.
Aglais urticae (Small Tortoiseshell butterfly); 14–15 mm.
Pieris brassicae (Large White butterfly); 16 mm.
(Figures from Knuth)

The tip of the proboscis is armed with numerous fine spines which are thought to be used for breaking open the tissues of nectarless flowers to release the sap. The evidence for this is that *Lepidoptera* are occasionally seen apparently sucking at these flowers (for example Common Centaury

(*Centaurium erythraea*)), but Kugler considers that there is still some doubt about it. However, an example is quoted by Knuth of *Lepidoptera* becoming pests of fruit in South Africa by boring the intact skin, and Francis Darwin (1875) has described the structure and action of the rasping tip of the proboscis of *Ophideres fullonica*, a pest of oranges in Australia. *Charaxes jasius* feeds especially on overripe fruit and Knoll (1922) once observed it boring through the intact skin of a tomato.

According to Knoll (1922) the hummingbird hawk-moth (*Macroglossa stellatarum*) can only drink solutions which are quite fluid; it strives in vain to suck up very thick syrupy solutions of sugar. The silver-Y moth (Plate 33), on the other hand, secretes saliva copiously and this enables it to dilute and suck up such materials.

Taking into account the records of Willis and Burkill, Bennett (1883) and Scorer (1913), it seems that the following flowers are probably those most visited by British *Lepidoptera*:

Silene	Campion
Lychnis	Campion or Ragged Robin
Lotus	Birdsfoot Trefoil
Prunus spinosa	Blackthorn
Rubus fruticosus	Bramble
Rubus idaeus	Raspberry
Hedera helix	Ivy
Valeriana	Valerian
Kentranthus ruber	Red Valerian
Eupatorium cannabinum	Hemp Agrimony
Solidago	Golden Rod
Achillea	Yarrow or Sneezewort
Senecio jacobaea	Ragwort
Centaurea	Knapweed
Cirsium	Thistle
Calluna vulgaris	Ling
Stachys spp.	Woundwort
Salix spp.	Sallows

Common Birdsfoot Trefoil (*Lotus corniculatus*) is pollinated by bees, and according to Müller *Lepidoptera* take the nectar without causing pollination.

In their study area in Scotland Willis and Burkill found that the flowers most visited by *Lepidoptera* were

Silene acaulis	Moss Campion
Rubus idaeus	Raspberry
Syringa vulgaris	Lilac (cultivated)
Thymus drucei	Wild Thyme

The scents of flowers pollinated by *Lepidoptera* are usually heavy and sweet, for example those of Honeysuckle (*Lonicera periclymenum*), Hyacinth

(*Hyacinthus orientalis*), Lilac (*Syringa vulgaris*), Wallflower (*Cheiranthus cheiri*) and Carnation (*Dianthus caryophyllus*). Carnation scent, found in various species of *Dianthus*, is very widespread in butterfly- and moth-pollinated flowers, something very like it occurring in such diverse plants as

Gymnadenia conopsea var. *densiflora* Fragrant Orchid	(family *Orchidaceae*) Europe
Narcissus poeticus Poets' Narcissus	(family *Amaryllidaceae*) S. Europe
Viburnum carlesii A sp. of Wayfaring Tree	(family *Caprifoliaceae*) E. Asia
Ribes odoratum Buffalo Currant	(family *Saxifragaceae*) N. America
Daphne cneorum, D. alpina, etc. Alpine spp. of Daphne	(family *Thymelaeaceae*) Europe
Petunia axillaris Petunia	(family *Solanaceae*) S. America

The scents of these sweet-smelling *Lepidoptera*-flowers resemble the scents sometimes diffused by the insects themselves, which probably play a part in the attraction of the sexes (see p. 117). However, one of the best-known butterfly flowers of gardens, *Buddleia davidii*, produces a rather curious fruity smell, which is perhaps related to its power of attraction for the butterflies of intermediate feeding habits (see pp. 96 and 128).

Butterflies are active in the daytime in fine weather. Delpino (1874) thought that the constant pursuit of the females by the males increased their usefulness as cross-pollinators. On the whole, butterflies favour *Compositae* and other small, tubular flowers grouped together, for example Teasel (*Dipsacus*) and Marjoram (*Origanum vulgare*). Some favour Thistles and many visit Bramble. Lederer (1951) found that the white admiral butterfly (*Limenitis camilla*), a species which feeds on tree sap and excrement as well as flowers, did not visit the *Compositae* and other flowers most favoured by butterflies. Nevertheless, it visited a few flowers apparently adapted to *Lepidoptera* (such as Privet (*Ligustrum vulgare*)) but it also went to Alder Buckthorn (*Frangula alnus*), Lime trees (*Tilia* spp.) and Sweet Chestnut (*Castanea sativa*), which have well-exposed nectar.

The flowers which are perhaps most *adapted* to butterflies in Britain are:

Certain *Caryophyllaceae*, e.g. Red Campion (*Silene dioica*) and Pinks (*Dianthus*);
Forget-me-not (*Myosotis*);
Pyramidal and Scented Orchids (*Anacamptis* and *Gymnadenia*, Plates 44 and 45);
Valerianaceae;
Some *Rubiaceae*;
Some *Compositae*, e.g. Hemp Agrimony (*Eupatorium cannabinum*);

Those *Labiatae* in which the corolla has no appreciable upper lip, e.g. Bugle (*Ajuga reptans*, Plate 28) and Wood-sage (*Teucrium scorodonia*).

These flowers are chiefly blue or deep pink and, as already mentioned, some, such as *Dianthus* and *Gymnadenia*, are strongly scented. (A Campion from the Mediterranean region (*Lychnis chalcedonica*) and many tropical and subtropical plants which are pollinated by butterflies have scarlet flowers.) The tubular part of the flower into which the proboscis has to pass is usually very slender. Consequently, even though the proboscis itself is also very slender, it does not easily avoid touching the stamens and stigmas. Some species of *Dianthus* have protruding and spirally twisted stigmas which are thought to be specially adapted to increasing the chance of their being touched by the insect's proboscis, while in *Anacamptis* there are ridges on the flower which guide the proboscis into the spur (p. 241).

In Willis and Burkill's work, butterflies were found to form only a small proportion of insect visitors, and only 13 species were seen to make visits to flowers. No doubt butterflies are more important in the south of Britain, and especially on calcareous soils, where they and the flowers they visit are numerous in species and individuals.

Some of the flowers visited by butterflies are also visited at night by moths, *Anacamptis* being an example of these. The different species of night-flying moths have their characteristic emerging times. The flowers that are specially adapted to pollination by them are mostly white, pale rose or pale yellow; they therefore show up better than flowers of other colours in poor light. Many are strongly scented. *Caryophyllaceae* and *Orchidaceae* again provide examples, such as White Campion (*Silene alba*), Night-scented Catchfly (*S. noctiflora*), Nottingham Catchfly (*S. nutans*) and Butterfly Orchids (*Platanthera* species.). These *Caryophyllaceae* open their flowers and produce scent at night, the petals of *S. noctiflora* being made inconspicuous during the day by their greenish exteriors and those of *S. nutans* by being rolled up. The Butterfly Orchids are also more strongly scented at night than by day. Orchids adapted for pollination by *Lepidoptera* have very slender spurs and species pollinated by moths have, in Britain, longer spurs than similar butterfly-pollinated species. Similarly, the night-flowering *Silene* species have a longer calyx than the day-flowering species. There is also a tendency among flowers adapted to nocturnal *Lepidoptera* to have the petals or other display organs long and narrow, or divided up into long narrow segments (Vogel, 1954); this is most marked in exotic plants but it can be seen among the British Catchflies and Orchids.

The hawk-moths (family *Sphingidae*) appear to be particularly effective as pollinators. They fly rapidly from flower to flower, and usually take the nectar without settling; their long tongues are no doubt helpful in enabling them to do this, and the longer-tongued species can reach deeply-seated nectar. Exceptions to the rule of feeding in flight have been noted in California, where hawk-moths have been seen to crawl into large trumpet-shaped flowers (Baker, 1961). Most hawk-moths fly in the evening, and a British flower specially adapted to pollination by crepuscular hawk-moths is the Honeysuckle (*Lonicera periclymenum*) (p. 215). The hummingbird hawk-moth (*Macroglossa stellatarum*), which migrates to Britain and is some-times common, flies by day, however, and visits a great variety of flowers of the types favoured by butterflies. In the Alps, Müller (1881, quoted by Knuth) observed individuals of this species differing in their choice of flowers at the same locality. Thus one visited *Primula integrifolia*, another visited three species of Gentian and *Viola calcarata*, while two others visited only *V. calcarata*. The rapidity of the visits is shown by the fact that one insect visited 106 flowers of the *Viola* in four minutes.

In Willis and Burkill's work, about 40 species of moths were recorded at flowers, though it is not stated whether evening observations were made in order to get a fair sample of moth visitors. Much the best represented family was *Agrotidae* (also known as *Noctuidae*); next came *Geometridae*, while a few other families were weakly represented. Hawk-moths were scarce, no doubt because members of this family are rare in Scotland where most of the observations were made. Among the most frequent visitors were:

Pyrausta spp.	(family *Pyralidae*)	
Celaena haworthii	(family *Agrotidae*)	Haworth's minor moth
Charaeas graminis	(family *Agrotidae*)	Antler moth
Hydraecia oculea	(family *Agrotidae*)	Ear moth
Plusia gamma	(family *Agrotidae*)	Silver-Y moth
Psodos coracina	(family *Geometridae*)	Black mountain moth
Dysstroma citrata	(family *Geometridae*)	Common marbled carpet moth
Calostigia salicata	(family *Geometridae*)	Striped twin-spot carpet moth
Glyphipterix fuscoviridella	(family *Glyphipterigidae*)	a 'micro-moth'

The families *Agrotidae* and *Geometridae* are the two largest families of British moths. Each has been broken up into a number of smaller families since the time of Willis and Burkill.

Certain families of moths are quite absent from their lists, and the reason for this in some cases is that the adults never require food. In these moths the proboscis is very short and non-functional, the reduced galeae usually not being held together.

One family that appears in Willis and Burkill's lists but which is said to have vestigial mouth-parts is *Hepialidae*. The ghost moth (*Hepialis humuli*) was recorded at a few flowers, and was said to be sucking honey at one of them. Probably it was mistakenly thought to be feeding, and the visits to flowers were presumably the result of an attraction which light-

a

PLATE 9. **a**, empid fly
(*Empis tesselata*) on
flowers of Hawthorn
(*Crataegus monogyna*);
see also Pl. 17**c**. **b**, wild
cockroach (*Ectobius
lividus*) on Sea Carrot
(*Daucus carota* subsp.
gummifer). **c**, hover-fly
(*Sphaerophoria ruepellii*)
on flower of Tormentil
(*Potentilla erecta*).

b

c

a

PLATE 10. **a**, common wasp *(Vespula* sp.*)* feeding at flowers of Ivy *(Hedera helix)*. **b**, blow-fly *(Calliphora* sp.*)* feeding at Ivy. **c**, drone-fly *(Eristalis* cf. *arbustorum)* on inflorescence of Hogweed *(Heracleum sphondylium)*.

c

coloured objects have for the males of this species. The family belongs to
the *Monotrysia*, one of the two small sub-orders mentioned on p. 96. This
sub-order also contains the yucca-moths, in which the mouth-parts are
specially modified for unusual functions (see Chapter 10). In the same
sub-order, the small moths of the family *Eriocraniidae* (several of which
occur in Britain) are of special interest in having small mouth-parts which
are believed not to be reduced but to represent an early stage in the
evolution of the typical lepidopteran proboscis (Fig. 19A). Each galea,

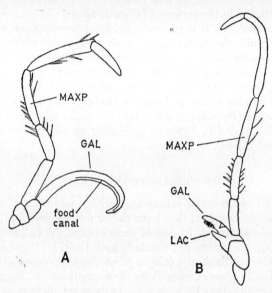

FIG. 19. A, The moth *Eriocrania*: maxilla (lacking lacinia); the galea
forms one half of the proboscis. B, the moth *Micropterix*: maxilla (compare
with Fig. 10C). After Tillyard (1923).

though considerably elongated, is nevertheless shorter than the five-jointed
maxillary palps, and is opposed to the corresponding galea of the other
maxilla. The galeae are softly chitinised, delicately ribbed transversely and
grooved on the opposing faces in the usual way. They end at the base in a
heavily chitinised segment. Above their bases is a wide soft labrum, and
below them a narrower hypopharynx, above which opens the mouth.
These two organs close over the bases of the galeae, and nectar sucked up
through the tube passes between them and into the mouth. There is a pair
of distinct mandibles which, however, are soft and cannot be moved.

The remaining sub-order of *Lepidoptera*, the *Zeugloptera*, is the most
primitive, its members having well-developed mandibles and unspecialised

maxillae (Fig. 19B) similar to those of allied orders with biting mouth-parts. There is only one family, the *Micropterigidae*, the adults of which feed on pollen and other powdery vegetable matter. The two mandibles are dissimilar and have a complicated arrangement of interlocking teeth. The muscles which work them also cause the compression of the highly specialised pouch-like hypopharynx, which is lined inside with a rasping surface. The epipharynx (i.e. under-surface of the labrum) has specialised brush-like developments, and the whole system is adapted to pushing the food between the mandibles and into the hypopharyngeal pouch, grinding it thoroughly in the process (Tillyard, 1923).

Micropterix calthella is a common British representative of this family (Plate 14a) and is the moth referred to on p. 53 as one of the characteristic visitors of Meadow Buttercup (*Ranunculus acris*). Willis and Burkill found it a frequent visitor to some of the flowers they observed, including Bog Asphodel (*Narthecium ossifragum*). They also recorded it both feeding on pollen and sucking nectar at Bramble (*Rubus fruticosus*), Meadow Butter-cup, Tormentil (*Potentilla erecta*) (Plate 9c) and Lady's Bedstraw (*Galium verum*), while M.C.F.P. has seen it feeding on the pollen of an anemo-philous plant (*Carex*).

Flower-visitors among other orders of insects

The most primitive insect orders are without any trace of wings. There are only four such orders; they have rather few species and the insects are mostly small or minute. In one of these orders, the *Collembola* or spring-tails, feeding on pollen is recorded, and Burkill (1897) saw them at flowers of Golden Saxifrage (*Chrysosplenium oppositifolium*) which grows in the damp shady places favoured by these small insects. Springtails could cause cross-pollination in such a plant because their power of leaping gives them sufficient mobility.

The winged insects are divided into those in which development is through a series of moults in which the insect becomes gradually more like the adult, and those in which the larva, usually very unlike the adult, enters a more or less dormant pupal phase in which the transformation to the adult form takes place. The orders of the former are of relatively slight importance in pollination. The earwigs (*Dermaptera*) use flowers commonly for hiding in, and they also devour them in the same way as the more destructive kinds of beetle, while some of them feed on pollen (Popham, 1961) or nectar (Corbet, 1970). Cockroaches (*Dictyoptera*) are apparently only occasional flower-visitors in Europe, although in Brazil a species has been observed which feeds on nectar. One of the three British species of wild cockroach, *Ectobius lapponicus*, was recorded apparently licking the nectar of flowers of *Spiraea* and *Filipendula* by Müller (1879); Plate 9b (p. 102) shows a flower-visit by another member of this genus in Alderney, where the insects were visiting the umbels in some numbers (M.C.F.P.). Knuth has one record, perhaps a chance one, of a member of the book-louse order (*Psocoptera*) at a flower, and visits to flowers by dragonflies (*Odonata*)* are doubtless also by chance, of no significance in connection with the feeding of the insect, and of very little in connection with pollination. Stone-flies (*Plecoptera*) have mandibles and some are said to feed on lichens and unicellular green algae. Pollen is similar in texture to the latter, and it is probable that those stone-flies that were observed by Willis and Burkill at the flowers of Marsh Marigold (*Caltha*), Rock-rose (*Helianthemum*), Rose and Lady's Mantle (*Alchemilla*), were feeding on pollen, as all these flowers produce it abundantly. Porsch (1957), however, saw stone-flies visiting the Twayblade Orchid (p. 235), feeding on its nectar and pollinating it; he also once saw a single individual of a different species taking nectar from an umbellifer.

The large order *Hemiptera* has more often been recorded at flowers. This order comprises aphides, leaf-hoppers and bugs. All species have piercing and sucking mouth-parts constructed similarly to those of the mosquito, and they use them for sucking the juices of plants or animals. Aphides occur casually in flowers and may cause pollination. Bugs of several families have been recorded at flowers, some as regular visitors (Porsch, 1957). Among these, *Capsidae* are medium or small-sized bugs, some of which visit various *Compositae*, *Umbelliferae* and other open flowers. The members of another family, *Anthocoridae*, are smallish, rather active bugs commonly seen on leaves and flowers, where they seek other insects on which they prey, but they also feed on the nectar of flowers similar to those visited by *Capsidae*, and also that of *Salix*. Species of these two families seen visiting flowers by Müller became dusted with pollen and appeared to him to be effective in pollination. Flower-visiting *Hemiptera* are usually about as reluctant to fly around as flower-visiting beetles.

The minute thrips (order *Thysanoptera*) have piercing and sucking mouth-parts and feed on the juices of plants; they are sometimes pests of crops, either through feeding on them or by transmitting virus diseases. They are often found in flowers, and sometimes cause pollination. They are said to be helpful in the pollination of beet (Imms, 1957), and Hagerup (1950) found that they were pollinators of Ling (*Calluna vulgaris*) (Plate 49), Cross-leaved Heath (*Erica tetralix*) and Common Catsear (*Hypochaeris radicata*). *Taeniothrips ericae* goes through its whole life-cycle in the flowers of *Calluna*, and often causes self- and cross-pollination; this is especially important in the Faroes, where bees and butterflies are very rare.

Apart from the *Hymenoptera* (which are described in Chapter 5), the most important of the orders in which a pupal phase occurs in the life history have already been dealt with, but three relatively primitive orders

* And by bush-cricket nymphs (*Orthoptera*) – seen by M.C.F.P.

of this kind, the *Neuroptera*, *Mecoptera* and *Trichoptera*, have also been recorded at flowers on a few occasions. Knuth gives a record of an alder-fly (genus *Sialis*) at an umbellifer, and green and brown lacewings (genera *Chrysopa* and *Hemerobius*) have also been recorded at *Umbelliferae*; these insects all belong to the *Neuroptera*, which have biting mouth-parts. However, some of them are said to feed on liquids such as honey-dew, so they might well feed on nectar when it is freely exposed. The scorpion fly (*Panorpa communis*, order *Mecoptera*) visits *Umbelliferae* and *Compositae*, including those in which the nectar is not freely exposed, and it visits abundantly the flowers of Jacob's Ladder (*Polemonium caeruleum*) (Pigott, 1958). It has biting mouth-parts, but often feeds on nectar and can reach the nectar at the base of small tubular flowers because the mouth is situated at the apex of a long snout formed from the head and mouth-parts. The mouth-parts of caddis-flies (*Trichoptera*) are adapted for licking up liquids, the mandibles being absent, and while some species are believed not to feed in the adult state, others feed at flowers, visits to Ivy (Sperring, 1933), Twayblade Orchid (Porsch, 1957), *Umbelliferae*, Common Valerian (*Valeriana officinalis*) and Yellow Water-lily (genus *Nuphar*) having been recorded. Caddis-flies are weak and infrequent fliers, and are therefore of slight importance in pollination. It is probable, however, that the very important *Lepidoptera* evolved from them.

THE SENSES AND BEHAVIOUR OF THE INSECTS INCLUDED IN PART 1

INSECT-POLLINATED flowers inevitably show adaptation to the general abilities of insects and also, in some cases, to their special responses and specialisations, some of which will have arisen in conjoint evolution with the flowers. A full knowledge of relevant insect behaviour is needed for an understanding of the observed structure and behaviour of flowers, and this knowledge can be obtained only from carefully planned experiments. Such investigations have been rather few so far, but in addition to the details of insect behaviour they show the kinds of problem that arise.

A general account of insect senses has already been made available in this series by Imms (1947). Tactile stimuli are received by hair-like organs scattered over the body with varying density. The perception of chemicals in the vapour state (sense of smell) is closely related with that of chemicals in solution (sense of taste), and it is not usually possible to say whether an organ is a tasting organ, a scent organ or both. The structures concerned with chemical sense are thin-walled discs in the cuticle, or thin-walled projecting or sunken hairs, and they occur on the tarsi of the legs, the antennae and the mouth-parts. The insect eye is a highly developed structure which can produce a detailed retinal image. Its faceted structure results in good ability to detect movement, but its focus cannot be altered, so that the range of clear vision is short. Most insects are sensitive to ultra-violet light but have little or no sensitivity to red, so that their visible spectrum covers the wavelengths 300nm to 650nm, compared with 400nm to 750nm in man (Burkhardt, 1964). Many insects possess colour vision, and much work on this faculty has made use of the insects' visits to flowers.

COLEOPTERA AND OTHERS

In beetles, the sense of smell plays a large part in feeding and in finding a place for egg-laying. Fritz Knoll, in his study of the plant *Arum nigrum* (1926), obtained evidence that dung-frequenting beetles found their way to its inflorescences by scent. The smell of this plant is produced by its spadix, and as the beetles flew past they suddenly turned in flight towards the spadices on coming within about 30 cm. of them. They then either flew straight towards them, with repeated brief reversals of direction, or flew round, gradually getting nearer. These flight paths are indicative of the insects' finding their way by scent; this becomes specially apparent when they are compared with the flight paths of insects finding their way to flowers by sight, as *Bombylius fuliginosus* does (see p. 112).

Some beetles definitely have colour vision, according to Schlegtendal (1934) who used a method of investigation independent of feeding reactions. She found that chrysomelid beetles could clearly distinguish yellow and orange from blue, and violet from green; dung-beetles (*Geotrupes*) could distinguish yellow, orange and violet from blue, and also yellow-green and light green from other colours. A hemipteran (*Troilus luridus*) and an orthopteran (*Dixippus morosus*) were found to be without colour vision, but some members of these orders do have colour vision (Burkhardt, 1964).

DIPTERA

Smell and Taste

In the blow-fly (*Calliphora erythrocephala*) Graham-Smith (1930) found three types of sense-organ on the labella of the proboscis: at the extremity of each pseudotrachea is a 'gustatory papilla', which as its name implies is believed to be an organ of taste; similar-looking sense-organs, which Graham-Smith suggested are both gustatory and tactile, appear on the surface between the pseudotracheae; the third type of sense-organ is the 'gustatory seta' or bristle, of which there are a number of rows on the outer surface of the labella just outside the pseudotracheal area.

An investigation of the antennal scent organs of flies in relation to mode of life was carried out by Liebermann (1925). The thin-walled olfactory hairs were found to be either solitary on the surface, or sunk, singly or in groups, into pits. The pits, when large, may be partially subdivided and they are always oblique, the openings facing towards the apex of the antenna. The olfactory bristles occur only on the terminal joint of the three-jointed antennae. In the species investigated, the total number of olfactory bristles on the two antennae ranges from 300 in the male of *Psila rosae* to 9260 in the male of *Calliphora vomitoria*. On the average there were more olfactory bristles on the males of dung- and carrion-frequenting species than on those of flower-frequenting species. The females were left out of this comparison because some species feed on flowers as adults but lay their eggs in filth. In these species the females have more olfactory bristles than the males. Liebermann thought that the explanation of these differences lay in the fact that vision cannot play such a big part in finding decaying matter as it can in finding flowers.

Liebermann made some observations which showed insects finding flowers by scent. An area 20 metres square round a large plant of Cowbane (*Cicuta virosa*, family *Umbelliferae*) was cleared of other flowers by mowing, the flowers of the *Cicuta* were covered with large leaves and the behaviour of flies in its neighbourhood was then observed. Those passing within two metres on the downwind side of the plant turned towards it and flew to and fro, then alighted and crawled under the leaves to the flowers; on the windward side of the plant, many flies flew within two metres of it but none was deflected towards it. In this way *Lucilia*, *Dexia*, *Pyrellia*, *Sarcophaga* (all *Calypterae*) and some other flies reached the flowers. The plant was presumably visible to these insects even if its flowers were not, but their behaviour in relation to wind direction gave strong indications of attraction by scent. In another experiment a plant of *Cicuta* was placed in a room behind a curtained window, which was then opened 5 cm. On the wall of the house near the window many flies were resting in the sunshine; soon after the window was opened those near it became restless and after some flying to and fro before the opening flew through it and on to the flowers. Flies were also attracted from a distance, first settling on the wall and thereafter behaving in the same way as the others. Here scent clearly played the main role. This experiment shows also that insects which are not actively seeking food may respond to stimuli normally associated with it.

Knoll (1926) found that the *Diptera* which were attracted to *Arum nigrum* were attracted by the scent, as were the beetles. They were species normally associated with human excrement.

Kugler (1951) investigated the reactions of green-bottle flies (genus *Lucilia*, family *Calliphoridae*) to scent. Captive insects were presented with yellow discs on each of which was a smaller disc of white absorbent paper moistened with sugar-water. Some of these imitation flowers were provided with the scent of Hawthorn (*Crataegus monogyna*); 62 per cent. of the flies' visits were to the scented and 38 per cent. to the scentless models. If the smell of ammonia was substituted for that of Hawthorn, the flies visited scented and scentless models equally. The smell associated with egg-laying habitats was therefore not instinctively associated with food by the flies. They could, however, perceive the ammonia, for if the flower models with this scent were not provided with food the visits to them soon became less frequent than those to the unscented models moistened with sugar-water, the insects having learnt to associate the smell with absence of food.

Kugler found that sight and scent were involved in the visits of *Lucilia* to the flower-heads of Scentless Mayweed (*Tripleurospermum maritimum* sub-species *inodorum*), which has a scented flower but scentless leaves. There were nearly twice as many visits to each of the daisy-like flower-heads of this plant as there were to each of the imitation heads nearby; 95 per cent. of approaches to real heads resulted in alighting, but only

66 per cent. of those to the models led to alighting. When, however, the central yellow disc of the model concealed the disc of a real flower-head and was pierced with a pin to let out the scent, the number of approaches and visits to the models was the same as to the natural heads. It thus appears that the insects were first attracted visually, and at close quarters sometimes chose not to visit those heads which were scentless. If a different flower scent was exhaled by the models, 75 per cent. of approaches led to visits, a result intermediate between that obtained with the accustomed scent and that obtained with no scent at all.

Kugler's experiments thus showed that scent plays a large part in the visits of *Lucilia* to flowers and that they are sensitive to the differences between scents. The observations also illustrate the distinction that can usually be made, when insects visit flowers, between the approach and the visit, which involves touching the flower in some way. The stimulus that elicits approach from a distance does not usually enable the insect to tell whether the flower is suitable for it. When it is close to the flower, different sensory perceptions usually come into play which determine whether a visit takes place. In the instance just described, the approach from a distance is purely visual, while at close quarters scent plays a part in determining whether or not the insect alights. The flies which Liebermann observed visiting concealed flowers were, on the other hand, both attracted from a distance and induced to alight by scent. Kugler found that *Lucilia* would occasionally fly direct to a flower, alight and then fly off without feeding; in such instances the long-range visual stimuli led to alighting as well as approach, and the short-range stimuli were not taken account of by the insect until it had settled.

Further experiments were carried out by Kugler (1956) using flies hatched in captivity. With *Lucilia*, *Calliphora* and *Sarcophaga* it was found that coloured discs were more or less ignored until scent was brought near them, whereupon visits and proboscis-reactions took place. If scent was actually applied to some discs, these were preferred to the unscented ones. Both fragrant and excremental scents were used, but they induced different colour-preferences (see p. 115).

Kugler also studied a hoverfly (*Eristalis tenax*) (1950). He added artificial carnation scent to some flowers that the flies were constantly visiting in a natural habitat; the result was that only 2 per cent. of approaches to them resulted in normal visits, while 46 per cent. resulted in short visits (alighting and flying away immediately) and 52 per cent. of approaches resulted in no visit at all. Kugler also used artificial flowers of the type used with *Lucilia*. When some yellow models were provided with sugar-water and flower scent, and others with a solution of common salt and no scent, both groups received the same number of visits. When blue flowers were used instead of yellow models and the scent was associated with the salt solution instead of with the sugar, 83 per cent. of visits were

PLATE II. Beetles *(Coleoptera)*:
a, *Oedemera nobilis* (male) on Mouse-ear Hawkweed *(Hieracium pilosella)*.
b, soldier beetles *(Rhagonycha fulva)* on Hogweed *(Heracleum spondylium)*.
c, long-horn beetle *(Pachytodes cerambyciformus)* on Hogweed.
d, garden-chafer *(Phyllopertha horticola)* on Hogweed. **e**, flower-beetles *(Meligethes sp.)* on Wall Rocket *(Diplotaxis muralis)*. See also Pl. 42.

PLATE 12. Two-winged flies *(Diptera)* : **a**, St. Mark's fly *(Bibio marci)* and a smaller species of *Bibio* on inflorescence of Alexanders *(Symrnium olusatrum)*. **b**, soldier fly *(Chloromyia formosa)* on Hogweed *(Heracleum sphondylium)*. **c**, bee fly *(Bombylius minor)* resting on Gorse *(Ulex europaeus)*. **d**, 'long-headed flies' *(Dolichopodidae)* on flower of Creeping Cinquefoil *(Potentilla reptans)*. **e**, long-tongued hover-fly *(Rhingia campestris)* feeding at a Dandelion *(Taraxacum officinale* agg.). **f**, hover-fly *(Helophilus pendulus)* on Goldilocks *(Aster linosyris)*. See also Pls. 7**a** and **c**, 17**a** and **c**, and 43**c**.

to flowers with sugar, the increased effect of scent being perhaps connected with the low stimulatory value of the colour blue (see p. 114). It
is an interesting fact that scent can have the effect of preventing a visit
from taking place. Several of Kugler's experiments showed learning by
experience; in one instance the effects of a repellent smell were partially
overcome by training.

Daumann (1932, 1935) studied the visits of the hoverflies, *Helophilus
pendulus* and *Syrphus balteatus*, to Grass of Parnassus (*Parnassia palustris*).
He found that the approach to the flowers was visual but that at very short
range the scent induced the insects to alight. He covered the nectaries,
which are the source of scent and are green (see Chapter 3), with small
glass discs, and then noted where the flies probed with their tongues after
alighting. It was found that the proboscis always touched the edge of the
glass or just outside the edge of it, suggesting that the flies touched only
where the scent was strongest, and that they found the nectaries by scent.
If the nectaries were cut out and placed in abnormal positions in the
flower, the flies touched them directly, or if they were covered with glass
the insects touched near the edges of the glass. They were thus not finding
the nectaries by their position in the flower. If pieces of green paper
shaped similarly to the nectaries were placed about the flowers the flies
ignored them and fed only at the nectaries, which suggests that the green
colour of the nectaries had no significance for them.

The behaviour of a species of bee-fly, *Bombylius fuliginosus*, at the flowers
of Grape Hyacinth (*Muscari atlanticum*) was studied by Knoll (1921). He
thought that two features of the flowers, the scent and the white rim of the
inky-violet perianth, had respectively a strong and a weak effect on the
behaviour of the visiting *Bombylius*. He investigated this by putting out a
row of *Muscari* inflorescences, some with open flowers, some with all the
flowers shrivelled and scentless, and some with flowers still open and
scented but with the white rim faded. Where only scentless flowers were
present, visits were very hasty, but where the flowers were scented but
without a white ring at the mouth the insects sucked the nectar.

In *Diptera* the taste organs are chiefly in the region of the mouth, but
they often also occur in the tarsi of the legs. When they taste food with
their legs insects automatically begin to lower the proboscis in order to
feed; it is, therefore, easy to investigate the sensitivity of the taste organs
in the legs. In this way it has been found that the legs of the blow-fly
(*Calliphora vomitoria*) are 100 to 200 times more sensitive to the taste of cane
sugar than the human tongue.

Sight

It seems probable that at least all the higher *Diptera* have colour vision.
The lesser house fly (*Fannia canicularis*) was included in Schlegtendal's

investigation (see p. 108) and was found to have similar colour vision to that of the chrysomelid beetles.

A detailed study of species of bee-fly, chiefly *Bombylius fuliginosus*, which is not found in Britain, was carried out by Knoll (1921) in southern Dalmatia. When feeding, the bee-fly flew quickly and straight from flower to flower in the characteristic manner of visual flower-seekers. Where flowers were scarce it made wide sweeps over the ground and might slow down near an attractive flower and then, if it was not suitable for it to feed at, take on speed again and continue its sweeps. If it became satiated it might still show signs of attraction to flowers, and approach or visit them hastily. When feeding it had the habit of passing from one plant to another

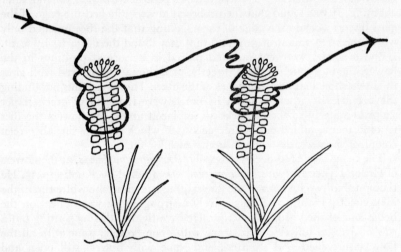

FIG. 20. Flight path of *Bombylius medius* visiting *Muscari comosum*. After Knoll (1921).

in a constant direction, even though at each plant it had gone all round the inflorescence. This and other behaviour showed that the flies could orientate themselves by the sun. These flight lines were quite independent of wind-direction, which made it clear that scent was not helping the insects to maintain them. If the inflorescences of Grape Hyacinth (*Muscari atlanticum*), the flower at which most of the observations were made, were covered by inverted test-tubes, the flies flew towards and pushed against the glass where it covered the flowers, showing that at close quarters as well as at a distance scent was not used for guidance, for in this experiment the strong scent of the flowers was emanating from the end of the tube some distance below the inflorescence. In this kind of experiment, the flies showed their normal behaviour in relation to *Muscari*, which was to approach very closely the light violet, sterile, nectarless flowers at the top of the inflorescence, and then to descend and feed at the lower, dark violet, fertile flowers (Fig. 20). Knoll carried out a number of experiments which showed that the violet colour of the *Muscari* could be distinguished from grey. In one, a board was prepared with sixteen contiguous squares, one of them blue-violet and similar in colour to the middle flowers of a *Muscari* inflorescence, and the others of various shades of grey. The flies were led towards the board by a line of flowering *Muscari* stems in water, the layout being set up in a place where the bee-flies were visiting this flower exclusively. The result was that the flies passing over the board regularly paused and hovered over the violet square but not over the grey ones.

In all his observations, Knoll recorded only three approaches to yellow flowers by *Bombylius fuliginosus*. Observations in different habitats showed that this fly readily approached flowers of all other colours, though it did not attempt to feed at some of them, and it did not encounter scarlet. However, Knoll put the rolled-up scarlet flowers of Corn Poppy (*Papaver rhoeas*) into two tubes and set them out, together with two other tubes containing rose-pink *Cistus* flowers, in a place where the flies were feeding at a white-flowered *Cerastium* species; six flies made seven visits to the pink flowers, and in doing so passed very close to the red ones which they completely ignored.

Knoll tried the reaction of *Bombylius fuliginosus* to the inflorescences of *Muscari comosum*, which has violet sterile flowers and brown scentless fertile flowers, by placing some of them in a habitat of *M. atlanticum*. The result was that the flies went to the sterile flowers and repeatedly pushed in amongst them but did not find the fertile flowers. The honey-bee behaved in the same way when *M. comosum* flowers were placed where the bees were visiting the blue flowers of Viper's Bugloss (*Echium vulgare*), but in another place where *M. comosum* grew naturally they were visiting its fertile flowers. The bees were thus slow to change their habits, rather than incapable of dealing with *M. comosum* effectively. Even in its natural stations, however, *M. comosum* was ignored by *B. fuliginosus*, but it was the favourite flower of *Bombylius medius*. This species conducted itself in the same way as *B. fuliginosus* at *M. atlanticum*, in that it flew first to the sterile flowers and then descended to feed at the fertile ones.

The drone-fly (*Eristalis tenax*) has had not only its olfactory responses experimentally investigated (p. 110) but also its reaction to colour (Ilse, 1949). The flies were caught on yellow flowers and placed in a greenhouse with a large Perspex panel let into it to admit ultra-violet light. The flies were tested with artificial flowers containing sugar-water. Yellow flowers were presented among 'grey' flowers which ranged in colour from black to white, and although there were five grey flowers to each coloured one many more visits were made to the coloured models. Similar results were obtained when the intensity of the yellow was varied. When yellow was offered with other colours, the flies visited it much more than red, green, blue and violet, and somewhat more than greenish yellow and two shades of orange. Some insects were fed only on blue flowers for four days, after which a choice of yellow and blue was offered. Blue now received at least as many visits as yellow. These results showed that *Eristalis* could distinguish yellow from grey regardless of intensity, and from various other colours, and that its feeding preference could be modified by training. In its general preference for yellow, *Eristalis* presents a contrast with *Bombylius*.

The same species of hoverfly was also investigated by Kugler (1950) with closely similar results. A very strong preference for yellow over all other colours was recorded, and deep yellow was preferred to pale. Periods of training on several other colours always modified the result of presenting a choice of colours, causing an increased proportion of visits to the training colour and to allied colours. After training on blue, it was possible to show that the flies distinguished blue from various shades of grey. However, only a slight (though statistically significant) preference for blue over grey was found, showing that blue has a low stimulatory effect on the flies. Kugler found that deep red was not at all distinguished from grey; the same applied to light violet and purple, apparently because the blue component of these colours was too weak.

As already mentioned, when Kugler added artificial scent to some naturally growing flowers, few visits to them occurred; however, the approach flights of the flies towards the flowers were perfectly normal, showing that the distant stimulus was visual. When the flat heads of Yarrow (*Achillea millefolium*) were covered with inverted glass beakers there was also no interference with the approach flights.

High degrees of flower-constancy were observed by Kugler in a habitat containing Hoary Alison (*Berteroa incana*) and Scentless Mayweed (*Tripleurospermum maritimum* sub-species *inodorum*). One individual of *Eristalis tenax* visited 47 Mayweed heads and two unopened heads of a yellow composite, and made nine approaches to other flowers, but none to *Berteroa*. Another individual of the same species showed a similar constancy to *Berteroa* and made no approaches to *Tripleurospermum*. Such degrees of constancy are found only when there is an abundance of productive flowers of the species concerned. Even when suitable flowers are scarce or more mixed, however, these flies are often useful as cross-pollinators, for they still tend to make successive visits to the same species of flower.

Parmenter (1958) counted the flies visiting the flowers of approximately equal-sized neighbouring patches of three *Compositae* (Yarrow (*Achillea millefolium*) which is white, Common Catsear (*Hypochaeris radicata*) which is yellow, and Hardhead (*Centaurea nigra*) which is pinkish purple). The

PLATE 13. Two-winged flies *(Diptera)*: **a**, hover-fly *(Baccha obscuripennis)* on Germander Speedwell *(Veronica chamaedrys)*. **b**, hover-fly *(Melanostoma* sp.*)* feeding on nectar of Germander Speedwell; the style and stamens are held against underside of abdomen. **c**, hover-fly *(Syrphus* cf. *luniger)* sucking nectar from disc of London Pride *(Saxifraga ×　urbium)*. See also Pl. 36**a** and **b**.

PLATE 14. Butterflies and moths *(Lepidoptera)* : **a**, *Micropteryx calthella* feeding on pollen of Creeping Buttercup *(Ranunculus repens)*. **b**, green hairstreak butterfly *(Callophrys rubi)* sucking nectar from flower of Creeping Buttercup. **c**, green-veined white butterfly *(Pieris napi)* sucking nectar from Ragged Robin *(Lychnis flos-cuculi)*.

a

b

c

two commonest hoverflies present were *Eristalis arbustorum* and *Helophilus parallelus*. The visits of the *Eristalis* were 132 to *Achillea*, 34 to *Hypochaeris* and 1 to *Centaurea*, and those of *Helophilus* were 3 to *Achillea*, none to *Hypochaeris* and 24 to *Centaurea*. The bees present visited only *Centaurea*.

Hoverflies are attracted to bright colours even when these do not indicate the presence of food. They have been seen hovering over patches of yellow paint and over yellow pears on barrows in London streets, and are also lured from their food plant by collectors by the placing of white sheets on the ground (Parmenter, 1958). The fact that bright colours can draw the flies away from their food suggests that this attraction is related to some part of the insect's life other than feeding.

As mentioned earlier (p. 111), Daumann found that the green colour of the nectaries of *Parnassia* was apparently of no significance to hoverflies feeding in the flowers. Kugler (1950), however, found what was apparently a short-range visual effect in *Eristalis tenax*. When the capitula of Ragwort (*Senecio jacobaea*) are old and have stopped producing pollen and nectar, the central florets turn brown, while the ray florets remain yellow for a time. Kugler observed that the flies which approached capitula with brown centres did not visit them, and he carried out experiments which strongly suggested that the flies discriminated between the old and the young heads visually.

Another aspect of flowers investigated by Kugler (1950) is their outline. *Eristalis* flies were presented with light yellow discs, 2 cm. in diameter, and eight-pointed stars of the same colour and similar surface area. On both types of model there were small discs of white absorbent paper, but only those on the stars contained sugar-water. Some insects visited both equally, some preferred the discs in spite of the absence of food, while others showed an increasing proportion of visits to stars. Thus the flies apparently had no inborn preference for one type of figure (unlike some other insects) but they were able to distinguish the two types of model.

The visual responses of the green-bottle (*Lucilia*) were also investigated by Kugler (1955a, 1956). The reactions of the flies to shape, to the colours yellow, blue and grey (in various shades), and the results of training to visit blue for food were the same as with *Eristalis*. In his later experiments Kugler (1956) found that *Calliphora*, hatched in captivity, preferred yellow to brown-purple in the presence of sweet scents, and brown-purple to yellow and white in the presence of excremental scents. A similar change in colour-preference occurred in *Sarcophaga*. These observations show that with the young flies scent is the first stimulus, but it is mainly excitatory, inducing the insects to seek and alight on objects of appropriate colours, yellow and white being the colours of flowers most commonly suited to them, and brown and dark purple being the colours of excrement and carrion. When the response of *Lucilia* to a disc and to a funnel of the same diameter was tested, the flies preferred the funnel.

The shiny rounded knobs on stalks in the flowers of *Parnassia* have long been regarded as 'false nectaries' whose function is to attract flies which, once lured to investigate the flower, may cause pollination and also find the true nectaries. Daumann found that the hoverflies that were visiting *Parnassia* flowers in nature entirely ignored the 'false nectaries', and that the removal of the latter made no difference at all to the behaviour of the flies in their visits to the flowers. He therefore concluded that the 'false nectaries' did not function in the way supposed. Kugler showed, however, that flies are in fact attracted by small shining objects. Some *Lucilia* were offered plain blue discs together with blue discs with drops of sugar-water on them; only the first few visits of each fly were counted, and it was found that the discs with drops received 66 per cent. of the visits. The flies were now presented with discs, some of which had sugar-water drops on them and others small aluminium discs, 4 mm. in diameter. The two types of disc were then visited about equally. Flowers of *Parnassia* were offered by Kugler to some *Lucilia* which could not previously have seen this type of flower; in their first visit to the flowers, five flies licked first the false nectaries and none the real ones. The experiment was repeated twice with the same flies, and in the first repeat two went straight to the real nectaries while six went first to the false ones; in the third test, four went straight to the real nectaries and four to the false. Thus, when inexperienced, the flies were strongly attracted to the false nectaries, but they soon learned where the nectar was to be found. A similar result was obtained with the hoverfly (*Sphaerophoria scripta*) and *Parnassia* flowers.

Kugler found that *Lucilia* were sometimes constant in their visits to flowers in nature, and he recorded an instance of constancy to *Parnassia*.

In general, the *Diptera* tend to visit white, pink, yellow and green flowers most readily. Visits to purple and blue flowers are commonly made only by certain genera of the *Brachycera* and *Cyclorrhapha*, for example in the *Brachycera*, some *Empididae* visit purple thistles and knapweeds (*Cirsium* and *Centaurea*), some *Dolichopodidae* visit blue *Myosotis*, and *Bombylius* visit the blue *Myosotis* and *Pulmonaria*, and the violet-coloured *Vinca minor* and *Limonium*; examples from the *Cyclorrhapha* of visits to blue and purple flowers are provided by some *Syrphidae* (*Volucella* and *Rhingia*), some *Conopidae*, and some *Larvaevoridae* (for example *Calirrhoe*). The flies concerned also visit flowers of other colours quite readily. All of them have longer tongues than their nearest relatives, and the blue and purple flowers they visit usually have more deeply seated nectar than flowers of other colours. Brown and brownish purple colours attract flies which feed on or breed in excrement and carrion, and in flowers these colours are found chiefly in species which deceive or trap such flies (see Chapter 10).

The work of Knoll, Kugler and Ilse has shown that flies can distinguish colours when feeding; the prevailing colour differences between the longer- and the shorter-tubed flowers therefore help the insects to find the flowers most suited to them. The colour preferences of the individual insect might either be acquired through learning or be inherited. Some of them, at least, are probably inherited in *Bombylius fuliginosus*, since this species showed an almost complete aversion to yellow flowers, and they appear to be inherited in a number of other flies also. Since they confer some advantage on their possessors, such innate preferences for certain colours would indeed be expected to have evolved once the association between particular colours and deep-seated or exposed nectar had become established.

LEPIDOPTERA

The reactions of *Lepidoptera* to colours and scents have also been given much attention. Müller observed in the Alps instances of butterflies preferring flowers similarly coloured to themselves; he thought that the colour-preference in choosing a mate must have become transferred to the choice of flowers (Knuth). Many *Lepidoptera* produce scent which attracts the opposite sex; these scents are often like those of flowers, and Müller thought that the scents of Lepidopterid flowers might have been evolved as an adaptation making use of the already existing attractiveness of these scents. A similar suggestion has been made to account for the scents of bat-pollinated flowers (p. 337).

Silver-Y Moth

One of the commonest British flower-feeding moths, the silver-Y (*Plusia gamma*), has been studied in Austria by Schremmer (1941a). In natural conditions the moths sometimes fly and feed during the day, but their main period of activity is in the evening twilight, continuing until after dark in moonlight or on clear starlit nights. If the moths are disturbed when at rest during the day they alight on the herbage and settle with their heads downwards and the eyes shaded from the sun. In feeding flights the moths always fly against the wind, if any; on calm evenings they fly about irregularly. The flight of the moths among flowers growing close together is slower than in passage from one group of flowers to another. The moths fly low when feeding, so that they often approach flowers from below and readily discover flowers hidden in the grass. Though the wings continue in motion during feeding, only occasionally stopping momentarily, the flowers are gripped with the legs, and in fact a silver-Y moth lacking fore-legs has great difficulty in feeding from a flower. When the nectar in a flower is nearly exhausted, or when the flower contains no nectar, the proboscis is repeatedly almost withdrawn from the flower and pushed in again. This probing is accompanied by a noticeable activity of the antennae, and is no doubt an attempt to ensure that no nectar has been overlooked. Schremmer noticed that the antennae were used for feeling around the entrances to the flowers of Red Clover (*Trifolium pratense*) while the proboscis was seeking entry; after the proboscis had entered the antennae were raised.

On one occasion, Schremmer found that the types of flower visited early in the evening were later neglected in favour of others; those visited first were a Pink (*Dianthus carthusianorum*) and a yellow-flowered Scabious (*Scabiosa ochroleuca*) while those visited later, when it was almost dark, were White Campion (*Silene alba*) and Bladder Campion (*Silene vulgaris*).

Varying degrees of constancy to a particular kind of flower were recorded; for example, one moth visited 68 Pinks and four heads of the mauve Field Scabious, meanwhile making many approaches to the yellow *Scabiosa ochroleuca*, none of which led to visits; in roadside habitats with a varied flora, however, the insects commonly visited, one after another, several different flowers with different structure and colour.

The feeding flights of the silver-Y moth against the wind suggest that it seeks flowers by scent, but Schremmer pointed out that to travel downwind at high speed would make it difficult for the moths to alight on a chosen flower. However, on one occasion he saw a moth after dark by the light of a pocket lamp visiting Bladder Campion and flying very slowly near the ground with a characteristic scent-directed flight.

The use of sight to the exclusion of scent was also demonstrated by Schremmer. He covered some *Dianthus* flowers with specimen-tubes open below and closed at the top. Moths feeding on *Dianthus* in the daytime alighted on the tubes at the level of the flowers and crawled around the flowers but did not find their way into the tubes, as they might have done had they followed the scent.

The uses of the senses of sight and smell by these moths were further investigated by Schremmer in laboratory experiments. The experiments were carried out in boxes large enough to allow the moths to fly about. Newly hatched silver-Y moths, when presented with variously coloured discs and artificial flowers filled with sugar-water, ignored them. The scented flowers of Phlox (*Phlox paniculata*) were then placed in the box and the insects soon found them, in one instance with an evidently scent-directed flight. Another insect was given yellow discs with drops of sugar-water on them; it discovered the sugar-water by accident and in the course of the evening learnt to fly straight to the yellow discs for food. The following night, however, it quite ignored yellow discs, but when it was given Phlox flowers it soon found them and fed from them. Newly hatched moths were also given scented flowers concealed in black containers with an entrance at the top; the moths approached the containers with much swinging from side to side and twisting, sometimes passing over them and then turning back, eventually alighted at the entrance and entered the containers. The distance at which the moths appeared to respond to scent was 10 to 15 cm. from the source.

Schremmer showed that the effects of previous experience can sometimes last from one day to the next. A moth which had on previous days been fed on Phlox was offered Phlox flowers and white discs, of the same diameter as these flowers, with drops of sugar-water on them. The moth fed from the Phlox and approached the discs but did not alight on them or even begin to unroll its proboscis, presumably being deterred by the absence of the scent to which it was accustomed. On the other hand, a newly hatched moth began feeding at the Phlox and then went on to feed at the white discs.

Tests on colour vision were made by presenting to the moths a grey board with discs on it of various colours in various intensities, a series of greys ranging from black to white being included. When violet artificial flowers containing sugar-water were also provided, the moths visited blue discs of two intensities and ignored red, yellow and the greys. A moth fed on sugar-water placed on yellow petals visited only the yellow discs, also in two intensities.

Schremmer found that the moths could be trained to visit only flowers with one particular scent. Wild-caught moths were fed from the flowers of Soapwort (*Saponaria officinalis*) concealed in black containers. Next evening six Soapwort flowers and four Phlox flowers were presented close together intermixed on a board. All the Soapwort flowers were visited and the Phlox flowers were repeatedly approached, once with unrolling of the proboscis, but the moths never alighted on them nor fed from them. Both types of flower were white and were very similar in size and outline. Other captured insects were not given any training and these, when offered Soapwort and White Campion in an alternating arrangement, visited both readily but showed a slight constancy, frequently making two or three successive visits to the same kind of flower. Training to the scent of Soapwort for three days induced a constancy to that flower, coupled with a neglect of White Campion, when the two types of flower were presented visibly and intermixed, or invisibly in black containers. A converse training to White Campion gave the converse result. Another insect trained to White Campion was given a mixture of this flower and Soapwort, but the Soapwort flowers were provided with sugar-water as usual and the Campion flowers were left empty. The moth persistently tried to feed at the Campion, sometimes flying away and returning to it; after a time, however, it started to visit the Soapwort with increasing frequency, though still making some visits to Campion. Later it made two approaches to Campion without alighting, and then made two more visits to Soapwort before going to rest. The original training was thus broken, though with some difficulty, and finally almost replaced by constancy to a new flower.

The possibility of training the moths to visit only flowers with one particular colour and one particular scent was also investigated, and Schremmer succeeded in training a moth to feed chiefly at transparent blue discs masking flowers of Soapwort, which were offered together with yellow-masked Soapwort flowers, blue discs and yellow discs.

Schremmer also tested the silver-Y moth with white discs and real flowers concealed but accessible a short distance away to act as scent sources. From his observations he concluded that when the moth smells the flower it follows the concentration gradient of the scent; if it then meets a visually conspicuous object it will usually approach it and sometimes visit it, but it will then continue to follow the scent until it reaches the source.

Schremmer's work, therefore, shows that scent plays an important role in the finding of food by experienced silver-Y moths, while in the newly hatched moth it is the only sense which leads it to food.

Hawk-moths

Important contributions on *Lepidoptera* have been made also by Knoll, who experimented with four species of hawk-moth and a butterfly. His first object of study was the hummingbird hawk-moth (*Macroglossa stellatarum*) which flies by day. This work (Knoll, 1922), like that on the bee-flies, was carried out in southern Dalmatia.

The hummingbird hawk-moth feeds in flight without the use of the legs. The food-seeking reaction can be recognised by the unrolling of the proboscis which occurs whether the stimulus is accompanied by food or not. When the insect is among flowers it flies directly, but fairly slowly, from one to another; when away from flowers it flies much faster, making wide sweeps over the ground as if searching, as *Bombylius* do.

Insects taken into captivity gave the feeding reaction to violet and yellow shades about equally. If they were fed on the yellow flowers of Common Toadflax (*Linaria vulgaris*) (p. 213) for a matter of days and were then offered the violet-purple flowers of Lesser Calamint (*Calamintha nepeta*), they ignored these, and vice versa. In nature, the flowers of both plants were freely visited by *Macroglossa*. A moth which had been fed on Common Toadflax was presented with a series of coloured and grey objects unaccompanied by food, and for two days it visited only shades of yellow and orange without distinction. On the third day it began to attend to violet and purple; it was then offered violet and purple coloured objects provided with sugar-water. The next day violet and purple (again without distinction) were preferred to yellow. Further experiments showed that red was not distinguished from black, and that the insects could be trained to feed at black or red, and also at dark blue when offered in company with light blue. Within the two main colour

groups which the insects readily distinguished, namely the blue group
(blue, indigo, violet and purple) and the yellow group (orange, yellow,
greenish yellow and yellow-green), no discrimination was observed, but
there is no indication that Knoll seriously tried to train the moths to make
distinctions within these groups. The colour discrimination for light from
a spectrum was the same as for pigmented surfaces and solutions. The
experiments included various shades of grey to ensure that the insects
were not being attracted merely to objects of a particular luminosity.
Newly hatched moths were presented with flowers of various colours and
with leaves, all covered with a sheet of glass to exclude effects of scent,
and with discs of yellow and violet on a grey background. The insects
visited the flowers and the coloured discs, but not the leaves. Their
behaviour was thus the same as that of experienced insects.

When yellow and violet objects were offered in three different intensities, the intermediate and darker shades were preferred to the lighter.
The feeding reaction was only elicited by objects similar in size to the
flowers visited by this moth in nature.

The direct flights to flowers which the insects made suggested that
scent was not being used in finding them. Observations in nature showed
no tendency on the part of the moths to fly upwind to find flowers. In
experiments, Common Toadflax flowers were placed well down in test-
tubes, and the insects attempted to visit the flowers seen through the glass;
they did not go to the mouths of the tubes where the scent was emerging. Scented flowers presented free, in test-tubes or between glass plates,
were all visited equally when offered in a 'flight-box' (30 or 50 cm.
cubed). Cutting off a moth's antennae did not affect its flower-visiting
behaviour.

The existence of the hummingbird hawk-moth's sense of smell was
shown in its egg-laying behaviour. Females in egg-laying condition were
attracted to green, yellow and orange-yellow objects, but only when
such an object was made to smell of Bedstraw (*Galium*) by the addition
of its juice did a moth give a full egg-laying reaction and lay an egg
on it.

The hummingbird hawk-moth can thus distinguish certain colours and
variations in intensity of colour. It can develop constancy to a particular
colour which in nature may be induced by the productivity of certain
kinds of flower, but Knoll states that constancy is also maintained by a
tendency of the insect to recognise its accustomed flowers by the shape of
the entry to them, which it perceives by tactile sense organs in the proboscis. In feeding, it apparently does not use its sense of smell either at
very close quarters or at short distances away from the flowers.

The flowers of Common Toadflax were much visited by *Macroglossa
stellatarum* late in the season in Dalmatia and Knoll used these yellow
Snapdragon-like flowers, which have a conspicuous orange area on the

palate, to test this moth's reaction to guide-marks.* Normally the proboscis drummed on the palate until it found the entrance, whereupon the
insect lunged forward and pushed the tip of the proboscis right down to
the tip of the spur. A variant form of the flower, which was plain yellow
without orange on the palate, was offered to the moth, and it fed from it
in a normal way. Another abnormal form had a normal orange palate,
but the mouth of the flower was wide open. With this the moths drummed
on the palate but rarely found the entrance which was some distance from
the palate. Thus the deep orange guide-mark, if present, guides the insect
but its presence is not essential to the finding of the nectar, at least for
experienced insects.

This subject was further investigated with the aid of the 'proboscis-trace
method'. In this, an object to be investigated is placed between glass
plates; the insect is allowed to feed on sugar-water from an adjacent
flower or model and, when it attempts to feed at the object behind the
glass, its proboscis leaves a trace of syrup on the glass. After the experiment
the traces are dusted with lead oxide, the surplus powder is knocked off,
and the glass is heated until the sugar is dry. The traces are then quite
hard, and the plate can be used to make a photographic contact print.
The special advantages of this method are that it shows exactly where the
proboscis touched the glass, and that it can be used in light too dim for
ordinary observation. Some of the objects placed between the glass plates
by Knoll are illustrated in Fig. 21. They were all light violet in colour
with dark violet markings. With A and B the traces were over the rings,
and with C they were over the convergence of the lines; with D the traces
were all over the ellipse, and with E they were concentrated over the
group of dots. Thus nearly all these artificial guide-marks had a directive effect, and circles were of major importance. The effectiveness of
converging lines in the absence of a ring (object C) varied with their
thickness.

The striped hawk-moth (*Celerio livornica*) was also investigated by Knoll
(1925). This moth, a rare visitor to Britain, has a proboscis 26 mm. long –
a similar length to that of the hummingbird hawk-moth. It hovers when
feeding, but the fore-legs hold on to the flower if it is within reach. In a
'flight box', 50 cm. cubed, in which experiments were carried out, the
moths rested by day and flew in the evening and morning twilight. They
continued to feed in light so dim that flowers of Sage (*Salvia officinalis*)
could hardly be seen by the experimenter. If the darkness became sufficiently deep the insects would fly against the window of the box; when
the light was artificially increased they returned to their feeding.

Wild-caught moths in the flight box were offered flowers of *Lonicera*
implexa which were artificially filled up with sugar-water; another flower of

* This term is here used to replace the traditional 'nectar guide', which is in-
appropriate when such markings occur in nectarless flowers.

this Honeysuckle (which is similar to our native *L. periclymenum*, p. 215 and Fig. 22A) was put in the flight box between glass plates. After repeated visits had been made to this flower, treatment of the proboscis traces showed that all the marks were on or near the broad, roughly rectangular upper lip. White discs 10 mm. in diameter, placed between the glass plates and accompanied by free *L. implexa* flowers, also received traces.

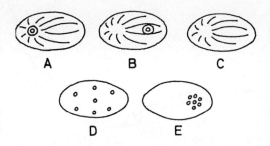

FIG. 21. Designs used by Knoll (1922) to test the effect of guide-marks on the hummingbird hawk-moth.

A striped hawk-moth was presented with a series of sixteen coloured discs (between glass plates) representing the entire spectrum and accompanied by free, blue-violet flowers of Sage (*Salvia officinalis*); proboscis traces showed that visits were chiefly to blue and, to a lesser extent, to blue-green, violet and purple. When the moth was similarly given Sage flowers and discs of two shades of blue, two of yellow, one of violet, and several of grey, it visited blue and violet very much more than yellow, and grey not at all. It was evidently able to distinguish colours of the blue group from other colours and from grey.

A moth was trained for three evenings to feed at yellow artificial flowers. Then a heavily scented white flower of Tobacco Plant (*Nicotiana*) filled with sugar-water was also presented, but it was four further evenings before the moth went to this flower directly. Similar experiments with another individual confirmed the possibility of training to a constancy which was only broken with difficulty, and again showed the apparent disregard of scent.

In fact Knoll was never able to observe any reaction by this moth to scent in his experiments. Moths hatched in captivity were offered the scented flowers of *Antirrhinum* and *Delphinium* and green parts of plants, all placed well down in test-tubes open at the top. The moths ignored the green objects but flew against the other tubes at the level of the flowers and not to the open tops of these tubes, showing that they were reacting to the visual stimulus. Nevertheless, this moth comes to collectors' 'sugar' to which it must be attracted by scent.

The type of tobacco flower used by Knoll has a nearly flat five-lobed 'limb' at the end of a long tube (Fig. 22B). He prepared two imitation 'limbs' from white paper; from one he cut out a central hole represent-

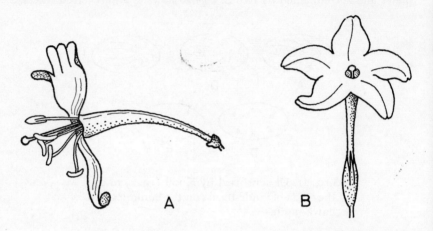

FIG. 22. A, flower of Honeysuckle (*Lonicera periclymenum*). B, *Nicotiana* flower of the type used in Knoll's experiments. B, after Knoll (1925).

ing the entrance to the tube, and from the other a hole was cut out off-centre. The holes were backed by grey paper. The proboscis-trace method showed traces distributed evenly all over the surfaces of both the paper stars. The moth, therefore, apparently does not find the entrance to these flowers visually, and in fact it frequently fails to find the entrance at all, being more attracted to the bright parts of the flower. It seems as though funnel-shaped flowers are better adapted to pollination by this insect, but the flowers of *Lonicera implexa* are probably also well adapted to it, as the proboscis strikes the upper lip and is easily deflected down into the tube. In nature the striped hawk-moth has been recorded at flowers of all these shapes.

Vogel (1954) has pointed out that moth-pollinated flowers tend to have narrower perianth lobes than related plants pollinated by day-*Lepidoptera*. The resulting floral shape should help the moths to find the entry, as successive probings of the bright perianth will tend to be aligned in a radial direction, and not in all directions as they might be on a more disc-like flower. Flowers of this star-like shape can combine a large diameter with a restriction of the probed area, and do not suffer the loss of bright-ness caused by the shadow inside a funnel.

Knoll briefly mentions the elephant hawk-moth (*Deilephila elpenor*) saying that it is the commonest of its genus at collectors' 'sugar' (Knoll, 1925) but that, when he subjected it to the proboscis-trace method with the fragrant flowers of Lilac (*Syringa*), he obtained no sign of attraction to the scent issuing from the sides of the glass plates (Knoll, 1927).

Knoll (1927) investigated also the convolvulus hawk-moth (*Herse convolvuli*) which flies and feeds in even dimmer light than the striped hawk-moth. Bred specimens do not fly much in captivity, so insects caught at white tobacco flowers were used, either in a 50 cm. cubed flight box or in a specially made flight tent. A pocket lamp provided the illumination which could be reduced gradually to zero. Even in extremely dim light proboscis traces were obtained over white tobacco flowers when accompanied by free white tobacco flowers, and over deep purple tobacco flowers and a violet paper star when accompanied by free deep purple tobacco flowers. Thus optical stimuli were entirely sufficient for finding the entrance to a flower and light colours were not necessary.

Experiments with coloured and grey discs on a grey background showed that there was a well-developed colour sense, similar to that of the hummingbird hawk-moth, and effective in very dim light. Training with blue-violet objects resulted in constancy to this colour, and by further training this constancy was replaced by constancy to white. A preference was shown for coloured circles 35 mm. in diameter over circles 14 mm. or less in diameter.

The tobacco flowers used for experiments on the convolvulus hawk-moth were strongly scented but there was no indication of attraction to scent, and all Knoll's efforts to demonstrate a reaction to scent in feeding were unsuccessful.

Knoll was surprised that the striped and convolvulus hawk-moths showed no reaction to scent, finding their way to the flowers in twilight entirely by sight, and that they had a colour vision which enabled them to find flowers by colour in very weak light, rather than merely by brightness; for most of the flowers believed to be adapted to the visits of such moths are very pale in colour and are heavily scented. Biologists thought formerly that a strong scent and pale colouring were obvious means of facilitating the discovery of the flowers in poor visibility, and in spite of Knoll's results it is still possible to regard the light-coloured flowers as adapted to visits by these hawk-moths, for it might be that in nature they would initially discover light-coloured flowers more easily than dark ones and develop a constancy to them.

Since many moth-pollinated flowers produce their scent mainly at night it is hardly possible to doubt that some moths respond to it.

Butterflies

The butterfly *Charaxes jasius* appears to depend entirely on scent in finding food, both at long range and at very short range (Knoll, 1922). In Dalmatia *Charaxes* ignored the flowers visited by other butterflies and fed on tree sap (from wounds and beetle holes), ripe figs and tomatoes, and rotting grapes. It was strongly attracted to wine and could be captured with the fingers while imbibing. Knoll put out some overripe and spotty grapes on some rocks that were similar in colour to the grapes, and watched the behaviour of the insects. *Charaxes* always arrived upwind, or if the wind was changeable or it was calm by zigzag flight. The butterflies alighted in the neighbourhood of the fruit, which was then invisible to them in the hollows in the rocks, and they walked or hopped around occasionally lowering the antennae. When a butterfly passed within a few centimetres of a grape, unrolling movements of the proboscis took place. The fruit was sometimes not found, even when only one centimetre from the insect. When a large board was put out with figs on it, the behaviour of the insects was exactly the same even though the fruit was now in full view.

Captive *Charaxes* were observed in a flight tent. Drops of plum juice in the immediate proximity of the insects induced feeding, but pure sugar-water (which is odourless) had no effect. Waxed paper 'flowers' coloured violet, yellow, grey or black and containing sugar-water produced no reaction. As soon as plum juice was added, the proboscis reaction was elicited, and the insects appeared to find the food with their feelers. The removal of both antennae suppressed the proboscis reaction, but the removal of one antenna did not. After the removal of the antennae, the proboscis reaction could still be obtained when the plum juice was made to touch the proboscis, and this led to feeding. The same result was obtained with sugar-water, so evidently there are taste organs in the proboscis.

Lederer (1951) investigated some other butterflies which feed exclusively or characteristically on foods other than flower-foods. He found that tree sap and excrement were always found by smell by the white admiral (*Limenitis camilla*), the purple emperor (*Apatura iris*) and allied species. The scent of the food had an activating as well as a directive effect. The newly hatched white admiral was also directed by scent in its first visits to flowers, but in later visits it was often guided visually; its behaviour was therefore like that of the silver-Y moth. The peacock butterfly (*Nymphalis io*) found tree sap by scent and flowers by sight, though flower scent had an activating effect. Lederer found that the distance at which the butterflies first reacted to scent ranged from 20 cm. to 30 m. (or, with a favourable wind, to 60 m.).

Butterflies have been extensively investigated by Ilse. Several species were kept in a greenhouse and were offered artificial flower-models containing sugar-water (Ilse, 1928). Newly emerged insects were first tested for inherited colour-reaction, a few coloured models being presented among a series of 'greys' ranging from black to white. Six species showed a statistically significant positive colour-reaction. Colour preference was tested by offering 2 or 4 colours with 16 or 18 grey models, or 16 colours, covering the whole spectrum, with 24 grey flowers. The results for each species were presented as a graph showing the number of visits to each colour, the colours being arranged in spectral order. The butterflies fell into two groups: the small tortoiseshell, the large tortoiseshell and the peacock (*Aglais urticae*, *Nymphalis polychloros* and *N. io*) showed strong peaks at yellow and blue, with the yellow peak higher than the blue for the large tortoiseshell and the peacock; three other species, the large white, the brimstone and the swallowtail (*Pieris brassicae*, *Gonepteryx rhamni* and *Papilio machaon*), formed the second group, and showed a strong preference for blue, violet and purple. The brimstone and the swallowtail made a few visits to red, orange and yellow. Green and blue-green were ignored in feeding by all these six species.

Substantial numbers of visits to colours other than the spontaneously preferred ones could be induced by means of training in the first group only. Even then the spontaneously preferred colours were never given up and received many visits; sometimes the result of training was not an increase of visits to the training colour but to the tint of the preferred colour most closely approaching the training colour. The Camberwell beauty butterfly (*Nymphalis antiopa*), which had no spontaneous preference for any colour compared with grey, was the most easily trained species of all.

It was found that in some species the full feeding reaction was elicited by coloured objects covered with glass. The resulting approach to food was called the *Pieris*-type of approach: the insect flew to the coloured object directly, with clearly aimed flight, and alighted on it or occasionally near it. The *Pieris*-type of approach characterised the typical flower feeders, and was also more or less developed in some species that divide their attention between flowers and other food sources (see Table 2).

In what was called the *Charaxes*-type of approach, the insects flew in a zigzag or swinging fashion, alighted some distance from the source of scent and then advanced by hopping or crawling, testing the ground with the antennae and often touching it with the proboscis; this was found fully developed in five species, as shown in Table 2. Other species showed features of the *Charaxes*-type approach only when in the closest proximity to the scent-source, or not at all.

When scent was sprayed into the greenhouse the non-feeding insects became restless, flapping their wings, drumming with their antennae and making unrolling movements with their proboscides. Some species then flew in a zigzag path until they came in sight of coloured 'flowers', when they flew directly down to them.

The experimental results are shown in Table 2, where they are compared with the insects' feeding habits. Some of the species included feed both at flowers and at other food sources. Those that feed only at flowers have a definite colour preference and make very little use of scent in feeding; the one species that never feeds at flowers is not attracted by colours, cannot be trained to them, and does not use vision in recognising and finding food.

Results of further investigations with the large white butterfly are also given by Ilse (1941). The colour-reactions of this butterfly were studied in all phases of its activity, and it was found that it could distinguish at least three groups of colours, as follows: (a) red to yellow, (b) green to blue-green, (c) blue to violet. According to Ilse, these results apply to the families *Pieridae* and *Papilionidae* in general.

Table 2. Food-seeking responses of butterflies

Butterfly	Natural colour preference	Trainable to other colours	Approach to food Visual	Approach to food Olfactory	Food
Small Tortoiseshell *Aglais urticae*	Yellow and blue	Yes	Yes	No	Flowers
Large Tortoiseshell *Nymphalis polychloros*	Yellow and blue	Yes	More or less	Yes	Flowers and sap
Peacock *N. io*	Yellow and blue	Yes	More or less	Yes	Flowers chiefly
Large White *Pieris brassicae*	Blue	No	Yes	Scarcely	Flowers
Brimstone *Gonepteryx rhamni*	Blue	No	Yes	Scarcely	Flowers
Swallowtail *Papilio machaon*	Blue	No	Yes	Scarcely	Flowers
Camberwell Beauty *N. antiopa*	None	Yes	When trained	Yes	Flowers and sap
Purple Emperor *Apatura iris*	None	No	No	Yes	Dung
Red Admiral *Vanessa atalanta*	?	?	Yes	Yes	Flowers, dung and sap

The effects of the size and the form of coloured objects which elicit feeding reactions in butterflies were earlier investigated by Ilse (1932). Using the silver-washed fritillary (*Argynnis paphia*), the peacock and the small tortoiseshell, she found that the larger the coloured surface the more attractive it was – a contrast to the results with the hummingbird hawk-moth; the largest surface tried was 30 × 50 cm. and the colours used were blue and yellow, each presented on a black background.

Using the peacock and the small tortoiseshell the effects of form were studied with blue objects on a black or white ground; a ring of external diameter 5 cm. was greatly preferred to a disc of diameter 3.5 cm. but with the same area of colour; similarly, a chequerboard 8 cm. square, divided into 8 blue and 8 black or white squares, was preferred to a blue area 5.6 cm. square (again the same area of colour). The preferred figures had a greater extension and a longer outline than the less favoured figures. When the figures to be compared were made equal in extension, the preference, though somewhat reduced, was still for the figures with a longer outline, although these now had a smaller area of colour.

These results did not hold when yellow objects (on a black background) were used instead of blue. To a yellow ring and disc of equal coloured area (dimensions as before) visits were about equal; when the ring and disc were equal in extension, the disc received two-thirds of the visits (compared with one-third when the colour was blue). When offered a choice of yellow and blue, the insects used usually showed a strong preference for yellow; which perhaps explains why, with yellow, area is more important than the length of outline. However, chequerboards showed that the longer outline was preferred if the difference in the length of outline of two figures of identical yellow area was made sufficiently great. Nevertheless, a limit to the increase in attraction caused by subdivision into chequerboards was reached when the small squares were reduced to 8 mm. square. From further experiments Ilse concluded that no particular form was preferred.

Bennett's field observations (1883) relate to some of the species used in Ilse's experiments. Species of *Pieris* visited many purple, violet and white flowers but no yellow ones, which accords with Ilse's results. The small tortoiseshell on the one occasion on which it was observed visited only yellow flowers of a single species, while the allied painted lady (*Vanessa cardui*) invariably visited only violet and purple flowers. Members of two other genera of butterflies were also observed; the meadow brown (*Maniola jurtina*) visited white and yellow flowers, and the common blue (*Polyommatus icarus*) went to yellow, pink and purple flowers.

Bennett many times observed single visits, or three or four successive visits, to one species of flower, but instances of marked constancy were also seen. For example, a meadow brown made seven visits to Bramble (*Rubus fruticosus*), a small tortoiseshell made many visits to Ragwort (*Senecio jacobaea*), a small white (*Pieris rapae*) made 11 visits to Hardhead (*Centaurea nigra*), another made 8 visits to Red Bartsia (*Odontites verna*), another 23 visits to Marsh Woundwort (*Stachys palustris*), while numerous painted ladies spent their time visiting both Hardhead and Greater Knapweed (*Centaurea nigra* and *C. scabiosa*), often flying a considerable distance between visits.

In various experiments described in this chapter insects were tested with flowers covered with glass vessels or plates, so that the effective scent source was separated from the flower, and in all cases the insects flew direct to the flowers. It is important to make it clear that these experiments merely show that once the insect has brought the flower within its range of vision it uses vision for its final approach. When one considers that the insects concerned are all strong fliers, highly dependent on vision for guidance, and that any moderately developed sense of sight must always be able to provide more precision than scent, the results of this kind of experiment are hardly surprising: vision would be *expected* to take precedence over any perceived scent. This precedence of vision is well shown by the case of the silver-Y moth (*Plusia gamma*) following a scent gradient and being temporarily diverted by a bright object on the gradient but not at the scent source. To demonstrate sense of smell the flowers must be out of sight, as in some other experiments with this moth. However, the seemingly improbable alternative of failing to employ the sense of sight is seen in *Charaxes*. Restriction of an animal's responses to only one of a number of available signals is a common feature of instinctive behaviour, often leading to absurd results.

Apparent anomalies in the behaviour of the nocturnal hawk-moths in Knoll's experiments (Knoll, 1925, 1927) can perhaps be explained with the aid of Schremmer's (1941a) observations on the silver-Y moth. Knoll found that insects allowed to drink, with little or no prior training, from scented flowers made proboscis traces over adjoining flowers sandwiched between glass plates. Thus they were not deflected to the edges of the glass plates where the scent could escape (which, as we have just seen, is hardly to be expected), nor were they put off by an absence of scent at the place where the flower was. This is like the behaviour of the newly hatched silver-Y moth which was presented with Phlox flowers and unscented white models and, being untrained, visited first the flowers and then the models. On the other hand, the hawk-moth which refused to visit a white flower despite its strong scent had been trained to visit scentless yellow models for the previous three days. Its behaviour was perhaps, therefore, analogous to that of another silver-Y moth which was trained to Phlox and then refused to visit unscented models. The hawk-moth's apparent disregard of scent may in fact have been an all too tenacious regard for it!

All that Knoll deliberately tested in the way of response to scent was the possibility of guidance at short range and this was ruled out by the unexpectedly good vision in dim light which his experiments revealed.

PLATE 15. Butterflies and moths *(Lepidoptera)*: **a**, comma butterfly *(Polygonia c-album)* feeding on nectar from Michaelmas Daisy; note the 'knee bend' in the butterfly's proboscis. **b**, noctuid moth visiting Nottingham Catchfly *(Silene nutans)* at dusk.

PLATE 16. Bees and allied insects *(Hymenoptera)* : **a**, sawfly on Hogweed *(Heracleum sphondylium)*. **b**, mason wasp *(Odynerus* [sensu lato] sp.*)* on Hogweed. **c**, solitary bee *(Andrena* sp.) on Dandelion *(Taraxacum officinale* agg.*)*. **d**, solitary bee *(Nomada* sp.*)* on Dandelion. **e**, solitary bee *(Prosopis* sp.*)* on Wallflower *(Cheiranthus)*. **f**, solitary bee *(Halictus* sp.) on Common Daisy *(Bellis perennis)*. See also Pls. 8, 24, 32 and 41.

Apart from this there is evidently nothing surprising about Knoll's results. He could not test long-range guidance because his flight boxes were too small, and he did not arrange any experiments on scent as a recognition mark for trained insects.

CONCLUSIONS

From among the many significant observations recorded in this chapter we may now pick out, by way of summary, those which best lend themselves to a comparison of one insect with another. We have seen that scent is the means of guidance from a distance of beetles approaching the trap flowers of *Arum*, and that it is the sole means of finding food in some non-flower-visiting butterflies. In the lives of other butterflies and some *Diptera-Calypterae* it can have the effect of activating the insects in a search for food, and it can be used by hoverflies (*Eristalis*), by the green-bottle fly (*Lucilia*) and by the silver-Y moth as a recognition sign. For some butterflies it can play a part in guidance also, but the hummingbird hawk-moth (*Macroglossa stellatarum*), other butterflies and the bee-fly (*Bombylius fuliginosus*) appear to make no use of scent in guidance to flowers.

Researches have also shown that though colour vision may be absent in members of some orders of insects less highly adapted to visiting flowers, it is widespread in the *Diptera* and *Lepidoptera*. Red-blindness is proved to exist in the hummingbird hawk-moth, the blow-fly (*Calliphora*) and in the drone-fly (*Eristalis tenax*), and it probably occurs in the bee-fly as well. It has been proved that guide-marks have some significance for *Macroglossa* and for the striped hawk-moth (*Celerio livornica*), but in an unmarked flower the striped hawk finds the entrance by more or less random probing. Trainability — which implies ability to develop constancy — to both colour and scent occurs in the drone-fly and the green-bottle fly (*Lucilia*) as well as in many moths and in some but not all butterflies. Some butterflies will visit yellow and blue but prefer yellow; the flies *Eristalis* and *Lucilia*, with their strong preference for yellow over blue, are similar. Other butterflies have a strict preference for blue and violet; they are untrainable and do not use scent in finding food. As *Bombylius fuliginosus* is likewise known to refuse to visit yellow in nature and does not use scent for guidance it seems similar to these butterflies and it would be interesting to know whether it lacks trainability. Butterflies of the genera *Aglais*, *Nymphalis* and *Argynnis* prefer a greater extension of objects and length of outline to a lesser, whereas *Eristalis* has no such innate preference. When the food object shows the less favoured colour, which Kugler characterises as having a low stimulatory value, it is found that outline becomes much more important for these butterflies, and scent likewise for *Eristalis* (see pp. 111, 114 and 129).

CHAPTER 5

THE INSECT VISITORS – II – HYMENOPTERA

THE *Hymenoptera* form a large and very diverse order of extreme biological interest, with about 6,200 species in Britain. The *Hymenoptera* are character-ised by having four wings (the fore- and hind-wings being held together by a system of hooks) and by the possession in the female sex of an ovi-positor or a sting. The order is divided into two sub-orders, the *Apocrita*, comprising gall-wasps, ichneumon-wasps, true wasps, bees and ants, and the *Symphyta*, a smaller and more primitive group.

SAWFLIES AND STEM SAWFLIES

The *Symphyta*, or sawflies, are distinguished from the other *Hymenoptera* by their lack of a narrow 'waist' (Plate 16a). They are somewhat wasp-like in appearance but have proportionately larger wings, usually soft clumsy bodies, and comparatively slow movements; the flower-visiting species range in length from about 5 to 22 mm. The larvae of most sawflies are like caterpillars in appearance and live externally on plants. The mouth-parts of the adults are unspecialised and bear a distinct resemblance to those of beetles. The jaws are well developed, while the maxillae are more or less joined to the sides of the labium to form what is known as the labio-maxillary complex (Fig. 23a and b). This is a tongue-like structure with lobes at the tip formed by the galeae and laciniae of the maxillae, and the glossa and paraglossae of the labium; the usual two pairs of palps are also present. Solid food is finely chewed by the mandibles, and liquids are licked up by the tongue. There are three main types of food: (1) insects, (2) moisture (rain, dew, cuckoo-spit, honey-dew and damaged ripe fruit) and (3) flowers and leaves, frequently including nectar, pollen, stamens and petals, the floral foods being taken chiefly by females (Benson, 1950). The mouth-parts of sawflies are very short – usually 2 mm. long or less – so that these insects can obtain nectar and pollen only from flowers in which these foods are well exposed. Sawflies are quite commonly seen at such flowers and must certainly be effective as pollinators, but their benefit to the plants is sometimes offset by the injuries which they do to the flowers.

According to Benson (1950) the flowers most favoured by sawflies are *Umbelliferae*, *Rosaceae* with large inflorescences, yellow *Compositae* and

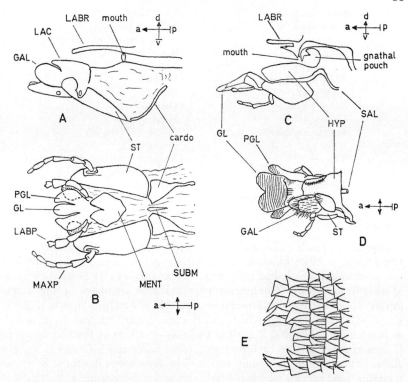

FIG. 23. A, labio-maxillary complex of a sawfly (*Cimbex* sp.), palps removed. B, the same from beneath. C, social wasp (*Vespula* sp.): section through mouth-parts, simplified. D, *Vespula*: labio-maxillary complex from above, right maxilla removed. E, *Vespula*: some of the flattened hairs of the glossa. (Abbreviations as in Fig. 10 & 11). A, B, after Bischoff (1927); C, D, E, after Duncan (1939).

Ranunculus. The records of Knuth and of Willis and Burkill well support this statement, but these authors did not find many sawflies visiting flowers. The larvae of sawflies are often closely restricted to certain food-plants, and there is a marked tendency of the adults to visit mainly the flowers of the larval food-plant. Examples of such flowers are Figwort (*Scrophularia*), Snowberry (*Symphoricarpos rivularis*), Scabious (*Scabiosa*), *Rubus*, Sallows (*Salix*). Such attachments are particularly strong with species feeding on *Salix*, some of them confining themselves entirely to the flowers of the species on which the eggs are laid, even when other *Salix* species are available (Benson, 1950 and 1959). Some other flowers visited by sawflies are:

some *Cruciferae*

Reseda lutea	Wild Mignonette
Geranium dissectum	Cut-leaved Cranesbill
Geranium sanguineum	Bloody Cranesbill
Acer campestre	Maple
Acer pseudoplatanus	Sycamore
Saxifraga hypnoides	Mossy Saxifrage
Sedum telephium	Orpine
Galium	Bedstraw
Valeriana officinalis	Common Valerian
Ajuga spp.	Bugle
Verbascum	Mullein
Polygonum viviparum	Alpine Bistort
Euphorbia cyparissias	Cypress Spurge
Cephalanthera longifolia	Narrow Helleborine

(Records given by Jones – 1945, Knuth, Kugler, Poulton – 1932, and Willis and Burkill)

One of the more striking flower-visiting sawflies is *Abia sericea* (family *Cimbicidae*), a large blackish insect with a green metallic sheen and a wrinkled abdomen; it has been recorded as frequent at Small Scabious (*Scabiosa columbaria*) (Chambers, 1947). Many members of the large family *Tenthredinidae* visit flowers; an example is *Tenthredo arcuata*, a black insect with yellow bands, which was the sawfly most commonly recorded at flowers by Willis and Burkill in their observations in the Cairngorms; it is a regular visitor to Buttercups (*Ranunculus* spp.) (Harper, 1957). Another member of the family which visits Buttercups is *Athalia bicolor*, a medium-sized species which is black except for the orange abdomen – a very common colour-pattern among sawflies. Another of the *Tenthredinidae*, *Dolerus picipes*, was once found visiting the Narrow Helleborine Orchid in England. When captured the insect was placed in a box with the Common Twayblade Orchid (*Listera ovata*); it proceeded to visit the flowers and removed from them four pairs of pollinia on its mouth-parts and one pair on its leg (Poulton, 1932). A somewhat unusual family of sawflies (*Cephidae*) are known as stem-sawflies because the larvae live in the stems of plants, including those of cereal grasses; they are small or medium-sized sawflies with long and very slender bodies, and in colour they are black with narrow yellow bands. The adults visit flowers, especially *Ranunculus* spp.

HYMENOPTERA-PARASITICA

The great majority of the *Hymenoptera* belong to the sub-order *Apocrita* and may be divided into those with stings (*Hymenoptera-Aculeata*) and those without (sometimes referred to as *Hymenoptera-Parasitica*, since most of them are parasitic). The stingless *Hymenoptera*, all so-called wasps of various kinds, are very numerous in species but are not of great importance in pollination. There are about ten families of them which contain flower-visiting species; rather few species have been recorded at flowers, but these are often numerous in individuals, and will be dealt with in three groups.

The first group contains the ichneumon-wasps (family *Ichneumonidae*); these are parasitic in the larval stage, usually on the larvae of other insects. Some of the adults are very small, but a good proportion of them are of moderate or large size (with a body length of 1–2 cm.). They are particularly slender insects with long antennae and long legs, and they are usually very active (Plate 8). They eat sap, honey-dew, nectar and pollen; flowers provide them with both nectar and pollen, but they also obtain nectar from extra-floral nectaries (p. 146) and eat pollen that has been trapped by dew-drops (Leius, 1960). The mouth-parts, even of quite large species, are usually under 1 mm. long; they are rather similar to those of sawflies but the glossa is larger than the paraglossae.

Some of the flowers characteristically visited by *Ichneumonidae* are given in the following list:

Saxifraga aizoides	Yellow Mountain Saxifrage	
Saxifraga hypnoides	Mossy Saxifrage	
Parnassia palustris	Grass of Parnassus	
Adoxa moschatellina	Moschatel	
Sedum acre	Wall-pepper	Nectar well exposed
Acer pseudoplatanus	Sycamore	
Umbelliferae		
Hedera helix	Ivy	
Listera ovata	Common Twayblade	
Ranunculus spp.	Buttercups	
Reseda lutea	Wild Mignonette	
Cerastium holosteoides	Common Mouse-ear	
Stellaria media	Common Chickweed	Nectar partly concealed
Potentilla erecta	Tormentil	
Potentilla palustris	Marsh Cinquefoil	
Rubus chamaemorus	Cloudberry	
Rubus idaeus	Raspberry	
Cochlearia officinalis	Common Scurvy-grass	
Valeriana officinalis	Common Valerian	
Compositae		Small, tubular flowers
Lycopus europaeus	Gypsywort	
Thymus drucei	Wild Thyme	
Mentha aquatica	Water Mint	
Salix herbacea	Least Willow	Catkins
Lamium maculatum	Spotted Dead-nettle	Large, tubular flowers

(Records given by Harper – 1957, Harper and Wood – 1957, Knuth, and Willis and Burkill)

Comparisons with the lists in Chapter 4 part 1 show that the *Ichneumon-idae* visit much the same plants as the *Diptera-Brachycera* (*Bombylius* excluded). The laboratory experiments carried out in Canada with three species of ichneumon-wasp showed that one of them would visit only *Umbelliferae*, among plants of several families offered, while the other two visited various flowers but made most visits to *Umbelliferae*. Such investigations are carried out mainly with a view to finding out the food requirements of *Ichneumonidae* used for biological control of insect pests, since failure of biological control projects might possibly result from a lack of suitable food for the adult wasps after their release. In general, any destruction of wild flowers tends to reduce the natural populations of these useful parasites of crop pests (Leius, 1960; van Emden, 1963).

The Twayblade Orchid (see p. 235 and Plate 42) is often given as an example of a rare class of flowers called ichneumon-wasp flowers, and Knuth gives records of six genera of these insects visiting this flower; it is, however, also visited to some extent by small *Diptera* and *Coleoptera* and other short-tongued insects. The *Orchidaceae* are the most advanced family of plants in floral structure, and it is interesting to find that some of them, such as the Twayblade, are at the same time adapted to pollination by such unspecialised insects as those just mentioned. Evidently they can be valuable pollinators to suitably adapted flowers, and Willis and Burkill's designation of the *Hymenoptera-Parasitica* as merely injurious is erroneous.

Among the more distinctive of the ichneumon-wasps is the genus *Pimpla*, the species of which range from about 6 to 20 mm. long, and are black with brown legs and a transversely wrinkled abdomen. The species of the genera *Ichneumon* and *Amblyteles* are also conspicuous, being over 10 mm. in length and black in colour, with bands and spots of cream, yellow and light red in various patterns. All these insects are common at *Umbelliferae*, especially Hogweed (*Heracleum sphondylium*) and Parsnip (*Pastinaca sativa*), while *Amblyteles uniguttatus* is one of the ichneumons recorded at the Twayblade.

An Australian species of ichneumon, *Lissopimpla semipunctata*, effects the pollination of four species of the orchid genus *Cryptostylis* by the process of pseudocopulation (p. 244). In this process, which is also known in other *Hymenoptera*, the males act as if mating with the flowers, which in form and scent resemble the female insect. The precise positioning and the movements involved in mating make it a suitable process for exploitation by Orchids, since their method of pollination (see p. 232), involving the transfer of pollen masses to the stigma, also requires accurate positioning of the insect. The males of *Lissopimpla* emerge from their pupae before the females and are powerfully attracted by a scent from the Orchids which is imperceptible to human beings. The insect alights on the Orchid with

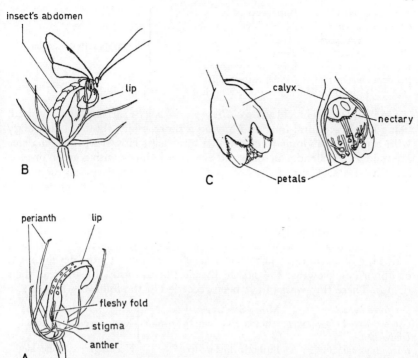

insect's abdomen

lip

B

calyx

nectary

C

petals

perianth   lip

fleshy fold

stigma

anther

A

FIG. 24. A, flower of the orchid *Cryptostylis*. The lip is brown with gleaming spots. B, ichneumon wasp *Lissopimpla* visiting flower of *Cryptostylis*. C, flower of *Cotoneaster integerrima*, side view and section; this is a dull brownish red 'wasp flower'. A, B, after Coleman (1928b); C, after Ross-Craig (*Drawings Brit. Pl.* **9**, 1956) and Knuth.

its tail towards the anther and stigma, and carries the pollinia on the tip of its abdomen (Fig. 24A and B). These observations are due to Mrs Edith Coleman (1927, 1928a, b, 1929a, b, 1930a, b, 1931, 1938).

The second group of flower-visiting *Hymenoptera-Parasitica* comprises several families consisting of small or very small insects, mainly 1–4 mm. in length. Three of these families, called chalcid-wasps, have been recorded at the following flowers:

Adoxa moschatellina	Moschatel	
Conopodium majus	Pignut	
Listera cordata	Lesser Twayblade	Nectar well exposed
Hammarbya paludosa	Bog Orchid	
Herminium monorchis	Musk Orchid	

Ranunculus acris	Buttercup	
Cerastium alpinum	Alpine Mouse-ear	
Geranium sylvaticum	Wood Cranesbill	Nectar partly concealed
Potentilla erecta	Tormentil	
Veronica chamaedrys	Birdseye Speedwell	
Veronica scutellata	Marsh Speedwell	

(Records given by Harper – 1957, Knuth, and Willis and Burkill)

One of these chalcid-wasps, *Brachymeria minuta*, is easily recognised, being about 4.5 mm. long and having a short, stout, shining black body with a pointed abdomen and swollen hind legs. However, most chalcids are moderately slender in form and are usually black with a green or blue metallic sheen. The members of another chalcid family live in the tissues of figs which they cross-pollinate in the course of their egg-laying activities (see Chapter 10).

Two other families of this group (*Cynipidae* and *Lamprotatidae*) are known as gall-wasps because, though some are parasitic on insects, most of them form galls on plants. The eggs of those that form galls are laid inside the plants and the presence of the larvae stimulates the development of abnormal growths; the larvae live inside the growths and eat their tissues. These tiny wasps have been recorded at the following flowers:

Saxifraga hypnoides	Mossy Saxifrage	
Chrysosplenium alternifolium	Alternate Golden Saxifrage	Nectar well exposed
Umbelliferae		
Ranunculus bulbosus	Bulbous Buttercup	Nectar partly concealed
Senecio jacobaea	Ragwort	Small, tubular flowers
Hieracium vulgatum	Hawkweed	
Campanula rotundifolia	Harebell	Large, tubular flowers

(Records given by Harper – 1957, Knuth, and Willis and Burkill)

Our third group of *Hymenoptera-Parasitica* is closely related to the true wasps. The only British family of this group to be noted is the *Chrysididae*. These wasps have brilliant metallic colouring: the prevailing pattern is blue or green on the head and thorax and red on the abdomen, giving rise to their popular name of rubytail-wasps. They range in size from about 3 to 10 mm. long, are rather scarce, and in habits they are parasitic on the larvae of solitary bees and wasps. In the British species the mouth-parts are short, and these wasps have been recorded in Britain almost exclusively at the flowers of *Umbelliferae*. However, the members of the European chrysid genus *Panorpes*, which is not found in Britain, are specialised flower-visitors with a proboscis 6 to 7 mm. long. In this genus the glossa is elongated and rolled at the sides to form a tube, while the galeae of the maxillae are also elongated and cover the glossa.

TRUE WASPS

The *Hymenoptera* which have stings are classed as *Hymenoptera-Aculeata*; they mostly build nests in which food for the larvae is stored by the adults, and they comprise the true wasps, the ants and the bees. In many families, however, there are species or whole genera which have the habits of the cuckoo, the adults laying eggs in the nests of other *Hymenoptera*, and the resulting larvae feeding on the food supply of the rightful owner; such insects are called inquilines. Apart from the inquilines, however, there is a group of wasp families regarded as the most primitive of the aculeates, which are entirely parasitic on other larvae. Members of four British families of this group (*Sapygidae, Methocidae, Mutillidae* and *Myrmosidae*), each with only a few species, have been recorded at the following flowers:

Parnassia palustris	Grass of Parnassus	⎫
Saxifraga sp.	Saxifrage	⎬ Nectar well exposed
Umbelliferae		⎭
Cakile maritima	Sea Rocket	Nectar partly concealed
Jasione montana	Sheepsbit	⎫
Chrysanthemum leucanthemum	Ox-eye Daisy	⎬ Small, tubular flowers
Thymus vulgaris	Thyme	⎭

(Records given by Knuth, and P.F.Y.)

Except in the family *Sapygidae* the females are wingless and do not contribute to cross-pollination. Related to these families is the non-British family *Scoliidae*. These are large, dark-coloured, hairy wasps, the larvae of which feed as external parasites on the subterranean larvae of certain beetles, and the genus *Campsoscolia* is chiefly noted as a pollinator of the Orchid *Ophrys speculum* by pseudocopulation (see Chapter 7).

The great majority of the true wasps capture insects or spiders on which they feed their young. Usually the prey is stung and stored in cells in the nest, one egg being laid on the food supply in each cell. Sometimes these predatory wasps feed on the juices exuded by their victims, but otherwise they have the same food as the *Hymenoptera-Parasitica*, the inquilines and the parasites just described, namely sap, honey-dew and nectar.

There are three families of predatory wasps, the *Pompilidae*, the *Sphecidae* and the *Vespidae*. The *Pompilidae*, comprising about forty British species, prey on spiders and are known as spiderhunting-wasps. They have very long legs and slender bodies from about 5 to 14 mm. long; they pursue their prey largely on foot, skimming the ground in short flights from time to time. In colour they are mostly black with red on the fore part of the abdomen, but some are entirely black or black with white spots or bands. Spiderhunting-wasps have been recorded at the following flowers:

Umbelliferae		⎫
Euphorbia	Spurge	⎬ Nectar well exposed
Potentilla	Cinquefoil	Nectar partly concealed
Compositae		⎫
Calluna vulgaris	Ling	⎬ Small, tubular flowers

(Records given by Chambers – 1949, Richards and Hamm – 1939, Knuth, Spooner – 1941, and P.F.Y.)

Like most aculeate *Hymenoptera* the *Pompilidae* require warm conditions, and this perhaps accounts for their absence from Willis and Burkill's records from the Cairngorms. However, even when they are at their commonest they are probably not very important as pollinators of the flowers they visit, being usually greatly outnumbered by *Diptera* and *Hymenoptera-Parasitica*. The tongues of *Pompilidae* are extremely short, like most of those of the next family, the *Sphecidae*.

The *Sphecidae* is a large family containing most of the solitary wasps and having about 100 species in Britain. The length of the body ranges from about 3 to 25 mm., and its shape is very variable, especially that of the abdomen, whose narrowed fore end can have many different conformations. The colour may be black (or black with red on the fore part of the abdomen) but often it is black and yellow, the yellow being confined to small spots or occurring in large spots or bands on the abdomen so that the insect resembles the layman's idea of a wasp.

The following is a selection of the flowers visited by *Sphecidae*:

Umbelliferae		⎫
Listera ovata	Twayblade	⎬ Nectar well exposed
Ranunculus spp.	Buttercups	⎫
Reseda lutea	Wild Mignonette	
Geranium spp.	Cranesbills	
Rubus fruticosus	Bramble	
Potentilla erecta	Tormentil	⎬ Nectar partly concealed
Bryonia dioica	White Bryony	
Symphoricarpos rivularis	Snowberry	
Epipactis palustris	Marsh Helleborine	⎭
Solidago spp.	Golden Rod	⎫
Senecio jacobaea	Ragwort	
Cirsium arvense	Creeping Thistle	
Jasione montana	Sheepsbit	⎬ Small, tubular flowers
Calluna vulgaris	Ling	
Thymus sp.	Thyme	
Mentha arvensis	Corn Mint	⎭

(Records given by Chambers – 1949, Knuth, Kullenberg – 1956a, Spooner – 1930, and P.F.Y.)

Most visits seem to be to *Umbelliferae* (see also p. 349), and in fact the tongue length of two fairly large species was measured by Kugler and found to be only 1.5 mm.; however, the large wasp *Ammophila sabulosa* has the labio-maxillary complex 3 mm. long and the sides of the glossa rolled back to form a tube. This wasp visits quite a lot of flowers, including Bramble, Thistles and Snowberry.

The *Sphecidae* is the third family we have now met in which pseudo-copulation is known, and in this case both the pollinating insects and the flower are British. The insects belong to the genus *Gorytes* and they pollinate the Fly Orchid (*Ophrys insectifera*) by this means, as described in Chapter 7. Like *Campsoscolia*, but unlike *Lissopimpla*, *Gorytes* carry the pollinia on their heads.

Taken as a whole the *Sphecidae*, though they are found at flowers more often than the *Pompilidae*, are probably also insignificant as pollinators compared with the other insects that visit the same flowers. No doubt chiefly on account of their need for warmth they are completely absent from the records of Willis and Burkill, but possibly their ability to avoid capture also played its part. As in the *Chrysididae* there is a Continental representative of the *Sphecidae* with an elongated proboscis 7 mm. long; this is *Bembex rostrata*, which is able to obtain nectar from and pollinate the pea-like flowers of Lucerne (*Medicago sativa*).

The members of the third family of predatory wasps, the *Vespidae*, fall into two groups – the remainder of the solitary wasps, and the social wasps. The first group, with the same life-history as the *Sphecidae*, contains the genus *Eumenes*, with one species in Britain, and the genus *Odynerus*, with twenty British species. They are all yellow and black (or rarely white and black) wasps, about 8 to 12 mm. in length (Plate 16b). In the second group each colony is founded by a queen, who lays eggs and feeds the offspring as they grow. These offspring are the workers and after they become adult the queen remains in the nest and lays eggs, while the workers bring in the food and extend the nest. These social wasps are represented in Britain by the hornet (*Vespa crabro*) and by six species of the genus *Vespula* which are the insects normally recognised by the layman as wasps (Plates 32, 10a).

The species of *Vespula* have slightly longer tongues than the majority of *Sphecidae* (Fig. 23c and d, p. 133). The mandibles are not used in feeding at flowers, though they are important in other activities. The glossa and paraglossae are covered with flattened hairs among which liquids can be imbibed (Fig. 23E); this hairy part of the tongue can be brought upwards into a chamber formed by the hypopharynx, maxillae and epipharynx, and suction can then be applied to draw the imbibed liquid into the mouth. A filter is formed by the fitting together of comb-like rows of hairs on the hypopharynx and on the galeae of the maxillae. Another filter is formed by a similar row of hairs across the narrow slit-like mouth, and this prevents all but microscopic particles from entering the mouth. Material filtered out by this second comb is temporarily stored in the gnathal pouch.

The adults of *Vespula* feed both themselves and the larvae on liquid foods, often obtaining these from flowers, although they feed the larvae chiefly on insects. It used to be thought that the larvae were fed only on animal food, but the transport of large quantities of a solution of sugar and honey to the nest was observed by Verlaine (1932b), and it is now known that sugary fluids are essential to the diet of the larvae (Brian and Brian, 1952). On the other hand, pollen is not a wasp food (Duncan, 1929).

The use by wasps of their sense of smell in finding food is common knowledge. Verlaine's (1932a) experiments with scented sugar-water showed that certain scents (heliotrope, violet, jasmine, bergamotte and aniseed) attracted wasps while others (cinnamon, lily-of-the-valley, creosote and turpentine) repelled them. A constancy to a particular scent was observed when a choice of attractive scents was offered, and this persisted when the strength of the preferred scent was greatly reduced. This behaviour is similar to that of *Diptera* and *Lepidoptera* described in Chapter 4 part 2.

Verlaine (1932b) found that when workers have discovered a good food source they display some excitement on returning to the nest and that when they leave again other workers follow them and may be led to the food directly; others also appear to be alerted and, though not able to follow the direct route to the food, they begin to search for it, probably knowing its scent from encountering it in the nest. An experiment was carried out with unscented sugar-water 25 m. from the nest. A wasp that was deliberately introduced to the liquid returned once to four times every ten minutes throughout the day, but no other wasp from the nest came to this food, presumably because it was unscented and rather far from the nest. Communication between wasps relating to food sources is thus of a rudimentary nature.

An investigation of the red-blindness of *Vespula rufa* was carried out by Schremmer (1941b), who found a wasps' nest with a conveniently situated entrance – a knot-hole 2 cm. in diameter in the side of a white-painted hut. He screened the hole with a white card, so that the wasps could emerge but, on returning, could not see the black entrance hole to the nest. Discs of various colours and 2 cm. in diameter were placed on the surface of the hut near the nest hole, and the behaviour of the returning wasps was observed. The wasps mistook the red and purple discs for their black hole; yellow, green and blue discs were ignored, however, evidently being distinguished from black. These results showed that to these wasps red appears as black. Similar methods have been used for some of the *Sphecidae*, and red-blindness has been found to occur among them also (Molitor, 1937a).

The small tubular flowers visited by *Vespidae* in Britain, as well as flowers with nectar well exposed, are much the same as those in the list given for *Sphecidae* on p. 140.

Of the flowers with well-exposed nectar, the *Umbelliferae* are the most visited, especially by the social species. In addition, Ivy (*Hedera helix*) is visited abundantly (Plate 10), and both the hornet (*Vespa crabro*) and the common wasps can be seen on it in October. They spend a lot of time taking nectar, though they may also be attracted by the presence of flies for prey.

Both solitary and social species of *Vespidae* visit the small tubular flowers. The solitary wasp *Eumenes coarctata* is found on heaths in Britain, and it probably feeds from the flowers of Ling (*Calluna vulgaris*). Its tongue is strongly elongated, being 4 mm. long, and the glossa is rolled back at the edges to form a tube, as in *Ammophila* (Kugler).

Individuals of *Vespula*, when visiting Heath and Cross-leaved Heath (*Erica cinerea* and *E. tetralix*), take the nectar through borings, the entrances to the flowers being much too narrow for them to get their heads in (Willis and Burkill). Such borings may be made by the wasps themselves (Block, 1962) but they are more often made by the shorter-tongued species of bumble-bee, and of course they by-pass the pollination mechanism.

The following is a list of the other types of flower visited by *Vespidae* in Britain:

Reseda luteola	Weld	⎫
Geranium phaeum	Dusky Cranesbill	
Rubus idaeus	Raspberry	⎬ Nectar partly concealed
Rubus fruticosus	Bramble	
Prunus spinosa	Blackthorn	⎭
Salix spp.	Sallows	Catkins
Lotus corniculatus	Common Birdsfoot Trefoil	⎫ Pea-type flowers
Vicia sylvatica	Wood Vetch	⎭
Scrophularia nodosa	Common Figwort	⎫
Scrophularia aquatica	Water Figwort	
Epipactis helleborine	Common Helleborine Orchid	
Epipactis purpurata	Violet Helleborine Orchid	
Cotoneaster spp. (in gardens)	Cotoneaster	⎬ 'Wasp-flowers' and similar forms
Berberis sp.	Barberry	
Ribes uva-crispa	Gooseberry	
Vaccinium myrtillus	Bilberry	
Vaccinium vitis-idaea	Cowberry	⎭

(Records given by Brian and Brian – 1952, Chambers – 1949, Spooner – 1930, Willis and Burkill, Yarrow – 1945, and P.F.Y.)

The solitary wasps of the genus *Odynerus* were the only visitors to the pea-type flowers, and they were feeding through borings.

The partiality of *Vespula* for the flowers of the Figworts (Plate 32) and the Helleborines (p. 232 and Plate 40b and c) is often most striking, the wasps visiting them persistently and in substantial numbers. In some way these flowers seem to be specially adapted to attract wasps, and they were placed by Müller in a special class of flowers called 'wasp-flowers', in which he also included the cotoneasters.

Vespula have been recorded at all the 'wasp-flowers' and similar forms listed, but the solitary species of *Vespidae* are only known to visit the Figworts in Britain, although they have been recorded visiting Gooseberry on the Continent. The flowers of the native British Figworts and of some European and American ones (Fig. 70, p. 217) are dark brownish red in colour and have an unpleasant odour (see also p. 217) (Trelease, 1881; Shaw, 1962).

The *Cotoneaster* species visited by *Vespula* have nodding globular flowers with dull brownish red petals (Fig. 24c). The remaining species of *Cotoneaster*, which are apparently not visited by wasps, have upright flowers with spreading, pure white petals. One June I saw several queen wasps visiting five species of *Cotoneaster* flowers in a mixed collection of the genus in Cambridge Botanic Garden. There was much bustling around the flowers, interrupted by short spells of actual visits to suck nectar. None of these wasps visited Cotoneasters with spreading white petals, although three species of this kind were being visited by honey-bees. The head of a queen wasp is too big to enter the globular *Cotoneaster* flowers, and only the mouth-parts and the lower part of the head are inserted.

The Snowberry (*Symphoricarpos rivularis*) is yet another member of Müller's group of 'wasp-flowers', and visits to it by *Vespidae* could probably quite easily be observed in Britain. So far, however, the only records are from the Continent where both social and solitary species visit it.

The 'wasp-flowers' generally have pinkish or dirty red flowers, or flower parts, forming horizontal or drooping cups into which a wasp can thrust its head, or at least part of it. These flowers are dull in colouring and probably appear even darker to the wasps because of their red-blindness.

Besides being attractive to wasps, some flowers of this kind are also attractive to bees, especially honey-bees. However, it seems that the Helleborines and the Figworts are very strongly adapted to wasp-pollination, and their late date of flowering, coinciding with the appearance of numerous worker-wasps, supports this view.

The Barberry, Gooseberry, Bilberry and Cowberry (all mainly visited by bees) have never been classed as 'wasp-flowers', but their points of similarity to them suggest that the wasps are attracted by the same means. The fact that pendent flowers discourage visits by short-tongued flies and encourage visits by bees has been mentioned in Chapter 3. The wasps appear to share with bees a readiness to cling underneath a flower to get nectar.

The ants, forming a single family, *Formicidae*, are great lovers of nectar and regularly collect it from flowers. Since the worker-ants are wingless and have to reach the flowers by crawling up the stems they are very unlikely to cause cross-pollination between different plants, and their method of entry to the flowers will in many cases permit them to take the nectar without effecting pollination at all. Dr L. C. Frost (unpublished) has discovered, however, that ants are the legitimate pollinators of Rupture-wort (*Herniaria ciliolata*). This is a plant of dry open places and it has stems which sprawl about closely pressed to the ground. Clusters of minute greenish flowers (each 2 mm. wide) appear on the stems. In the mass the flowers produce a perceptible scent, like 'peppery honey'. Nectar is produced and the ants eagerly feed on it, passing quickly from flower to flower. In Cornwall the main pollinators were found to be *Lasius niger* and *Formica fusca*. In the course of their visits the ants become covered with pollen. Another plant believed by Dr Frost to be ant-pollinated is Honewort (*Trinia glauca*). This member of the family *Umbelliferae* is found in dry limestone grassland in south-west England and reaches a height of from 3 to 20 cm.; the flowers have a strong scent of honey.

It is interesting that both these plants have obligate outbreeding systems: *H. ciliolata* (in Cornwall and in Guernsey at least) is self-incompatible, while Honewort is dioecious. It is self-evident that such systems cannot come into being unless a reliable pollinator exists, and the evidence thus indicates that in some circumstances ants can serve as reliable pollinators. The coincidence of these outbreeding arrangements in two ant-pollinated plants is rather striking and suggests that ants are not always sufficiently reliable *cross*-pollinators, even for such a low-growing plant as the Rupture-wort, and that internal outcrossing mechanisms in the plant have developed because of this.

The visit of an ant to another low-growing plant of south-west English cliffs, Portland Spurge, is shown in Plate 8b.

Except for such plants as these, ants are liable to be harmful visitors. The adaptations of plants to exclude such harmful visitors have been treated in detail in a little book, *Flowers and their Unbidden Guests*, by A. Kerner, published in English in 1878. The two main ways by which plants exclude ants are the formation of impassable barriers between the ground and the flowers, and the provision of additional nectaries away from the flowers to act as decoys. Impassable barriers are found in Teasel (*Dipsacus*) in the form of pools of dew and rainwater around the stem, held by the united bases of each pair of leaves, and in Nottingham Catchfly (*Silene nutans*) in the form of sticky zones on the upper parts of the stem. Decoy nectaries occur on the stipules at the bases of the leaves of some Vetches (*Vicia* spp.) and near the base of the leaf-blades in the Cherry Laurel (*Prunus laurocerasus*) which is commonly grown in gardens.

In many tropical plants nectar is specially provided outside the flowers to attract ants. The ants, being powerfully equipped for biting and stinging, then protect the plant from various kinds of attack, including nectar-robbery by corolla-piercing. Among the tropical plants which secrete nectar on the leaves, bracts or calyces, are some which are pollinated by the large and powerful *Xylocopa* bees. Since the floral mechanism makes it difficult to obtain entry to the internal nectar, these bees are often tempted to pierce the corollas from the outside. The 'ant-guard', however, effectively deters them from doing so. The flowers seem to be provided with a chemical means of keeping the ants away from the inside of the corolla (Van der Pijl, 1954).

BEES

General Biology and Flower Preferences

The *Hymenoptera* dealt with so far are of slight significance as pollinators when compared with the one remaining group of this order, the bees. These are sometimes assigned to several families, and sometimes to only one, the *Apidae*. Unlike the wasps they are complete vegetarians, the adults (males and females) feeding on nectar (and sometimes pollen), and the larvae on both nectar (after its conversion to honey) and pollen, the larval food being collected by the female adults. Nectar consists of a solution of one or more sugars, the three sugars that may occur being sucrose, glucose and fructose (see Chapter 3). After collection, the nectar is carried in the bee's crop and converted into honey. The vast importance of the bees as pollinators arises from the fact that the larvae, unlike all other insect larvae, have large quantities of flower-food brought to them or stored up for them in the nest by the female adults. Many flowers are specially adapted to pollination by bees, and many others, less specialised, benefit from their activities.

There are about 240 species of bees in Britain; of these 20 are social and the rest are solitary. A considerable proportion of the species are in-quilines, laying their eggs in the nests of other bees; their larvae feed on the food provided for the larvae of the host species. The life-history of the solitary bees is much like that of the solitary wasps but a great many bees appear in March or early April, whereas in Britain very few solitary wasps emerge before about the end of May. Most of the species of social bees belong to the genus *Bombus*, the bumble-bees. These have exactly the same life-cycle as the social wasps, with fertilised queens founding colonies in spring, and being later helped by their worker offspring. Some species of the genus *Halictus* are social and have the same system as *Bombus*, but two or three females may combine to start a nest in spring, and the colonies are much smaller than those of *Bombus*. The single remaining social species, *Apis mellifera*, the honey-bee, is different from all other British social bees and wasps in that the colony remains in being all the year round, and new colonies are founded by queens which are accompanied by part of the worker population of the parent colony.

Probably there are more honey-bees in Britain than all the other bees put together, but this situation is highly unnatural. The distribution of honey-bees is also unnatural, being governed by the provision and movement of hives (see Chapter 11). In addition they have a longer season of activity than any other kind of bee. When left to themselves they nest chiefly in hollow trees and cavities in buildings. Bumble-bees also have a rather long season, but their numbers are disproportionately small early in the season; they nest in grass tussocks or underground (usually in ready-made holes) according to species. The solitary bees, together with the social species of *Halictus*, burrow in the ground or make nests in beetle holes in wood, in hollow stems, in rotten wood, in snail shells, in holes and crevices in masonry, or under stones. The situation chosen depends on the kind of bee concerned, and the occurrence of solitary bees is thus somewhat restricted by availability of nesting sites. The ground-nesting species may also be restricted in occurrence by their soil preferences for nesting purposes. The season of activity of many species is short but many genera contain both early and late species, while some species can be seen over almost as long a period as the honey-bee, this being achieved by having two or three broods during the year so that there are periods of absence or scarcity between broods.

In Britain, the first bees to appear in spring are honey-bees (Plate 18b), and these are followed by the first queen bumble-bees (Plates 18a and 27) and, among the solitary bees, the early andrenas (Plate 35a) and two species of *Anthophora*. Honey-bees have medium-length tongues and visit practically any available flower. Bumble-bees have long tongues but their choice of flowers in early spring is very limited, the chief natural flower for them at this time being Sallow catkins (*Salix* spp.). *Andrena* is our largest genus of bees with over sixty species, all of which nest in the ground. Most of them are very short-tongued. It is the larger species of the genus which emerge first, and these range in size from rather larger than the honey-bee to rather smaller. The thorax of these larger species, and sometimes the abdomen also, is densely hairy. Some of these bees could be mistaken for the honey-bee but others are beautifully patterned and coloured. The andrenas active in March and April visit chiefly Sallows and the early yellow *Compositae*, Coltsfoot and Dandelion (*Tussilago farfara* and *Taraxa-*

cum officinale) (Plate 16c). The early species of *Anthophora* (*A. retusa* and *A. acervorum*) are rounded furry bees rather like small bumble-bees; the two species are much alike and in each the males are mainly brown while the females are entirely black except for their rust-coloured pollen brushes. Their movements are extremely quick and they are more easily frightened by the human presence than the bumble-bees. Their tongues are very long and they visit large tubular flowers such as Primrose and Cowslip (*Primula vulgaris* and *P. veris*), Deadnettles (*Lamium* spp.), Solomon's Seal (*Polygonatum* spp.) and the Borage family (though in spring the Borages are usually only available in gardens in Britain). When they first emerge in March they may visit Sweet Violet (*Viola odorata*) and Sallow catkins.

In April the first *Halictus* species appear; these early ones are among the smallest of their genus – tiny blackish bees with rapid oscillating flight. In gardens in April these small bees are often seen round Spanish Bluebell (*Endymion hispanicus*) which has large bell-shaped flowers which they can crawl right into. The smaller andrenas now emerge and these are similar to the small *Halictus* though a little larger and with more flattened abdomens. Both these groups of small bees visit yellow *Compositae* and Birdseye Speedwell (*Veronica chamaedrys*). A common bee in gardens in April is the reddish-brown *Osmia rufa* (Plate 17b), belonging to a genus with eight species in Britain and known as mason-bees. It is a furry bee about 12 mm. long with its pollen brush on the undersurface of its abdomen instead of on the hind legs as in most bees. It has an extremely long tongue but visits a wide variety of flowers, including the wide open fruit blossom flowers with partly concealed nectar, which also attract the larger andrenas at this time.

In late April and during May further species of *Andrena*, *Halictus* and *Osmia* appear, as well as representatives of some other genera. These include *Sphecodes*, which are black with a red band on the fore part of the abdomen, and *Nomada* (Plate 16d), which are banded with black and yellow, or brown and yellow, and look like wasps. In both genera the hair clothing is sparse, and both are inquilines, *Nomada* usually on *Andrena* and *Sphecodes* on *Halictus*. The blossoms of Hawthorn (*Crataegus* spp.) are much visited by *Andrena* and *Halictus* during May.

Representatives of most of the remaining genera appear in June. Among these are *Prosopis*, small shiny black bees, called in Germany 'mask-bees' because the males have the face pale yellow and the females often have a yellow mark on the face beside each eye. *Prosopis* have few body hairs and no pollen brush (Plate 16e); the pollen required for the larvae is carried in the crop with the honey. The proboscis is very short, so that only more or less exposed nectar can be obtained. A wide variety of flowers is visited, but the genus is particularly attracted by Wild Mignonette and Weld (*Reseda* spp.). *Colletes* also appear in June; these resemble medium-sized andrenas and the common garden species visits chiefly Yarrow (*Achillea millefolium*), cultivated *Achillea* spp. and Tansy (*Chrysanthemum vulgare*), all of which are *Compositae* with heads of closely-packed short tubular florets. Friese (1923) notes that species of *Colletes* restrict themselves to a smaller selection of plants than *Prosopis* and *Halictus* and that their emergence is closely related to the flowering of their favourite plants. *Panurgus* may also appear in June; the two British species are blackish brown with orange pollen brushes on the hind legs. They nest in sandy soil and are particularly fond of the yellow ligulate-flowered *Compositae*, such as Hawksbeard (*Crepis*), Hawkbit (*Leontodon*) and Hawkweed (*Hieracium*). Also in June appear the three remaining genera which collect pollen on the underside of the abdomen. These are *Megachile*, the leafcutter-bees (which favour *Compositae*, *Campanulaceae* and *Leguminosae*), *Chelostoma* (which favour *Campanulaceae*) and *Anthidium* (of which the only British species, *A. manicatum*, the carder bee, visits chiefly the large tubular flowers of *Scrophulariaceae* and *Labiatae*, as well as frequenting *Leguminosae*). *Anthidium* is unusual in that the males are larger than the females instead of smaller; they are large brown bees, with yellow on the face and yellow spots on the abdomen. Other bees that emerge in June are *Halictus leucozonius* (Plate 16f), one of the larger species of the genus, with similar flower-preferences to *Panurgus*, and *Anthophora quadrimaculata*, with similar preferences to *Anthidium*.

In July the species of *Colletes* that visit Ling (*Calluna vulgaris*) emerge, as well as various species of *Andrena* found on sandy heaths, some of which favour Ling and Bell Heather (*Erica cinerea*). *Andrena marginata* also emerges in July, when its favourite flowers, Scabious (*Knautia arvensis*, *Scabiosa columbaria* and *Succisa pratensis*) and Knapweed (*Centaurea*) are out. *Macropis labiata*, which for pollen visits almost exclusively Yellow Loosestrife (*Lysimachia vulgaris*), also emerges in July (see Chapter 3).

Owing to the bees' need for pollen, which is usually easily accessible, general compilations of flowers visited by the various species of bees do not show much correlation between the tongue-length of the insect and the accessibility of the nectar. Kugler (1940) reported, for instance, that species of *Halictus* readily collect pollen from flowers whose nectar they cannot reach. However, tongue-length (Table 3) must limit a bee's choice of flowers for nectar-gathering except where borings are made, as they often are by bumble-bees, some of which appear to be more resourceful than the solitary species in getting food in unconventional ways. The particular preferences of bees for certain flowers already mentioned often concern flowers which provide both nectar and pollen; most of the family *Compositae* have flowers which provide good supplies of both, and the same applies to many *Leguminosae*. These two families are particularly suited to the bees with abdominal pollen brushes as with few exceptions they present their pollen from below and the bees can scrape it directly into the brush. Sometimes bees will take pollen from some flowers and nectar from

Table 3.   Lengths (in mm.) of proboscis and body of some British bees
(both sexes unless otherwise stated)

Species	Proboscis length (mainly from Knuth)	Body length
Colletes daviesana	2.5–3	8
Prosopis communis	1–1.25	6.5
Halictus morio	2.5	5.5
,, *leucozoniu*	4	7–9
,, *rubicundus*	4–4.5	10
Sphecodes reticulatus	2	8.5
Andrena argentata	2.5	8
,, *gwynana*	2.5	7.5–9.5
,, *marginata*	3.5–4	8–9.5
,, *pubescens*	3.5	9.5–13.5
Melitta leporina	3.5	11
Panurgus calcaratus	3	7.5–9
Dasypoda hirtipes	5	14
Anthophora quadrimaculata	8	9–10.5
,, *acervorum*	13	14
Melecta luctuosa	11	14
Epeolus cruciger	4	7–9
Nomada goodeniana	4	9–11.5
Megachile centuncularis	6–7	9–11
Coelioxys elongata	4.5	12
Anthidium manicatum	9–10	11–16
Stelis punctulatissima	5.5	9
Osmia caerulescens	8	6.5–9
,, *rufa*	7–9	9–11
Chelostoma campanularum	3	5
Bombus agrorum (queen)	11–14	15–17
,, *pratorum* ,,	10–12	16
,, *jonellus* ,,	12	16
,, *sylvarum* ,,	14	16
,, *lapponicus* ,,	13	17
,, *hortorum* ,,	18–19	20
,, *lapidarius* ,,	14	22
,, *ruderatus* ,,	22	23
,, *terrestris* ,,	10	24
Psithyrus rupestris (female)	11–14	22–24
Apis mellifera	6	13

others; this is commonly the case with the social species, but it occurs
also with solitary bees; for example, *Anthophora acervorum* visits Peonies
(*Paeonia* spp.) in gardens for pollen only.

However, it seems that some long-tongued bees never visit certain
flowers with rather easily accessible nectar, for there are no records of

PLATE 17. **a**, hover-fly
(Rhingia campestris) sucking
nectar from flower of Red
Campion *(Silene dioica)*.
b, solitary bee *(Osmia rufa)*
sucking nectar from Wall-
flower *(Cheiranthus cheiri)*.
c, male and female empid
flies *(Empis* cf. *pennipes)* on
flower of Herb Robert
(Geranium robertianum).

a

b

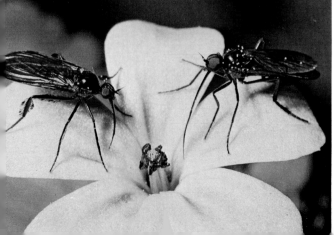

c

PLATE 18. **a**, bumble-bee *(Bombus agrorum)* visiting flower of Bluebell *(Endymion nonscriptus)*. **b**, honey-bee stealing nectar from Bluebell. See also Pl. 34.

visits of *Megachile* or *Anthidium* to Ragwort (*Senecio jacobaea*) according to Harper and Wood (1957). Even more restricted in their habits are *Macropis labiata* and *Andrena marginata* (see p. 149), together with *Andrena praecox*, a very early bee which takes pollen exclusively from sallows (Chambers, 1946). *Andrena bicolor* visits a great variety of flowers in its first brood, while in certain localities the second brood rarely visits anything but *Campanula* and *Malva*, and in some other localities nothing but Dandelion (*Taraxacum officinale*) and Bramble (*Rubus fruticosus*) even when *Malva* is available. Such situations are rather frequent in *Andrena* and they are paralleled in their inquilines, *Nomada*, which may actually specialise in the same flowers as their hosts (Friese, 1923). Bees that visit only one or a few species of flowers for food are described as oligotropic, while those showing a similar restriction for pollen supplies are called oligolectic (see also p. 372). Whereas oligolecty appears to be rather rare in British bees it is a common and striking phenomenon in some parts of the world and examples from America are quoted in Chapter 12. In some of these cases the flowers concerned are the meeting place of the sexes; this is so with the British bee *Macropis labiata* in which the females leave a special scent on the *Lysimachia* flowers which makes these attractive to other females (even when all pollen has been removed) and especially to males (Kullenberg, 1956b). In a particular experiment concerning four species of bumble-bee, with varying lengths of tongue, Brian (1957) found that when gathering nectar in competition with each other these species tended to restrict themselves to the flowers most appropriate to their tongue length, although at least the longest-tongued species was not so selective when there was little competition.

Chambers (1945, 1946) noticed that andrenas collected pollen from certain wind-pollinated trees, and honey-bees are particularly noted for visits to a variety of wind-pollinated plants, including even a gymnosperm, yew (*Taxus baccata*) (Hodges, 1952). Wind-pollinated flowers are inconspicuous but they usually produce very large quantities of pollen which, however, is not sticky. This does not seem to make it difficult for *Andrena* to collect it, although these species do not moisten pollen with honey as do some bees, including the honey-bee. Since many wind-pollinated flowers are unisexual the female flowers are not visited by the pollen collecting insects, which thus fail to pollinate the plants.

The scents of bee-pollinated flowers are sometimes indistinct to man but those which he can clearly perceive are rather varied. The honey-bee and short-tongued bumble-bees freely visit many of the flowers visited by *Diptera*, some of the scents of which are described on p. 95. The flowers that are more specially adapted to bees have generally a sweet scent which, even when strong, is more delicate to the human nose than the heavy scents common in *Lepidoptera*-pollinated flowers (p. 101), but as with that group, the same scent may be found in many unrelated flowers.

For example, the scent of Sweet Violet (*Viola odorata*) is like that of *Iris reticulata* (a commonly cultivated Middle Eastern species which, incidentally, flowers at about the same time as this violet and is similar to it in colour). Very similar to these, but more strongly scented, is the cultivated Mignonette, *Reseda odorata*, from North Africa. Honey-like scents are common in *Leguminosae*, examples being White Clover, (*Trifolium repens*) Tree Lupin (*Lupinus arboreus*, a Californian plant introduced into Britain) and Spanish Broom (*Spartium junceum*, from the west Mediterranean region). In the last two the scent is very powerful; both species, like Gorse (*Ulex europaeus*), which smells of coconut, have pollen flowers. Other recurring scents of bee-pollinated flowers are the plummy scent of the Grape Hyacinth (*Muscari atlanticum*) and Oxlip (*Primula elatior*), the disagreeable smell of *Helleborus foetidus* (see p. 54), and the pleasant scent of the garden Pansy (*Viola × williamsii*) which is to be found in species of Michaelmas Daisy from both Europe and North America (for example, *Aster sedifolius* and *A. puniceus*, which are visited mainly by bees and hoverflies) and in some other flowers.

Many species of bees are involved in the relationship of pseudocopulation with Orchids (*Ophrys* spp.) in the Mediterranean region; the bee genera usually concerned are *Andrena* and *Eucera* (the latter genus being represented by only two species in Britain) (see Chapter 7). Other bees are involved in the strange pollination processes of certain tropical American Orchids (see Chapter 10).

Mouth-parts

The mouth-parts of bees are adapted for both nectar-collecting and nest-building. All the main parts found in the more primitive insects can be recognised in bees, where they are less drastically modified than in the highly advanced *Diptera* or *Lepidoptera*. These parts can be seen in the diagrams of the mouth-parts of *Halictus* and the honey-bee (Figs. 27d and 29, p. 155 and 157). Like the primitive mouth-parts of *Cimbex* and *Vespula* (Fig. 23, p. 133) those of bees have a single glossa. It is a flexible structure with a springy rod running along it; muscles attached at the base of this rod can induce a great deal of movement in the glossa, and there is in addition always a mechanism for drawing the base of the glossa some way back into the gutter-shaped mentum. The secretion of saliva takes place on the mentum near the attachment of the glossa. At least in the honey-bee the mouth-parts also contain the taste organs.

Some bees have a long glossa and some have a short one. Those with a short glossa belong to the genera *Prosopis*, *Colletes*, *Andrena*, *Halictus* and *Sphecodes*. The first two of these are the shortest-tongued of all bees and are peculiar in having the glossa bilobed (Fig. 25). According to Demoll (1908) the detailed structure of the glossa and paraglossae in these two

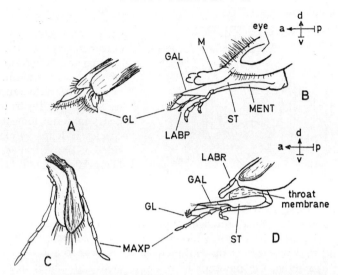

FIG. 25. The short-tongued bees. A, *Colletes daviesana*: view of tips of galeae with glossa protruding from beneath them. B, the same: side view of extended mouth-parts. C, *Prosopis pectoralis*: view of the maxillae (galeae and palps). D, the same: side view of extended mouth-parts.

genera is adapted to the job of lining the nest with regurgitated mucus. The glossa is a short structure with a fringe of hairs which presumably absorb nectar; by drawing the glossa back into the space above the mentum and beneath the galeae of the maxillae, pressure and suction can probably be applied to it so that nectar imbibed by the hairs can pass into the mouth. Demoll (1908) believed that the nectar first passes through a very narrow passage between the glossa and the paraglossae, in which case the lapping action just described might not be necessary. In any case the nectar then travels backwards through a tube formed by the mentum and maxillae which overlap one another (Saunders, 1890). These structures are linked towards the base by a membranous bag, the 'throat-membrane'. In the upper side of this are two folds which connect the edges of the maxillae to the labrum (Figs. 25D, 26 and 27C, D) forming a covered passage which links the mouth with the basal ends of the maxillae, so that fluid passing back from here is unable to escape. The folds are probably kept together by tension when suction is taking place, and narrow rods present in their edges may help to keep the gap closed. The underside of the membranous bag continues back under the head where it lines the cavity into which the mouth-parts are retracted when not in

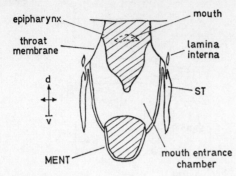

FIG. 26. Diagrammatic transverse section through mouth-entrance of a bee; the epipharynx hangs down in front of the mouth. (The lamina interna is not visible in the other illustrations of bees; in some long-tongued species it helps to close the sides of the entrance chamber.) After Demoll (1908).

use. Embedded in it are the sclerites which join the mouth-parts to the hard parts of the head and which project and retract the mouth-parts. The sclerites and the way in which they move can be seen in Figs. 29 and 31A and D. The cardines move the entire apparatus to and fro while the lorum can move the mentum in relation to the maxillae; in order to project the mouth-parts the cardines have to swing forwards and downwards and this great movement is made possible by the skin of the throat-membrane loosely linking the hardened joints together; the lowered position of the mouth-parts when nearly fully projected can be seen in Figs. 25D and 27C. When the mouth-parts of bees are retracted, the galeae, the end of the glossa and the palps fold downwards and backwards (Fig. 31A).

Bees are considered to be on a higher evolutionary level than wasps, but the primitiveness of *Prosopis* and *Colletes* is indicated by the resemblance of their mouth-parts to those of *Vespula*, all three genera having a short bifurcated glossa.

In the remaining genera with a short glossa this organ is pointed at the tip. In *Andrena* it forms a sort of scoop (Fig. 27A); the proboscis is again very short and constructed much as in *Colletes* and *Prosopis*, but the galeae are large, horny and opaque, instead of partly translucent, and have a pronounced apical fringe of coarse hairs. The paraglossae are well-developed. A very few species of *Andrena* have a considerably longer proboscis than the majority of the genus. In one of these (*A. marginata*) (Figs. 27B, c) all the parts are somewhat elongated in comparison with those of other species; the scoop-like appearance of the glossa is still evident at its base but the rest of the glossa is convex, tongue-like, and covered in long hairs towards the tip. In *Halictus* the ratio of the proboscis-length to the body-size is about the same as in *Andrena marginata*, but the proportions of the parts are different, the stipes (the basal part of the maxilla) being relatively longer (Fig. 27D). The mentum is also longer as it has to match the length of the stipes, while the hairy convex glossa is

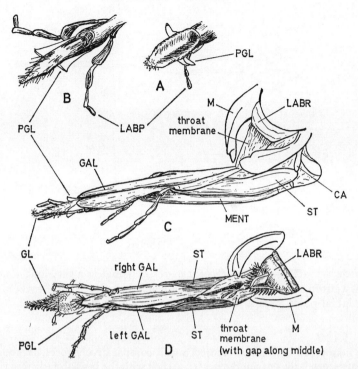

FIG. 27. A, a typically short-tongued *Andrena* (*A. tibialis*): scoop-like glossa. B, an unusually long-tongued *Andrena* (*A. marginata*): glossa. C, the same: general view of mouth-parts. D, *Halictus rubicundus*: general view of mouth-parts, from more directly above than C, so that mentum is hidden.

short. *Sphecodes*, which are mostly inquilines of *Halictus*, have similar mouth-parts but the cardines, labium and maxillae are shorter.

Although *Halictus* have a comparatively long proboscis, even longer proboscides are found in most of the bees with a long glossa. Evolution has undoubtedly progressed in the direction of greater proboscis length, but this process is to be thought of as specialisation rather than improvement. The gaining of access to deep-seated nectar is accompanied by decreased ease of collecting fully-exposed nectar, as observed by Kugler (1940) in bumble-bees.

The extended proboscis of a bee with a long glossa is shown in Fig. 28, while the mouth-parts of another are shown separated in Fig. 29A. The proboscis works in much the same way in all the bees with a long glossa, though there are some variations in construction. As is to be expected,

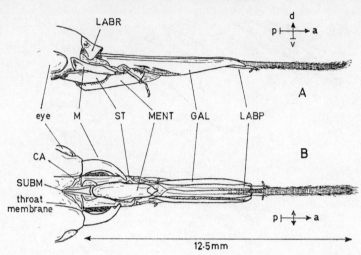

FIG. 28. Mouth-parts of *Anthophora acervorum* (female); a very long-tongued bee. A, from side. B, from below. The elongated galeae and basal joints of the labial palps form a tube round the very long glossa.

the honey-bee is the best studied and a comparatively recent account of its structure is that of Snodgrass (1956). As with the bees already described, the stipites of the maxillae, together with the labium, form a food channel, while food is licked up by the glossa; however, the glossa is very long and the food channel is continued forward by the great development of the galeae of the maxillae and the labial palps (Figs. 28 and 29D). The glossa is able to move to and fro in this front part of the food channel by the coiling and uncoiling of its base into and out of the front of the mentum, as already described (p. 152). The glossa itself is rolled at the edges to form a tube with a slit along the underside, which is closed by dense fringes of hairs; its surface is transversely ringed, and each ring is the seat of a row of stiff hairs (Fig. 29B). The shape and position of the glossa rod is seen in Fig. 29E; towards the base it has the possibility of some movement to and fro within the glossa, and as a result it is able to tighten or slacken the skin of the glossa in such a way as to cause the hairs of the glossa to spread or to lie down. When the glossa is fully extended the hairs spread and can thus absorb liquid between them, and as it is drawn back they close up and release the liquid into the food channel. The glossa rod, although activated only from the base, is also able to produce bending movements at the tip of the glossa which assist in licking up nectar. These movements usually take place inside flowers, but I have seen them

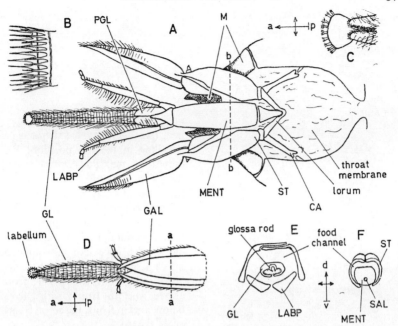

FIG. 29. Mouth-parts of the honey-bee (worker). A, extended and splayed mouth-parts from below. B, short length of glossa showing two rows of hairs, further enlarged. C, labellum from below and tip of glossa with ventral groove closed by hairs; the lower surface of the labellum is convex; same scale as B. D, distal part of extended mouth-parts, not splayed, from above. E, cross-section through a – a of D, showing overlap of galeae. F, cross-section through b – b of A, on reduced scale. A-E, after Snodgrass (1956); F, inferred.

in an *Anthophora* visiting a small *Crocus*, and Friese (1923) observed them in captive male bumble-bees drinking from little containers; he described how the galeae and palps were firmly united around the glossa, which moved to and fro and swept the fluid into the tube. The proboscis may be extended fully if this is necessary to reach the nectar (Müller), but when the food has been imbibed among the hairs of the glossa, the proboscis has to be slightly retracted; this brings the raised parts on the bases of the maxillae back into contact with the lobes on the underside of the labrum. The parts fit neatly together and close over the food channel right back to the mouth, just as the throat membrane does in *Halictus*, etc. (Fig. 30A).

So far we have seen that the food channel lies between two concentric tubes in its distal part (Fig. 29E) and in a tube formed by the maxillae

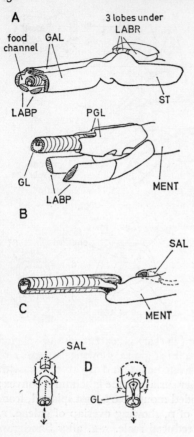

FIG. 30. Mouth-parts of the honey-bee (worker). A, basal parts of maxillae in feeding position. Removal of maxillae would reveal the parts shown in B, which shows the natural position of the paraglossae. If these are removed – C – the tip of the mentum with its salivary pouch, and the base of the glossa, are revealed. D, course of saliva from salivary pouch to interior of glossal tube; left, from above and in front; right, from below and in front. Based on Snodgrass (1956).

overarching the mentum in its proximal part (Fig. 29F). Another part of the system is formed by the large flattened paraglossae, which ensheath the apex of the mentum and the base of the glossa (Fig. 30B); their shape is such that the saliva, produced from the salivary pouch near the tip of the upper surface of the mentum, is carried downwards on either side and passes into the tube formed by the glossa, entering through an opening on the underside (Fig. 30c and D). The saliva emerges at the extreme tip of the glossa where there is a little flattened disc called the labellum (Fig. 29c and D). There are divergent views as to whether food enters the interior of the glossal tube.

Mouth-parts of a slightly more primitive kind than those of the honey-bee are found in several genera of British bees with a long glossa. In *Panurgus*, for example, the glossa is not particularly long and the labial palps are very slender, apparently playing no part in the formation of the

PLATE 19. **a**, bumble-bee *(Bombus* cf. *terrestris)* feeding at flower of Bilberry *(Vaccinium myrtillus)*. **b**, honey-bee on winter-flowering heather *(Erica mediterranea = E. carnea* auct.).

PLATE 20. **a**, honey-bee gathering nectar from *Trachystemon orientale*. **b**, honey-bee gathering pollen from *Trachystemon orientale*. **c**, bumble-bee *(Bombus lucorum* worker*)* collecting pollen from Bittersweet *(Solanum dulcamara.)*

a

b

c

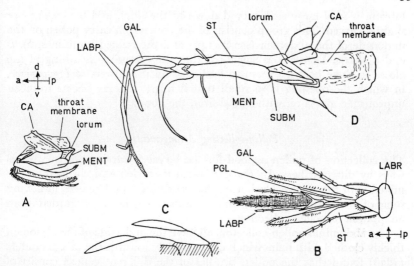

FIG. 31. A, mouth-parts of the honey-bee (worker), folded and seen from the side. B, mouth-parts of the bee *Panurgus*, from above, splayed. C, glossal hairs of the bee *Anthophora*, attached to a section of the glossa. D, mouth-parts of the bee *Osmia coerulescens* (male) seen obliquely from below; mouth-parts in this position about 6.5 mm. long, body 8 mm. long. A, after Snodgrass (1956); B, after Saunders (1890); C, after Demoll (1908).

food channel, while the maxillary palps resemble those of the bees with a short glossa (Fig. 31B). The galeae are long and strongly tapered, so they appear well adapted to sucking the slender tubular florets of *Compositae*, which are the favourite flowers of *Panurgus*. Similar mouth-parts are found in *Dasypoda*, and in *Melitta*, which has a specialised arrangement of long curved hairs at the tip of the glossa, making it look rather like a bottle brush. A closer resemblance to the honey-bee is found in the mouth-parts of *Nomada* and especially *Epeolus*, both of which genera are inquilines.

The mouth-parts of *Anthophora* are shown in Fig. 28; the overall length is much greater than in the honey-bee, and the proportions are rather different. When the galeae are folded back out of use they project well back under the thorax of the bee. In *Anthophora* and the closely similar *Eucera* the hairs of the glossa have oar-like flattenings which increase their surface area for absorption (Fig. 31C). The bumble-bees (*Bombus*) are rather similar in their mouth-parts to the bees just described.

The remaining bees all have rather similar mouth-parts (Fig. 31D). The glossa is long and is sheathed by the galeae for a greater proportion of its length than in other bees with a long proboscis (Demoll, 1908). The labrum is very large and curved down at the sides, and the galeae are slightly curved. This group consists of the bees which carry pollen on the underside of the abdomen (see p. 149) and the related inquilines, *Stelis* and *Coelioxys*. Demoll has suggested that the extensive sheathing of the glossa in these bees fits them to feeding from the flowers of *Leguminosae*, in which the opening is so small that it might squeeze nectar from an unprotected glossa on withdrawal from the flower.

Pollen-collecting Arrangements

The collection of pollen as food for the larvae is carried out in various ways by different bees. In *Prosopis* pollen is carried entirely in the crop mixed with the nectar. Curiously, they resemble all other bees in having some of their hairs branched, which is a feature generally regarded as an adaptation to pollen collection.

The abdominal pollen-collectors all have the underside of the abdomen thickly clothed with hairs which curve slightly towards the tail. Saunders (1878) found that the pollen brushes of the different genera consist of different kinds of hairs; thus in mason-bees (*Osmia*) and leafcutter-bees (*Megachile*) the hairs (unlike those elsewhere on the body) are unbranched, those of the latter being spirally grooved, whereas in *Chelostoma* the hairs are flexuous and branched. The legs are used to gather up the pollen and transfer it to the abdominal brush. They are adapted to this by bearing bristles (used in *Prosopis* only for cleaning the body) which are developed into a stiff brush on the inside of each meta-tarsus (compare Fig. 32A) and into a comb on the inner edge of each of the fore- and mid-tibiae towards their tips. These stiff hairs may be used both to collect pollen which is adhering to the body of the insect and to remove it from the flowers. While leafcutter-bees are flying from flower to flower their legs hang down and are scraped together, the pollen being passed back to the hind-legs. These are then raised and the pollen transferred to the abdominal brush. However, pollen can sometimes be transferred to this brush straight from the flower (see p. 149), with or without the assistance of the hind-legs.

The bees which carry home the pollen on their legs resemble the abdominal collectors in having stiff pollen-gathering hairs on the insides of the metatarsi, but these hairs sometimes extend to other joints of the tarsi also. *Andrena* carry home the pollen on the main joints of the hind-legs and also on parts of the thorax (Fig. 32 and Plate 16). The hind tibiae are densely clothed with branched and unbranched hairs, which form a large pollen-carrying brush. On the other hand, the hind femora carry part of their pollen in a brush on the front surface and most of it on the lower surface in a basket-like structure formed by fringes of branched hairs. The trochanter, one of the two small joints between the

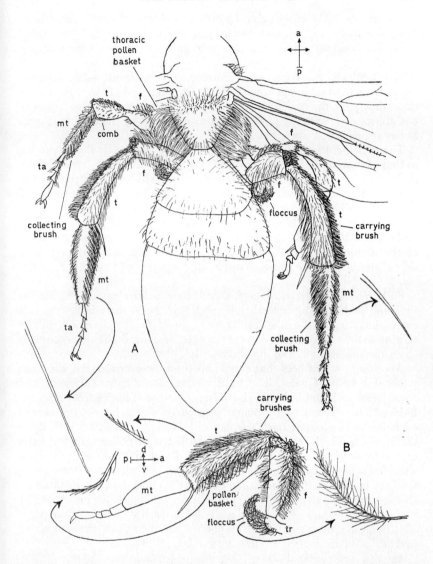

FIG. 32. Pollen-collecting apparatus of *Andrena denticulata*. A, the insect seen from above (front legs, other parts of body, and certain details omitted). B, hind leg, as seen when insect is viewed from side (coxa and details of tarsus and metatarsus omitted). f, femur; mt, metatarsus; t, tibia; ta, tarsus; tr, trochanter. The parts of the body shown in A measure 9.5 mm. long. Individual hairs shown at greater magnifications.

femur and the body, carries a group of beautiful long plumose hairs which descend and then curve rearwards, constituting the floccus. These hairs themselves enclose pollen and also help to close in the baskets of the femora.

Like the leafcutter-bees, andrenas commonly carry out pollen-packing movements with their legs when in flight from flower to flower, but sometimes they pack the pollen before leaving the flower. In addition, some, if not all, species make the pollen easier to transport by moistening it with honey regurgitated from their crop.

The pollen-collecting apparatus in *Colletes* and *Halictus* is closely similar to that of *Andrena*. In both genera, however, the hind femora carry all their pollen in a basket on the underside, and the branched hairs corresponding to the floccus of *Andrena* arise at the base of the femur. The branches of these hairs diverge widely and overlap to form a mesh; similar hairs clothe much of the thorax of *Colletes*. *Halictus* has no thoracic pollen baskets, but it collects some pollen in the dense hairs beneath the front of the abdomen. In both these genera the curved hairs on the lower side of the hind tibiae show a fan-like development of branches near the tip.

Both *Halictus* and *Colletes* pack their pollen before leaving the flower, brushing each fore-leg several times very rapidly on the middle leg, and each middle leg on the hind-leg. These movements take place only on one side at a time, but both hind-legs can be employed simultaneously to clean the under surface of the abdomen.

Another group of bees that carry pollen on their legs have a less complicated system of hairs. The hind metatarsi are more enlarged, and these joints and the hind tibiae have large brushes of backwardly directed hairs which carry most of the pollen. *Panurgus*, however, have some branched hairs on the femur, but their capacity is slight compared with those of the tibia, which is unusually well clothed with pollen-carrying hairs on the inner surface. These hairs are waved and pinnate with numerous short branches. *Dasypoda* is similar to *Panurgus*, but it has enormously long feathery hairs over the entire surface of the tibia and the large metatarsus, the last-mentioned joint having no stiff combing bristles on the inner surface. On account of its enormous pollen-carrying capacity *Dasypoda* is illustrated in almost every book on pollination, though the more complex arrangements in *Andrena* are much more interesting.

In the remaining bees of this group (*Anthophora, Eucera, Melitta, Macropis* and *Rophites*) pollen-carrying brushes are confined to the outer surfaces of the hind tibia and metatarsus. The genera *Eucera, Melitta* and *Macropis* are known to moisten their pollen with honey.*

A third and final group of bees which collect pollen on their legs

* This account of pollen-collecting apparatus has been based so far on the work of Müller, Braue (1913) and Kugler, supplemented with observations by P.F.Y.

consists of the social bees, *Apis* and *Bombus*. In these the collection of pollen is similar to that of other bees in its first stages but quite different in its final stages.

In the honey-bee (*Apis mellifera*) (Fig. 33) pollen is scraped off the head and fore-part of the thorax by the antenna-cleaners and the combs of the fore-legs. The middle legs clear pollen from the hind part of the thorax and the hind-legs clear it from the abdomen. All this pollen is worked into the brushes on the inner surface of the metatarsi of the middle legs, and is moistened by regurgitated honey. After this each middle leg in turn is placed between the two hind-legs and then drawn forward; this action scrapes the pollen into the metatarsal brushes of the hind-legs. When sufficient pollen has accumulated on the hind metatarsi, it is transferred to the pollen baskets (or corbiculae) on the outside of the hind tibiae (Plate 54c). This is made possible by a structure known as the pollen press, which is found only in the social bees and is, perhaps, the most striking adaptation to pollen collection found among all bees. As can be seen from Fig. 33c and D, the hind tibiae and metatarsi are flattened and greatly widened compared with those of the other legs. The metatarsi, however, are attached to the tibiae only at the front by a narrow joint (Fig. 33c); this allows the press to be opened by the bending down of the metatarsus and closed by its bending up. The proximal end of the metatarsus is produced (except at the actual joint) into an outwardly directed flange which slightly overlaps the apex of the tibia; this flange is called the auricle (Fig. 33E). Two actions are required to transfer pollen from the inner surface of the metatarsus of, say, the right leg to the pollen basket of the left leg. In the first, the rake of spines on the apex of the left tibia is pushed downwards through the hairs of the right metatarsus. This scrapes pollen into the space between the rake and the left auricle. The second movement is a bending upwards of the metatarsus which closes the gap between the auricle and the rake (Fig. 33E). The pollen is thereby forced upwards and outwards into the pollen basket (Fig. 33F) as a compact and sticky mass. Pressure applied by the springy auricle and its bristles causes successive masses of pollen to be plastered one upon the other. A solitary bristle is present on the surface of the tibia which forms a pin through the pollen mass and is apparently important in holding it in position. The tarsi of the middle legs shape the pollen mass, which is eventually kidney-shaped. The efficiency of the pollen press is probably greatly dependent on the moistening of the pollen, which also makes it possible for the rather sparse unbranched hairs of the pollen basket to carry a large quantity of compacted pollen. Honey-bees which collect pollen accidentally when concentrating on nectar-gathering sometimes retain it but sometimes discard it. In discarding pollen they make movements similar to those made when packing pollen, but the position of the legs is different and probably the press is kept closed, so that the pollen just drops away from the rake. Both processes are normally carried out while the insect is in flight.

The bumble-bee (*Bombus*) is almost identical in its pollen-packing apparatus, but it differs in small details.*

*   *   *

The study of the senses of bees (touch, taste, smell and sight) has been chiefly confined to the honey-bee, which has been the subject of several books summarising these researches (von Frisch, 1950 and 1954; Ribbands, 1953; Butler, 1954). Smell and sight are clearly the senses most relevant to a study of pollination.

Sense of Smell

The olfactory organs of bees are similar to those of other insects and are confined to the antennae where they are mixed with tactile hairs. In the worker honey-bee the organs of smell are absent from the first four joints of each antenna and present in the remaining eight. These bees can detect dilutions of scents ten to one hundred times weaker than those just perceptible to man, and they are also very good at discriminating between slightly different mixtures of scents (Ribbands, 1955). It was found by Lex (1954) that the different parts of a flower often smell differently to human beings. She then used flowers cut up into these parts for experiments with bees and found that the honey-bee could easily distinguish between the scents of the parts. The presence of the organs of smell in the mobile antennae enables bees to explore the exact distribution of the smell of an object. They could thus easily use the scented guide-marks of a flower, which often coincide with the visible guide-marks, to assist them in finding the nectar. Bolwig (1954), however, found that honey-bees did not follow linear scent traces made on coloured models, although coloured objects with scent at one end only were visited chiefly at the scented end.

The role of scent in the approach of honey-bees to food was investigated by von Frisch (1954), who trained bees to feed from a blue cardboard box containing jasmine scent, two similar but empty and uncoloured boxes being presented at the same time. The bees were then shown one plain empty box, one plain jasmine-scented box and one blue empty box. They approached the blue box directly from a distance, but on reaching the entrance hole they appeared startled and roamed around outside instead of going in. If they chanced to come within a few inches of the jasmine-scented hole they went in there. Observations show again and again that

* The discovery of the method of pollen-packing in *Bombus* and *Apis* is due to Sladen (1911, 1912) and descriptions of it in *Apis* are given by Hodges (1952) and Snodgrass (1956).

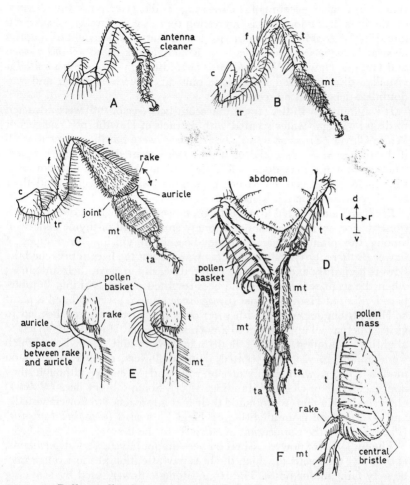

FIG. 33. Pollen-collecting apparatus of the honey-bee. A, left fore-leg, front view. B, right mid-leg, back view. C, right hind leg, turned forward, back view; arrows show opening and closing movements. D, both hind legs, back view; tibial rake of left leg is about to be pushed through metatarsal brush of right leg. E, two views of left pollen press, similar to those seen in D; hairs of pollen basket omitted in left-hand drawing. F, loaded pollen, basket. c, coxa; f, femur; mt, metatarsus; t, tibia; ta, tarsus; tr, trochanter. A - C, E, after Snodgrass (1956); D, F, after Hodges (1952).

vision is important in guiding bees to a food source from a distance, but that scent is taken account of at close range and exerts a powerful influence on the bees' behaviour. (It is interesting that exactly similar behaviour is found in the Diptera.) Kugler (1940) found that Halictus could distinguish accurately between flower-heads of Rough Hawksbeard (Crepis biennis) and Greater Hawkbit (Leontodon hispidus), although they are very similar visually; discrimination took place only at very short range and was doubtless dependent on scent.

It was found by Butler (1951) that scout-bees (see p. 176) were attracted to dishes of sugar-water scented with extracts of Hawthorn (Crataegus) or White Clover (Trifolium repens), but the bees were hardly attracted at all if the dishes were unscented or if they were scented with Spiraea arguta. The experiments were done before these plants had come into flower so that these young bees could not be familiar with the scents, and their reactions were presumably inborn.

Bees can be prevented from visiting a flower if an unaccustomed scent is present. For example, dishes of strongly smelling oil-of-thyme concealed among the plants prevented honey-bees from visiting three types of flower (Butler, 1951). While the scent was present the bees approached the flowers hesitantly and frequently alighted nearby to clean their antennae; when the scent was removed the bees resumed normal visiting. Bumble-bees feeding at Houndstongue (Cynoglossum officinale) were offered a plant of Houndstongue scented with rose; only half of the approaches led to visits, whereas all approaches to a normally scented Houndstongue led to alighting (Manning, 1956b). Both bumble-bees and honey-bees, which were trained to visit rose-scented artificial flowers, refused to alight on models scented with a mixture of rose and lavender, although they approached them closely (Manning, 1957). Some solitary bees (Halictus) observed by Kugler (1940) made their first approaches to flowers visually and were put off from alighting on Field Bindweed (Convolvulus arvensis) (Plate 21a) by the application of clove oil to the flowers.

The majority of flowers visited by bees do not have a very strong scent, and in fact the scents to which the bees pay attention at short range may be very faint to our noses. The Houndstongue flowers used by Manning and some of the flowers which Lex found to have internal scent differences are not normally thought of as being fragrant. The use of such faint scents by bumble-bees was clearly demonstrated by Kugler (1932). The method was first tested using the Sweet Pea (Lathyrus odoratus), which is strongly scented. After being trained to visit red Sweet Pea flowers, the bees were tested with the real flowers, with unscented paper models and with paper models in which Sweet Pea flowers were concealed and which smelt fairly strongly of the flowers. The scented models, although very imperfect visually, received as many visits and proboscis reactions as the real flowers, whereas the unscented models received fewer. Three types of weakly scented flower were then tested in a similar way. All visits to real flowers were accompanied by proboscis reactions; unscented models received few visits and no proboscis reactions; scented models received an intermediate number of visits, accompanied by proboscis reactions in some cases. The flowers were those of the Duke of Argyll's Tea Tree (*Lycium halimifolium*), Viper's Bugloss (*Echium vulgare*) and Common Toadflax (*Linaria vulgaris*), and the results were the same with each. Kugler showed in further experiments that the bees were responding to the specific flower scents and not merely to the smell of vegetable matter or flower scent in general.

Among the Louseworts (*Pedicularis* spp.) of North America, two species with little or no scent to the human observer have been shown, even when concealed, to be attractive to their bumble-bee pollinators. Although the petal colours could not be seen, the bees unerringly alighted on that portion of the small muslin cage closest to freshly matured flowers. Moreover, individual bees went only to the cages containing the species they had been visiting previously, showing that they could distinguish the Louseworts entirely by their smell. A third species of Lousewort was found to have a much stronger scent, and it was suggested that this was related to its shady habitat, in which scent might be more important than in the open (Sprague, 1962).

Among honey-bees scent plays an important part in the communication of information about sources of food. Returning foragers bring into the hive the scent of the flowers from which food has been obtained, and this scent is used by other bees to find the same kind of flower. Von Frisch demonstrated this in the following way: a bowl of fragrant Cyclamen flowers was put out near the hive, the flowers having been filled with sugar-water. Not far off, another bowl of Cyclamen flowers was put out side by side with one containing Phlox flowers which are also fragrant, but none of these flowers had food added. Some bees, presumably alerted by the finders of the original bowl of Cyclamen, found the other two bowls, whereupon they ignored the Phlox and persevered in searching for food in the Cyclamen flowers. If the original Cyclamens were replaced by Phlox flowers containing sugar-water, bees interested in Phlox began to appear at the site of the other two bowls of flowers. Von Frisch also discovered that when a rich source of food lacking scent is found the alerted bees visit scentless flowers. Furthermore, he found that the bees in the hive could learn a flower-scent either from another bee's body or from the nectar she had collected, which is normally passed round among the bees in the hive. Sometimes, however, the scent on the body is lost during the flight back to the hive, so that the second method is the more reliable. Honey-bees are also able to produce a scent themselves, and they sometimes use this when they are on a good food source to attract other bees in the neighbourhood towards them.

Sense of Sight

It was explained earlier (see Chapter 4 part 2) that the eye of an insect is faceted, has a short range of clear vision and is particularly sensitive to movement; in fact, each facet at least in the honey-bee, is able to perceive changes taking place ten times more often than the fastest that can be noticed by the human eye.

The reactions of honey-bees to two-dimensional shapes were closely investigated by Hertz. Honey-bees are very poor at distinguishing shapes but, like butterflies (Chapter 4 part 2), they are quite sensitive to differences in length of outlines. Tests were made by Hertz (1935) with coloured figures on a white ground (the colour being immaterial) and it proved impossible to train the bees to visit a circle for food when a cross of equal surface area was also offered. In fact there was found to be an innate and persistent preference for figures with longer outlines. A further property for which the bees had an inherited preference was unsteadiness of outline. Thus a group of small crosses was more attractive than a group of small circles, even though their total outline was of identical length. In addition, if leaves of various shapes were put out on a white background and covered with glass, the honey-bees were attracted and alighted over the leaves, visiting the more compound shapes and ignoring the others; if the more compound shapes were removed the others were then visited, and the bees concentrated on any teeth or lobes present and on the tips and stalks of the leaves.

Manning (1956a) offered bumble-bees coloured paper shapes with a diameter of about 12 cm. – large enough for him to follow the bees' reactions to different parts of the pattern. In a test with uniformly coloured models in the form of a circle, a six-pointed star and a 'flower' with six rounded 'petals', nearly all the visits were to the edges of the models, where the bees hovered, dipped down to the surface, and sometimes alighted and walked round the edge. Having approached the centre of the 'flower' along the side of one petal, they then followed the margin of the next petal and so receded from the centre again. Manning also noticed that some natural flowers are large enough to reveal this preference for colour boundaries; for example, bumble-bees and honey-bees visiting *Magnolia* repeatedly reacted to the edges of the petals, and some of the bees failed to find the stamens and flew off.

Having found out the reaction of bumble-bees to plain models, Manning now introduced 'flowers' of similar size, but with nectar guides. He used blue or yellow 'flowers', with a thin line of the other colour along the middle of each 'petal', reaching nearly to the centre. He now found that the bees made $1\frac{3}{4}$ times as many dips down to the centre as to the edge of a model, and often the bees reached the centre by flying along one of the lines. A bee visiting a model would make several dips over it, and it was found that the first dip was much more often at the edge than at the centre. On the other hand, there were many more subsequent dips to the centre than to the edge. What was apparently happening was that the bees, on sighting a model from a distance (usually not more than two feet), went straight to its edge without perceiving the guide-marks. The effect of the guides was therefore exerted only at very short range.

The effect of three-dimensional models on honey-bees was studied by Hertz (1931), and it was found that white structures against a white background were attractive in proportion to the depth of shade in their darkest parts. The reactions of bumble-bees to similarly coloured hollow cones and flat discs were compared by Manning (1956a). It was found that when the bees made clear choices of target from a distance of two or more feet there was no discrimination, and that dipping in flight after arrival was much more frequently to the centre than to the edge of the cones (which had their vertices downwards and were tilted at 30° to the vertical). Bees visiting the cones often flew right into them and quite frequently alighted near their centres. In addition a flat disc that was coloured more intensely towards the centre was found to receive more dips at the centre than a uniformly coloured one.

Kugler (1943) found that hollow cones received more visits from bumble-bees than convex structures. For this experiment the bees were in small flight-boxes, so that they must have made their choices at short range, and this doubtless enabled them to discriminate between the different types of model.

In Chapter 4 part 2 it was seen that most *Diptera* and *Lepidoptera* can distinguish colours of the blue group from those of the yellow group, and that many of them have a preference for one group or the other. It was also shown that some butterflies react to red as a colour of the yellow group, whereas others are red-blind. Red-blindness also occurs in all bees that have been investigated. That it exists in *Prosopis* was established by Knoll (1935), using the method that was used for *Vespula* by Schremmer. The only other European bees that have been investigated for colour vision are the honey-bee and the bumble-bees. In the early part of this century honey-bees were trained by von Frisch to feed on a paper of a particular colour among a series of grey papers of varying shades, and it was found possible to train the bees to blue and to yellow but not to red, which they confused with the greys. Later work by von Frisch and his associates showed that the honey-bee can also perceive blue-green and ultra-violet as distinct colours. A flower that reflects all light wavelengths but ultra-violet (a white flower to man) therefore appears coloured to a bee, but a flower that reflects all wavelengths including ultra-violet appears white to a bee too. 'Bee-white' flowers are rare, and it is difficult to train bees to this colour (von Frisch, 1950, 1954). The bumble-bees have similar colour-vision to the honey-bee (Kugler), and ultra-violet perception has also been confirmed in several solitary bees in America (Grant, 1950c).

A thorough investigation into the colour vision of honey-bees has been carried out by Daumer (1956) using an elaborate specially built 'spectral colour-mixing apparatus'. The equipment permitted the comparison of 'bee-white' light with various coloured lights of varying intensity.

The colour discrimination of bees was investigated by training the bees to a particular colour and testing their ability to distinguish it from colours near to it in the spectrum and from those distinct from it. It was found that for bees there were three main colour groups: yellow, blue and ultra-violet. These groups were almost totally distinguished from one another at all intensities used. Within each group some power of discrimination was also found, but it was not very accurate in the yellow and blue groups, and in both these groups it fell off with decreasing intensity. In the yellow group the colours distinguished were orange, yellow and green; in the blue group they were blue and blue-violet. The ultra-violet group was different in that two light-wavelengths quite near together were distinguished at all intensities. Between the yellow and blue groups the bees saw a colour (blue-green) which they could distinguish accurately from the main groups and, similarly, they could distinguish a colour (violet) between the blue and ultra-violet groups (Fig. 34). The situation in man is exactly comparable: there are three main ranges: red, green and blue; and two intermediate regions: yellow and blue-green. Man can see a further colour (purple) which is produced by mixing light from the opposite ends of the spectrum (red and blue). In bees, likewise, Daumer found that by mixing light from the opposite ends of the bee-visible spectrum (yellow and ultra-violet) another colour can be produced which is distinct to the bees ('bee-purple').

For man, when red is removed from white light the remaining light is a mixture of blue and green, and is seen as blue-green. If pure blue-green is added to red the effect is white, and the colours are therefore described as complementary. Similarly, for bees, white light with ultra-violet removed appears as blue-green, and if blue-green is added to ultra-violet the effect is 'bee-white'. Blue-green and ultra-violet are therefore complementary for bees. The complementary colour to yellow is violet for bees, and the complementary colour to blue is 'bee-purple' (Fig. 34).

Blue-green for man, besides being a spectral colour, can also be produced by mixing blue and green, while yellow can be produced by a mixture of red and green spectral light. The colours of the two transitional zones of the bee spectrum can similarly be produced by mixing the spectral lights on either side of them. Daumer found that when these transitional colours were produced by mixing it was possible to make different mixtures which were distinguishable by bees. Thus, three easily

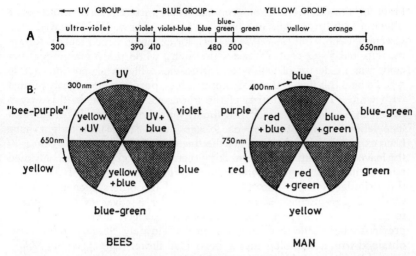

FIG. 34. A, spectrum of bee-visible light. Figures show wavelength. B, systems of complementary colours for bees and man. Opposite segments are complementary and when combined produce 'white'. Main colour groups shaded. nm=nanometres. After Daumer (1956).

distinguishable violet colours were produced by mixing blue and ultra-violet in different percentages. Similarly, with 'bee-purple', which has no pure spectral counterpart, different mixtures of yellow and ultra-violet produced two easily distinguishable colours. The sensitivity of bees to ultra-violet is brought out by the fact that a mixture of 2 per cent. ultra-violet and 98 per cent. yellow was distinguished by the bees from yellow, whereas 50 per cent. yellow had to be added to ultra-violet to make it different from pure ultra-violet to the bees. Mixtures of colours within the main colour groups also give intermediate colours; for example, a mixture of orange and green is confused by bees with yellow.

It has been shown that the eye of the honey-bee, like that of the blow-fly, has three types of light-sensitive cells, though their spectral sensitivities are different, there being a numerous type sensitive to ultra-violet, and two less numerous types sensitive respectively to blue and yellow. A hypothesis to explain how these give rise to the colour circle shown in Fig. 34B, and to account for the acute sensitivity to changes of wavelength in the transitional zones between the main colours (evident from Fig. 34A), is given by Burkhardt (1964).

Daumer followed his remarkable investigation of the honey-bee's colour-discrimination by an equally outstanding study of the colours of flowers and the reactions of bees to them (Daumer, 1958). Two hundred kinds of flowers were photographed, each through three interference filters: one ultra-violet, one blue and one yellow. This showed that Creeping Cinquefoil (*Potentilla reptans*), for example, which looks yellow to us, reflected 7 per cent. of 'bee-white' light, while the remaining 93 per cent. was made up of yellow and ultra-violet in the proportion 94.5 to 5.5. To bees this flower therefore appears 'bee-purple', very slightly diluted with white. The constitution of the flower-colours found in this way is shown in Table 4. Daumer also found that the bees could distinguish 'bee-yellow' flowers from green foliage more easily than might have been expected since, on average, the 'bee-white' element in the colour of the leaves is six times as great, making them appear greyish. He confirmed part of his analysis of flower-colour by experiments with bees. When three similar flowers (one 'bee-yellow' and the others each a different shade of 'bee-purple') were presented under a cover of glass opaque to ultra-violet, the bees confused them completely, although they had previously been able to distinguish them accurately. Similar results were obtained with a 'bee-blue' and a 'bee-violet' flower (two species of *Scilla*).

Table 4. Flower-colours

Actual flower-colour	Colour to bees	Colour to man	Example
yellow+ultra-violet	'bee-purple'	yellow	Creeping Cinquefoil (*Potentilla reptans*)
yellow	yellow	yellow	Cowslip (*Primula veris*)
green	yellow	green	Stinking Hellebore (*Helleborus foetidus*)
green+blue+red	blue-green	white	Wild Cherry (*Prunus avium*)
blue	blue	blue	Wood Forget-me-not (*Myosotis sylvatica*)
blue+red	blue	purple	Ling (*Calluna vulgaris*)
blue+ultra-violet	violet	blue	Birdseye Speedwell (*Veronica chamaedrys*)
blue+ultra-violet+red	violet	purple	Purple Loosestrife (*Lythrum salicaria*)
ultra-violet+red (*rare*)	ultra-violet	red	Corn Poppy (*Papaver rhoeas*)
red (*very rare*)	black	red	(non-British) *Nonnea pulla*

Many flowers which have some ultra-violet reflection show patterns caused by the absence of ultra-violet from certain regions. These patterns, invisible to man, act as guide-marks to the bees in a similar way to the nectar guides that we can see. An extensive survey of such patterns was carried out by Kugler (1963, 1966). He found that, whereas only 30 per cent. of the flowers investigated had patterns visible to the human eye, a further 26 per cent. could be added by including ultra-violet patterns. These 26 per cent. thus have patterns visible only to an eye sensitive to ultra-violet. However, many flowers with patterns visible to man also have patterns formed by ultra-violet absorption. No types of ultra-violet pattern were found that were not also found in the 'visible' range. The frequency of patterning in general increases with the complications of exploiting the flower, and it is thus higher among zygomorphic than actinomorphic flowers and, correspondingly, among bee-visited flowers. However, among these the frequency with which the pattern is formed *only* by ultra-violet absorption is much below average. This blending of ultra-violet absorption with absorption in the 'visible' range may perhaps lead to more varied colourings than are found in the simpler types of flower. Tests on the significance of these patterns were carried out by Daumer (1958) using petals with an ultra-violet-free spot at the base. The petals were arranged in the form of a flower, either the natural way, with the spots central, or with the spots at the periphery. The position of the spots made little difference to the approaches of honey-bees from a distance, but after alighting the bees made proboscis movements over the spots, wherever they happened to be. A similar result was obtained with honey-bees from an experimental colony which had never seen flowers or coloured flower models, and this showed that the attraction of the spots was inborn.

Foraging honey-bees navigate by the sun and they can tell the direction of the sun even if the sky is completely overcast, provided there is no obstruction other than the clouds. It has also been found by von Frisch (1950) that they can make use of the polarisation of light from a blue sky to orientate themselves.

Powers of Communication

One of the most remarkable powers of the honey-bee is its ability to inform members of its colony of the direction, distance, abundance and nature of a valuable source of food, the direction being indicated only if the food source is more than about twenty yards from the hive. We have already mentioned that information about the nature of the food is conveyed by its scent. The rest of the information is conveyed by dances executed by the bees that have found the food. This behaviour is well-known and has been described by its discoverer, von Frisch (1950, 1954), and also by Butler (1954).

Sense of Time

Honey-bees have a highly developed time sense. If they are fed at certain times of the day in a certain place or places, they will appear at the feeding places at the appropriate times, and if, as an experiment, food is withheld they will remain searching the accustomed area during the period when food is normally provided. In addition, if honey-bees are fed on one day at a site in a particular direction from the hive, they will set off in the same compass direction the following morning, even though the hive has been moved overnight to a new and entirely unfamiliar place and faces another way. Although this does not bring the bees to the food, it shows their skill in navigating by the sun and allowing for its position according to the time of day (von Frisch, 1954).

Foraging Behaviour

The constancy of the individual bee to a particular species of flower is an important factor in cross-pollination. Little is known about foraging habits among solitary bees, but Chambers (1946) investigated the pollen loads of some species of *Andrena* that visit fruit blossom and found that *A. varians*, *A. haemorrhoa* and *A. armata* were very constant in collecting fruit tree pollen (cultivated plum, cherry, pear and apple, and wild blackthorn (*Prunus spinosa*)), *A. varians* particularly so. *A. armata* later changed its constancy to Sycamore (*Acer pseudoplatanus*) when this came into flower. Evidence of appreciable and often substantial constancy has been found from pollen analysis of the loads of *Andrena*, *Halictus*, *Anthophora* and *Megachile* in America (Grant, 1950c). Direct observations on constancy in *Halictus* have been given by Kugler (1940). In one instance, one of these bees consistently visited the yellow flowers of Greater Hawkbit (*Leontodon hispidus*) without paying the slightest attention to various violet and purple flowers growing in the same place. Another *Halictus* repeatedly visiting Dandelion (*Taraxacum officinale*) occasionally alighted on flowers of Meadow Buttercup (*Ranunculus acris*), but immediately flew off and continued to collect food from the Dandelions; both kinds of flower are 'bee-purple' according to Daumer, though yellow to our eyes. At a place where Greater Hawkbit was growing with a similar yellow Composite (*Crepis biennis*) one *Halictus* fed only at the Hawkbit and another only at the *Crepis*. In addition to behaviour of this type, Kugler saw *Halictus* visiting alternately two species of flower on the same flight, but here also there were generally a few successive visits to the same type of flower.

As an example of an investigation on the foraging behaviour of the honey-bee we may quote that of Ribbands (1949). He studied the foraging of individual bees on crops specially planted for the purpose. These consisted of five species of flower planted together, three of them separately in long contiguous rows, and all five separately nearby in large or small rectangular beds, some contiguous, and some separated by distances of a few feet. The bees were marked and watched continuously for long periods. A bee which worked numerous flowers of Californian Poppy (*Eschscholzia californica*) ranged over a large part of the available crop, and generally progressed steadily in one direction during each foraging trip; she often made successive visits to newly opened flowers or returned to them after visiting one or two other flowers. Evidently the young flowers were supplying more pollen than those that had been open longer. Another bee, collecting pollen from Shirley Poppy (*Papaver rhoeas* cultivars), showed a remarkable constancy to a single very productive bloom, visiting it on each of nineteen successive trips, usually visiting it first on each trip, and on many trips collecting from it all, or nearly all, her load. This showed a very exact awareness of the position of this flower. (The Shirley Poppies were particularly variable in their pollen production, several of them being attractive, though less so than the one just mentioned.) These two bees provided examples of a very extensive and an extremely restricted foraging area. Cases of moderate restriction were provided by bees collecting pollen from another of these specially planted flowers, the Nasturtium (*Tropaeolum majus*); the bees all confined themselves to a few square feet of the available crop, though each had a different foraging area.

The three flowers mentioned so far supplied the bees only with pollen, whereas nectar (and a little pollen) could be obtained from the remaining two kinds of flower in the experiment. A bee that was working *Limnanthes douglasii* for nectar began work one day at 6.30 a.m., collecting pollen from Shirley Poppy. Then, at about 9.0 a.m., it gradually transferred its attention to the *Limnanthes*, which were by that time producing nectar, and abandoned the Shirley Poppy for the day. The bee's time sense, combined with its previous experience, were presumably responsible for its not attempting to visit *Limnanthes* early in the morning, though these factors are not infallible guides because the time of greatest discharge of pollen or nectar is affected by the weather. The value of this use of time sense is increased when alternative crops are far apart, as it reduces the number of unsuccessful journeys to the second crop.

Another bee was found on two successive days to be collecting both pollen and nectar from *Limnanthes*, and pollen from Californian Poppy. During periods when the Poppy was not productive, the bee visited it occasionally until pollen production began, when it increased its visits to this flower.

Ribbands observed that bees working a crop that was rapidly becoming unproductive became increasingly restless, and began to move hastily over the whole foraging area or even beyond it, instead of moving short distances from flower to flower. It is rather curious that, although some bees will change to an alternative crop when one is exhausted for the day, others return to the hive until it is time for their favoured crop to be productive again (von Frisch, 1954; Butler, 1954; Free, 1963). Bees may work an adequately productive crop for a long time before changing over to a better one that has been available all the time. They do not appear to spend time sampling all accessible crops and choosing the best, so that excessive competition on the best crop is avoided.

There were fairly frequent changes in the crops worked by the bees observed by Ribbands, but he noted an exclusive attachment of 12 days by a single bee to a single pollen crop, and one of 21 days to a single nectar crop. Bees that worked only one crop at a time invariably changed from a less fruitful to a more fruitful pollen crop, or from a pollen to a nectar (or nectar plus pollen) crop. Thus it seems that nectar crops are more attractive to the older bees – bees having a foraging life of only a few weeks – in spite of the fact that a bee has to do more work to collect a load of nectar than a load of pollen (in certain other studies, however, no evidence for such a foraging sequence has been found (Free, 1963)). The number of flower visits needed to make up a load of pollen was found by Ribbands to be 1 to 27 for Shirley Poppy, 66 to 178 for Nasturtium, and intermediate for Californian Poppy. The time required ranged from a minimum of three minutes for Shirley Poppy to a maximum of 18 minutes for Nasturtium. Only two loads consisting purely of nectar were fully observed; each required over a thousand flower visits and a time of more than an hour and three-quarters. Bees collecting both pollen and nectar from *Limnanthes* took a minimum of 250 flower visits and 27 minutes to complete a load.

Various factors influence the number of visits required for a particular type of load. For example, some bees are more thorough than others in collecting pollen and therefore complete a load with fewer flower visits and in a shorter time. Furthermore, an increase in the number of visits required to complete a load of pollen or nectar may be caused either by a decrease in production (which often takes place late in the day) or by competition from other bees. The time required to collect a load is also increased by a drop in temperature, which lowers the rate at which the bees can work on the flowers.

Though most honey-bees show strong crop-constancy there are always some bees in a colony that actively explore – at least for part of their lives – coloured and scented objects, and in this way discover new sources of food. These bees are known as scout bees. If they find a rich source of food they remain constant to it for a short time and communicate their knowledge to other members of the hive. Scout bees may be recognised from their behaviour, and it has been found that they are proportionately more numerous early and late in the season when there are fewer types of flower available (Butler, 1954).

The foraging behaviour of bumble-bees has been studied by Manning (1956b) with most illuminating results. The special interest of his investigation arises from his choice of extreme flower types, Houndstongue (*Cynoglossum officinale*), with its inconspicuous flowers (see p. 187), and Foxglove (*Digitalis purpurea*) (Plate 31, p. 238), with very showy ones. The reaction of the bees to each of these flowers (mainly nectar-producing) was investigated separately, but in each case the plants were in a natural habitat, some in a group and others widely scattered in the neighbourhood, being ten feet or more away from the group and each other. Some of the plants were growing naturally but others were grown in pots and put out in positions which completed the required arrangement. The potted plants made it possible to alter the arrangement and then to restore it to its former state.

The Houndstongue flowers have a short tube and a concave five-lobed limb; they are dull brownish red when young and purplish later, but bumble-bees tend to avoid visiting the purple flowers. On each flowering branch there are usually two flowers in the red stage and two in the purple; normally the flowers are nodding, but the first few to open face upwards. In both the seasons when Houndstongue plants were observed there was an interval of eleven to twelve days between the opening of the first flower and the visit of the first bumble-bee. During this time the flowers held copious uncollected nectar and, unlike the later flowers that received visits from the bees, they set no seed. The bumble-bees began to visit the Houndstongue when one of them, passing within two feet of a flower, alighted. Visits were at first rather tentative but were soon repeated and became more purposeful. After a few visits the bees became conditioned to the form of the Houndstongue plant as a whole; this was shown by the fact that they visited Houndstongue plants without any flowers, as well as other plants with a similar growth habit, sometimes searching in the leaf axils, where in the Houndstongue the flowering branches are produced. Manning found that a Houndstongue plant with no flowers was reacted to at a distance of six feet by these conditioned bees, whereas a plant with flowers but no leaves was reacted to at not more than two feet. The strong mousy smell of the plant did not seem to help the bees to find it. Bees moving from one plant to another in the main group used a slow, apparently exploratory flight and kept reacting to other similar species, until they gradually learnt the approximate positions of the Houndstongue plants. Each bee, having visited all the plants in the group, flew off fairly slowly, making wide sweeps over the ground 1½ to 2 feet above it in an exploration flight. The bees now reacted as before to Houndstongue plants and similar species, ignoring plants with conspicuous flowers. In this way all the Houndstongue plants were found, the farthest away being 25 yards from the central group and some being so hidden in bracken that they could be seen only at distances of less than two feet. Bees leaving each of the isolated plants made orientation flights and after a few visits had a very exact knowledge of their positions. Orientation flights are important in the lives of all nest-building *Hymenoptera*, and are used when foraging. The insect flies round the object in which it is interested in gradually increasing circles, and may fix its position by quite distant objects in the landscape (Butler, 1954). The bumble-bees re-visiting isolated Houndstongue plants usually followed a particular course, and flew fast and directly. One bee, which had visited a Houndstongue only once before, returned to the spot on three occasions, although the plant was taken away after the first visit.

One difference between the behaviour of bees at the central group and that of those at the distant plants was observed by removing a plant in each situation. In the central group, some bees did not appear to miss the absent plant, while others visited its site diminishingly in the course of half an hour. The removal of a distant plant, however, did not reduce the number of bees approaching its site, and they searched the site for much longer. This behaviour reflected the bees' approximate knowledge of plants in the central area and their exact knowledge of the positions of the distant plants.

In the Foxglove experiment the bees started to visit the plants as soon as the flowers showed colour, sometimes forcing their way into the flowers before they were fully open and before any nectar had been secreted. The bees quickly found all the distant plants, showing only slight signs of sweeping exploration flights and flying directly to plants up to twelve feet away. Very few orientation flights were seen round the distant plants, and only one of these was a perfect performance. The bees could presumably see the Foxgloves some way off, so that an awareness of their approximate positions was all they needed to be able to find them again. It thus seems that bees must estimate the conspicuousness of a plant and have the power to determine how much orientation is required. The removal of a plant from the central group reduced the number of visits to its site, as in the Houndstongue experiment; bees visiting the sites of distant plants that had been removed, however, searched for them over a wider area than in the Houndstongue experiment, being less sure of their positions. Bees visiting flower-less Foxglove plants spent a negligible time searching them, and other plants with a similar growth habit were not searched at all, although the bees would visit Foxglove flowers lying on the ground. It is to be expected that plants intermediate in conspicuousness between Houndstongue and Foxglove will elicit intermediate behaviour.

It was found that the several species of bumble-bee observed in these experiments, together with a few honey-bees that worked Houndstongue, all behaved in the same way.

Many insects, and particularly bumble-bees, when confronted with an inflorescence of columnar form, such as that of the Foxglove, visit first the

a

PLATE 21. **a**, solitary bee *(Halictus* sp.*)* sucking nectar from Field Bindweed *(Convolvulus arvensis)*. **b**, hover-fly *(Syrphus balteatus)* feeding on pollen of Field Bindweed; note passages to nectary between bases of the stamens.

b

PLATE 22. **a**, flower of
Gentian *(Gentiana kochiana)*
with half of corolla removed,
lit from behind to show
translucent corolla-tube.
b, flower of *Gentiana septemfida*
lit from behind to show
translucent corolla tube.

a

b

lowest available flowers and then work upwards. Bee-pollinated plants with this type of inflorescence show a corresponding behaviour. The lowest flowers open first, but there is an overlap of anthesis between adjoining flowers, and the flowers are protandrous. This means that the bee first visits the oldest flowers which are in the female stage, and any pollen deposited in them is likely to come from another plant. As the bee moves upwards it acquires pollen from the younger flowers in the male stage; after visiting the highest available flower the insect must fly off to another inflorescence (which usually means going to another plant) in order to avoid revisiting flowers that have just been depleted. The protandry of flowers in such an inflorescence thus constitutes a highly efficient outbreeding mechanism. In Rose-bay (*Chamaenerion angustifolium*), which has an inflorescence of this type, the visitors cling to the projecting and slightly drooping stamens and style and because of their weight assume an almost upright posture. Benham (1969) has suggested that this could be an arrangement to ensure that the insects move in the upward direction and that the occurrence of such arrangements may have led to upward movement becoming instinctive. A further feature of this plant which might reinforce the tendency to start at the bottom is the fact that nectar secretion is greatest during the female stage.

Foraging bumble-bees can often be seen rejecting certain flowers of a type on which they are working; this may be partly based on age-changes in the flower, such as a looseness among the petals of *Leguminosae*, withered petals in general, or colour changes in older flowers (which are quite pronounced in the Horse Chestnut (*Aesculus hippocastanum*), for example). Sometimes, however, one sees bumble-bees rejecting apparently suitable flowers, which suggests that they may be able to recognise recently visited flowers, possibly by means of scent. Some positive evidence that tropical *Xylocopa* bees select the flowers they visit in this way has been obtained by Van der Pijl (1954).

Kugler (1943) pointed out that many of the outstanding bumble-bee flowers (mainly long-tubed or funnel-shaped) have features which make the nectar difficult to find. These often take the form of physical barriers which have to be pushed aside; such barriers are found in Toadflax (*Linaria* spp.), Louseworts (*Pedicularis* spp.), Antirrhinum, Delphinium, Comfrey (*Symphytum officinale*) and *Leguminosae* (see Chapter 6). The bees are not particularly clever at finding their way into flowers, but their good memory enables them to repeat with ease their first successful visit to a flower. This explains the importance of flower-constancy to bees, as it enables them to continue working a flower to which they have become accustomed.

When all the flowers regularly visited by bumble-bees are listed it is found that those which have a reputation for being especially adapted as bumble-bee flowers represent a comparatively small proportion. This is because we get our idea of the bumble-bee flowers mainly from those preferred by the long-tongued species. In a study of four species of bumble-bee in a limited area Brian (1957) reported marked differences in their behaviour. The short-tongued species *B. lucorum* visits a wide range of short-tubed and fully open flowers. In addition it collects honey-dew and bites holes in flowers as a short cut to the nectar (Plate 34b). In the type of flower it visits and in its opportunistic behaviour it resembles the honey-bee. At the opposite extreme is the very long-tongued *B. hortorum*; most of the flowers it visits are too deep for the other species to reach the nectar; it does not collect honey-dew, nor bite flowers, although Brian found that its jaws are strong enough for it to do so. The two other species studied by Brian were intermediate in tongue-length and behaviour, *B. pratorum* being more like *lucorum*, and *B. agrorum* (Plate 18) more like *hortorum*. Leclercq (1960) studied *B. terrestris*, showing it to be closely similar in habits to *lucorum*. Leppik (1953) found that bumble-bees have a strong preference for irregular (zygomorphic) flowers over radially symmetrical ones, while honey-bees show the reverse preference.

Brian (1957) compared the tongue-lengths of the bumble-bees with the tube-lengths of the flowers they visited, and found that on average the tubes were a few millimetres shorter than the tongues that drained them. Each species therefore had different preferences, but no bee normally had to extend its proboscis fully. Presumably if a bee constantly had to forage at the limit of its reach it would find it more laborious and be less efficient.

Like the honey-bee, bumble-bees will sometimes work two or more kinds of flower simultaneously (Hulkkonen, 1928), and they can be simultaneously trained to visit models with either of two scents, which they then prefer to models containing scents to which they have not been trained (Kugler, 1932). Foraging at two kinds of flower is not at all the same as promiscuous visiting; it is probably induced by insufficient abundance of any one species of plant. When a crop of Foxglove came into flower, the bumble-bees which began working it did not at first confine themselves to it as there were not enough flowers open to keep them busy; later, when there were more Foxglove flowers, some of the bees restricted themselves to them, though some others occasionally visited Red Campion (*Silene dioica*) (Manning, 1956b). Kugler (1943) saw a bumble-bee working in an area where there were three yellow-flowered species of different genera growing together; the bee was working only two of them, and though it was attracted by the third it never collected any food from it. In addition, both honey-bees and bumble-bees are well known to visit different colour forms of the same species growing together, as they often do in gardens (Grant, 1950c). It therefore seems that bees continue to investigate flowers coloured differently from the one on which they first found food, and that the discovery of the right scent induces them to alight.

Bumble-bees were believed by Müller (1881) to show a strong preference for blue and purple over white and yellow flower colours. Circumstantial evidence obtained by Kugler (1943) and Brian (1957) suggests that there is no inherited preference in this respect, and the preference noted by Müller probably therefore reflects the prevailing colours of the flowers adapted to pollination by the longer-tongued species. The colour preference of honey-bee scouts, which is probably inherited, was investigated by Butler (1951), using white, green, pink, blue and yellow papers; blue and yellow were almost equally attractive, and much more so than the other colours.

Bumble-bees, like honey-bees, may restrict their foraging to a small area in a large crop (Free and Butler, 1959) and, also like honey-bees, they may persist in working one crop without giving attention to others, with resulting neglect of superior crops (Kugler, 1943, p. 302). A very important difference from honey-bees is shown by bumble-bees in their almost complete lack of communication of information about food sources. The only trace of communication arises when a bee sees others on flowers; then it may be induced to alight on the same flower, or one nearby (Brian, 1957).

According to a review article by Brian (1954) bumble-bees in general work more quickly than honey-bees, but the different species have different rates of working which increase in proportion with tongue-length.

* * *

Although bees are on the whole extremely reliable pollinators, their efficiency, especially that of the social species, appears to have gone beyond what is best for flower-pollination. The biting of holes in flowers and other kinds of illegitimate visiting are obviously bad for the plants (Plates 18b and 34d). In addition, the frequent concentration of social bees on very limited foraging areas means that, considering the number of flower-visits made, they contribute rather meagrely to outbreeding among plants. The organisation of the honey-bee colony makes for the ruthless exploitation of every food source. Social life makes possible communication, co-operative effort and division of labour. The honey-bee seems more or less to have taken charge of its floral environment, and there is perhaps here a faint parallel with the control of the environment achieved by man. The evolution of an almost total dependence on flower food by bees means that flowers in general must provide entirely for the nourishment of these insects. Of this nourishment, moreover, each species of bee-pollinated plant must provide a supply worth exploiting, if it is to retain its pollinators.

If the proboscis lengths of the *Hymenoptera* as a whole are compared with those of *Diptera*, a very similar picture is found in each order. In both, there is a large number of quite unspecialised short-tongued visitors to flowers and other food sources, some of these insects being occasionally induced by special devices to visit highly specialised flowers. Then in both orders there are groups which feed exclusively on flowers, and they mostly have tongues of moderate length, a few having really long ones. However, long-tongued insects, it should be noted, are much commoner among the bees than among the flies.

Our extensive knowledge of the structure and behaviour of flower-visiting insects will make it possible to appreciate the functioning of the advanced types of flower to be described in the next chapter.

PLATE I. Guide-marks on flowers. **a**, Forget-me-not, *Myosotis sylvatica*. **b**, Buxbaum's Speedwell, *Veronica persica*. **c**, *Cistus ladaniferus*. **d**, Common Toadflax, *Linaria vulgaris*. **e**, Dune Pansy, *Viola tricolor* subsp. *curtissi*. **f**, Eyebright, *Euphrasia montana*.

PLATE II. British species of *Ophrys*. **a**, Fly Orchid, *O. insectifera*, Wiltshire, June. **b**, Early Spider Orchid, *O. aranifera*, Dorset, May. **c**, Bee Orchid, *O. apifera*. Both pollinia have fallen from the anther; one has swung into contact with the stigma. Dorset, June.

CHAPTER 6

INSECT-POLLINATED FLOWERS - II

In Chapter 3 we described a number of flowers which are believed to be primitive, both in their structure and in their pollination mechanism – flowers with free sepals and petals and with freely exposed nectar or pollen, often pollinated rather indiscriminately by a variety of insects. We included with them some less primitive flowers, with united sepals and petals, which show the same kind of pollination mechanism. In the present chapter we shall consider the flowers which are much more specifically adapted in form and structure to particular pollinators – in most cases the long-tongued insects which themselves have become specialised flower visitors.

One's first impression on surveying these more specialised flowers is of exuberant, and indeed bewildering, variety. Many classifications of flower types have been proposed in the past. Such classifications are helpful in arranging descriptive accounts of flower pollination; they are also a useful tool in studies of floral ecology. The culmination of nineteenth-century efforts in floral classification was the system of Knuth, based on the earlier classification of Delpino and Müller. In this system, some groups were defined according to the accessibility of nectar and others according to the groups of insects to which the flowers were specially adapted, while, in addition, one group was defined by the arrangement of flowers in the inflorescence and another by the size of the insect visitors. Evidently, many plants will fall in more than one class in such a system (for example, most 'bee flowers' will also be 'flowers with concealed nectar'), though the way the system is set out implies that the classes are mutually exclusive. Clearly, if one wishes to classify by several different criteria, several different and complementary classifications will be needed. Flowers pollinated by a particular group of insects certainly tend to have features in common (the same applies to, say, wind-pollinated or bird-pollinated flowers). However, flowers may be very differently adapted to the same pollinator (It is a common principle in Nature that the same result can be attained in different ways), and in any case, a flower is often pollinated by more than one agent. It is more useful to think of a *syndrome* of characters commonly associated with a particular mode of pollination than to try to set up clear-cut pollination classes (see Van der Pijl, 1961). Even in a purely morphological classification there is no single set of criteria that will accommodate and display all that is significant and interesting in the diversity of flower form, and there are always flowers that are unique and will not fit satisfactorily into any classification.

FLORAL TUBES AND THE 'CONCEALMENT OF NECTAR'

Several broad trends of adaptation to specialised pollinators have evidently operated very widely among the families of flowering plants. One of the most important of these was described by Knuth in terms of 'concealment of nectar'. Leppik (1957) described another facet of essentially the same process in terms of development from two-dimensional 'actinomorphic' and 'pleomorphic'* flowers to three-dimensional 'stereomorphic' flowers. The nectar is so placed that it can be reached only by insects with a tongue of some length, and these flowers can be worked quickly and effectively only by an insect with considerable powers of perception of three-dimensional form.

Although union of perianth parts is especially characteristic of flowers pollinated by long-tongued insects, adaptation to these visitors has been achieved by a wide variety of plants in which these parts are free from one another. In the Wallflower (*Cheiranthus cheiri*) the sepals are elongated and pressed firmly together, forming a long narrow tube (Fig. 35). The petals are modified accordingly, having an expanded 'blade' and a slender 'claw'. This flower is related to the Common Winter-cress (*Barbarea vulgaris*) and Wild Candytuft (*Iberis amara*) described in Chapter 4, and it is pollinated by *Lepidoptera* and long-tongued bees. In the Larkspurs (*Delphinium* spp.) the sepals are petaloid and one of them is produced into a long hollow spur, while in the Columbines (*Aquilegia* spp.), which are in the same family (*Ranunculaceae*), a long spur is formed by each of the five petals. A fairly common way in which the flower tube is lengthened without union of the sepals or petals is deepening of the receptacle so that it forms a tube itself. In the Wild Cherry (*Prunus avium*) the numerous stamens add to the depth of the receptacle, which is only three or four millimetres long (Fig. 36). The receptacles of the Blackcurrant (*Ribes nigrum*) and the Gooseberry (*Ribes uva-crispa*) are even shorter, and their depth is slightly augmented by the small erect petals of these flowers. In the Raspberry (*Rubus idaeus*) the petals and stamens set round the edge of the nearly flat receptacle combine to achieve the effect of a short tube (Plate 7b). Very deep receptacles are found in Mezereon (*Daphne mezereum*) (Fig. 37) and Purple Loosestrife (*Lythrum salicaria*); in Mezereon,

---

* Leppik uses the term 'actinomorphic' in a narrow sense for flowers with stamens, pistil and nectar at the same level, with radial symmetry, and with relatively large and variable numbers of floral members: in 'pleomorphic' flowers the floral parts (especially the petals) appear in definite numbers, a fact which Leppik considers important in the recognition of these flowers by insects.

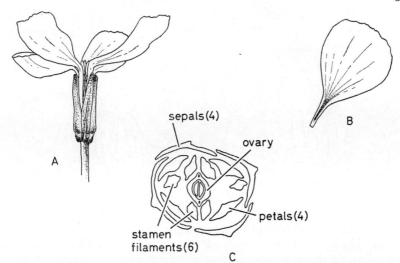

FIG. 35. Wallflower (*Cheiranthus cheiri*). A, whole flower. B, petal. C, transverse section of flower-tube, showing that there are two nectar passages.

which belongs to the family *Thymelaeaceae*, there are no petals and the receptacular tube and the sepals are petaloid.

There are two important groups of 'long-tubed' flowers in which the sepals are united into a tube while the petals remain separate. One of these is that part of the family *Caryophyllaceae* comprising the Pinks and their allies. An example is the Cheddar Pink (*Dianthus gratianopolitanus*) (Fig. 38), which is similar to the Wallflower apart from its united sepals and greater number of floral parts. The Pinks, with their slender-tubed flowers, are adapted to pollination by *Lepidoptera*, whereas the allied Red Campion (*Silene dioica*) (Plate 17a), with a shorter and wider tube, is adapted to bees and long-tongued flies. The second group comprises the Vetches, peas and their many allies in the family *Leguminosae*. These are like the Pinks and Campions in having five petals which are clawed so that they fit into the tubular calyx, but they differ from them in the marked inequality of their petals, and they will be considered later (p. 199).

Examples of flowers with united petals as well as united sepals are the Primrose (*Primula vulgaris*) (Fig. 74, p. 295) and the Periwinkles (*Vinca* spp.) (Fig. 40). The corolla here consists of a slender tube within the calyx and a flat 'limb' or disk on which visiting insects can alight. There are many flowers of this form in the Olive family (*Oleaceae*), for example

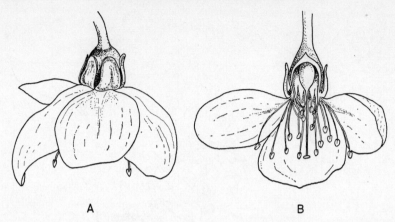

A   B

FIG. 36. Flower of Cherry (*Prunus avium*). A, side view. B, section showing sepals, petals and stamens attached to the rim of the cup-like receptacle, of which the inner surface secretes nectar.

Jasmine (*Jasminum* spp.), Lilac (*Syringa vulgaris*), and in the Gentian family (*Gentianaceae*), for example Centaury (*Centaurium erythraea*) and Spring Gentian (*Gentiana verna*). The Borage family (*Boraginaceae*) show great variation in the relative proportion of the tube and limb. The hanging, tubular bumble-bee flowers of Comfrey (*Symphytum officinale*) (Plate 34c) are at one extreme; at the other are the bright blue 'rotate' flowers of Green Alkanet (*Pentaglottis sempervirens*) and Borage (*Borago officinalis*), with a very short tube and a broad flat limb. Between these extremes lies the

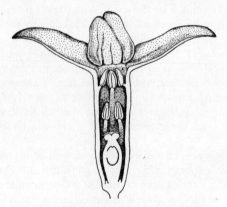

FIG. 37. Mezereon (*Daphne mezereum*).

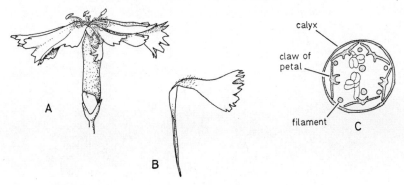

FIG. 38. Cheddar Pink (*Dianthus gratianopolitanus*). A, side view of flower.
B, single petal. C, transverse section of flower-tube about half-way up the
calyx. The channel on the claw of each petal is probably a proboscis
guide; it fades out below. The stamens elongate in succession from the
bottom of the tube; five cut filaments and two anthers just below the cut
are shown in C, but three even shorter stamens are omitted. The nectaries
are at the bases of the stamens.

Houndstongue (*Cynoglossum officinale*) with its dingy purple cups surmounting a short corolla tube (see p. 177).

A variation on these tubular flowers is seen in those species with broad
corollas contracted to a narrow mouth. Flowers with 'urceolate' corollas

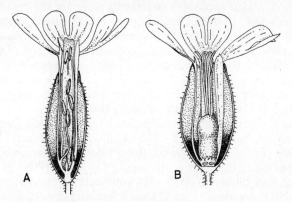

FIG. 39. Red Campion (*Silene dioica*). A, male flower.
B, female flower. Notice the rudimentary gynoecium
in the male flower and the ring of small knob representing rudimentary stamens in the female flower.
Exeter, Devon.

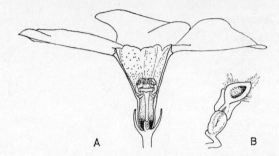

A   B

FIG. 40. Lesser Periwinkle (*Vinca minor*). A, side view
of flower with two corolla-lobes and the corresponding
part of the tube removed. B, stamen. The stamens and
style are very specialised in structure. The pollen is
shed in coherent masses on the non-receptive top
surface of the style. It will adhere to the tongue of a
visiting insect only after this has been made sticky by
the secretion from the receptive zone which encircles
the widest part of the head of the style.

of this kind are common in the Heather family (*Ericaceae*). The small
flowers of Bell Heather (*Erica cinerea*) are visited by a variety of insects with
slender mouth-parts, but probably only butterflies can reach the nectar in
intact flowers, and bumble-bees commonly perforate the corollas. Cross-
leaved Heath (*Erica tetralix*), with slightly larger flowers, appears to be
visited mainly by bees, but there is some doubt as to how many of these
can reach the nectar or bring about pollination; Hagerup and Hagerup
(1953) considered that pollination by *Thrips* or autogamy is the rule.
Most of the larger flowers of this type are more or less pendulous and
visited principally by bumble-bees, for instance Bilberry (*Vaccinium
myrtillus*) and St Dabeoc's Heath (*Daboecia cantabrica*).

Sometimes, flowers with united petals have the sepals free or much
reduced. Examples are Field Madder (*Sherardia arvensis*) (Fig. 41),
Honeysuckle (*Lonicera periclymenum*), the Bellflowers (*Campanula* spp.) and
the florets of many members of the family *Compositae* (Figs. 71–73). It is
noteworthy that these are all flowers with inferior ovaries, and all are
flowers which (for various reasons) have a less evident need than many for
extra support at the base of the corolla.

There are some interesting instances of tubular flowers among the
Monocotyledons. The Bluebell (*Endymion non-scriptus*) (Fig. 42) has six
free perianth segments, which are pressed tightly together to form a tube
about a centimetre long. The flowers are pendent, and are especially

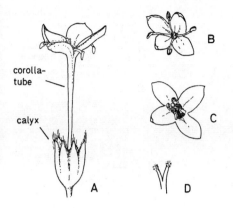

corolla-
tube

calyx

A

B

C

D

FIG. 41. Field Madder (*Sherardia arvensis*). A, side view of old flower with stamens recurved. B, face view of same flower. C, face view of young flower with stamens erect. D, slightly divergent style-branches of an old flower. Cambridge.

visited by bumble-bees (Plate 18a); another pollinator is the long-tongued hoverfly *Rhingia campestris*. Honey-bees, on the other hand, sometimes alight on the outside of the flowers and push their tongues between the perianth segments, thereby evading the pollination mechanism (Knight, 1961) (Plate 18b). In the Nodding Star-of-Bethlehem (*Ornithogalum nutans*), the stamen filaments are very broad, and overlap to form a tube similar in shape to the flower of the Bluebell (Fig. 43), while the perianth segments are held well away from the staminal tube and serve for display. The long-tongued bee (*Anthophora*) and the bumble-bee (*Bombus pratorum*) have been seen visiting the flowers in Cambridge Botanic Garden (P.F.Y.). Related plants in which the perianth segments are united are the Grape Hyacinths (*Muscari* spp.), and Solomon's Seal (*Polygonatum* spp.) (Fig. 44), which has flowers functionally similar to those of the Bluebell.

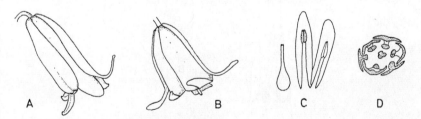

A

B

C

D

FIG. 42. Bluebell (*Endymion non-scriptus*). A, side view of young flower. B, older flower. C, an inner and outer tepal, each with attached stamen, and the pistil. D, cross-section of flower near the mouth; the stamen attachments are shown, although some of them are below the plane of the section. Cambridge.

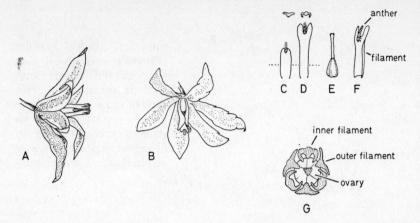

FIG. 43. Nodding Star-of-Bethlehem (*Ornithogalum nutans*). A, side view of flower. B, face view of flower. C, external view of outer stamen and cross-section taken at the level of the dotted line, outer surface uppermost. D, the same, inner stamen. E, pistil, side view. F, oblique view of inner stamen from the inside. G, cross-section of flower-tube taken above the level of the ovary.

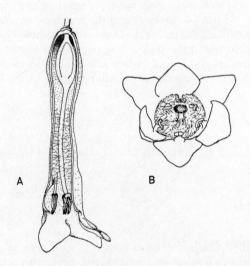

FIG. 44. Common Solomon's Seal (*Polygonatum multiflorum*). A, flower with half of perianth removed. B, face view of flower.

PLATE III. Bird-pollinated flowers. **a**, *Columnea microphylla*, (Gesneriaceae, Costa Rica). **b**, *Lobelia cardinalis*, (Lobeliaceae, eastern N. America). **c**, *Abutilon megapotamicum* (Malvaceae, S. America). **d**, *Strelitzia reginae* (Musaceae, S. Africa). **e**, *Manettia inflata* (Rubiaceae, S. America). **f**, *Callistemon citrinus* (Myrtaceae, Australia).

PLATE IV. Carrion-flowers and fly-traps. **a**, *Stapelia variegata* (Asclepiadaceae, S. Africa). **b**, *Aristolochia elegans* (Aristolochiaceae, Brazil). **c**, *Paphiopedilum villosum* (Orchidaceae, Burma). **d**, *Arisaema elephas* (Araceae, Yunnan).

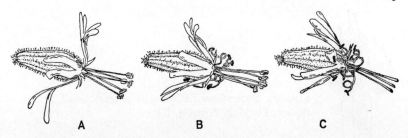

A   B   C

FIG. 45. Nottingham Catchfly (*Silene nutans*). A, flower on first night of opening; the filaments of five stamens have elongated and the anthers are beginning to dehisce. B, flower on second night of opening; the stamens which dehisced the previous night have now recurved against the corolla, and the filaments of the remaining five stamens have elongated and their anthers matured. C, flower on third night of opening; the filaments are now all recurved and the anthers empty, and the styles have elongated and are receptive. Moth pollinated; compare Plate 15b and Honeysuckle, p. 215–16. Branscombe, Devon.

Generally speaking, the form of these tubular flowers makes for greater precision in the behaviour of visiting insects than was the case with the flowers described in Chapter 3. However, as they are radially symmetrical the orientation of the visiting insect in relation to the flower matters little. Apart from the special case of pendulous flowers (visited largely by bumble-bees), the flowers are variously disposed on the plant – often erect, or inclined more-or-less indiscriminately at various angles. The position of the stamens and stigmas in relation to the tube varies from genus to genus and even from species to species – completely enclosed in some (for example Wallflower, Primrose, Periwinkle), and projecting to a greater or less degree in others (Nottingham Catchfly (Fig. 45), Purple Loosestrife, Nodding Star-of-Bethlehem). Borage (*Borago officinalis*) is representative of a rather striking group of (largely unrelated) flowers with rotate or reflexed tubular corollas, and a prominent anther-cone surrounding the style in the centre of the flower. Borage produces abundant nectar (like the related early-flowering 'Abraham, Isaac and Jacob' (*Trachystemon orientalis*) (Plate 20)), and is freely visited by honey-bees, which probe between the filaments to reach the nectary. In so doing, they prise apart two adjacent stamens and release a small amount of the pollen previously shed into the anther cone by the inwardly dehiscing anthers. The flowers are also visited by bees collecting pollen. Bittersweet (*Solanum dulcamara, Solanaceae*) has a rather similar mechanism, but the pollen is shed from pores at the tips of the anthers. It produces little nectar, and has evidently become specialised as a pollen flower. It is freely visited by bees, which release pollen by rapid vibration of their wings as they hang from

FIG. 46. Bittersweet (*Solanum dulcamara*). Half-section of flower. Dawlish Warren, Devon.

the anther cone (Fig. 46). Macior (1964) has observed precisely similar foraging behaviour on the similarly formed flowers of the American *Dodecatheon meadia* (*Primulaceae*). The same flower-form recurs in Cranberry (*Vaccinium oxycoccos*, *Ericaceae*), which is apparently little visited and usually selfed, and in the Australian genus *Dianella* (*Liliaceae*).

TRUMPET AND BELL-SHAPED FLOWERS

A rather different line of development is shown by the trumpet and bell-shaped flowers which an insect must crawl inside to feed and to bring about pollination. Here the emphasis is on adaptation to the body-form of the pollinator rather than to the length of its mouth-parts, though often there are also adaptations preventing short-tongued insects from reaching the nectar. Of course, many of the cup or bowl-shaped flowers described in Chapter 3 are obviously adapted in a general way to the size of their usual pollinators, and no completely sharp line can be drawn between these flowers and those with a deeper trumpet-shaped or bell-shaped corolla. Nevertheless, among the Dicotyledons most of the flowers of the kind we are considering here have gamopetalous corollas. Among the Monocotyledons this generalisation does not hold, and trumpet and bell-shaped flowers include (among other examples) Lilies and Fritillaries with free perianth segments.

A common trumpet-shaped flower is the Field Bindweed (*Convolvulus arvensis*) (Fig. 47A). The short-lived white or pinkish flowers are about 2 cm. long, and the corolla, which forms a narrow tube at the extreme base, flares out to a diameter of up to 3 cm. The five stamens closely surround the style so that, in effect, the style and stamens form a short column in the centre of the flower. Nectar is secreted at the base of the ovary, but can be reached only through five narrow passages between the broad bases of the stamens. The anthers dehisce outwards, so that their

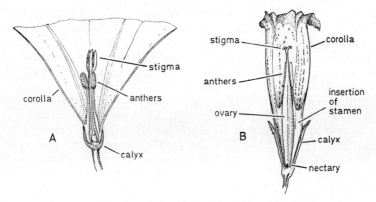

FIG. 47. A, flower of Field Bindweed (*Convolvulus arvensis*) with half of calyx and corolla removed. Alderney, Channel Is. B, flower of Gentian (*Gentiana kochiana*) with half of calyx and corolla removed. The corolla is divided at the base into five deep tubes by the insertions of the stamens. Compare Plate 22a.

pollen immediately comes into contact with the body of a visitor; the two stigma lobes project beyond the stamens, so that an insect bearing pollen from another flower will readily bring about cross-pollination as it enters the corolla. Bees and a variety of flies are the chief pollinators. The degree of variation in corolla colour and other characters suggests that much cross-pollination must take place.

The structure and mechanism of the two common larger bindweeds (*Calystegia sepium* and *C. silvatica*) are similar. It has widely been held that *C. sepium* is adapted to pollination by hawk-moths, especially the convolvulus hawk-moth (*Herse convolvuli*), but this supposition seems to have no more than the most tenuous foundation. None of the great nineteenth-century floral biologists ever seems himself to have observed the moth visiting the flowers. Müller writes (p. 425): 'Delpino mentions *Sphinx convolvuli* as a fertiliser of *C. sepium*; he tells me by letter that one of his friends catches this insect in numbers, standing by a hedge overgrown with the plant, holding thumb and forefinger over a flower and closing its orifice when the insect has entered!' This statement is surprising on two counts. It is commonly assumed that all hawk-moths habitually suck nectar while hovering, like the day-flying (and therefore familiar) hummingbird hawk-moth. Also, the convolvulus hawk-moth is a bulky insect in relation even to the large flowers of the Bindweed. However, it would at least be consistent with Baker's observations and photographs of the visits of the American hawk-moth (*Phlegethontius sexta*) to the similar

large white trumpet-shaped flowers of *Datura meteloides* (Baker, 1961).
Certainly the large size and pure white colour of the flowers suggests a
crepuscular pollinator, and daytime visitors are relatively scarce considering how conspicuous the flowers are. However, the flowers open soon
after sunrise, and are visited by considerable numbers of insects, especially
bees and hoverflies; on sunny days all the pollen is often removed by mid-
morning. Some flowers close in the evening, others remain open through
the night and close the following day (Stace, 1965). Observations of visits
by smaller moths seem to be conspicuously lacking. There can be no
doubt that in the British Isles the flowers are effectively pollinated by
hoverflies (Baker, 1957) or bumble-bees, particularly *Bombus agrorum*
(Stace, 1965; I have seen this species visiting the flowers in Devon and
Dorset), and that failure to set seed is due to self-incompatibility within
vegetatively reproduced populations (Stace, 1961) rather than lack of
pollinators.

Flowers of essentially the same type, though with a more bell-shaped
corolla, are found in the Alpine trumpet Gentians or 'stemless Gentians'
(*Gentiana kochiana* (Fig. 47B), *G. clusii* and their allies); our own Marsh
Gentian (*G. pneumonanthe*) is essentially similar. These are bumble-bee
flowers. Unlike the Bindweeds, the Gentians open for a number of days
in succession, closing at night and in dull weather, and they are protandrous, which further favours cross-pollination. The interior of the throat
is whitish, in contrast to the deep blue of the lobes at the periphery, so
that the interior remains light as a bee enters the mouth of the flower.
This feature, in various forms, is characteristic of many flowers with tubes
or bell-shaped corollas. In *Gentiana pneumonanthe*, and even more strikingly
in the cultivated *G. sino-ornata*, translucent 'window panes' form prominent
stripes up the side of the tube (see pp. 232, 303 and 306).

The Bellflowers (*Campanula*) have much in common with the flowers we
have just considered, but show some interesting differences in detail.
Many, like our common Harebell (*C. rotundifolia*) (Fig. 48), have pendulous flowers; this in itself favours the agile and specialised bees as pollinators, and bees of various kinds are much the most frequently observed
visitors. The flowers are protandrous, and functionally similar to those of
the Bindweeds or trumpet Gentians, with the nectar concealed in the same
way by the expanded bases of the stamens. However, the pollen is not
transferred direct from the stamens to visiting insects. In the young bud,
the anthers closely surround the hairy tip of the style, on to which they
shed their pollen before the flower opens. In the newly opened flower the
slender filaments of the stamens have already shrivelled, and the pollen
adhering to the tip of the style is ready to be picked up on the body of a
bee crawling into the bell. After some days the style branches diverge,
exposing the three receptive stigmas which take up the position hitherto
occupied by the pollen. Before the flower withers, the stigmas curve back

a

PLATE 23. **a**, bumble-bee *(Bombus hortorum)* sucking nectar from flower of Yellow Flag *(Iris pseudacorus)*. **b**, hover-fly *(Rhingia campestris)* sucking nectar from Yellow Flag.

b

PLATE 24. **a**, solitary bee (cf. *Andrena* sp.) visiting flower of Common Melilot (*Melilotus officinalis*). **b**, solitary bee *(Halictus* sp.) on White Clover *(Trifolium repens)*. **c**, bumble-bee *(Bombus* sp.) on Red Clover *(Trifolium pratense)*. **d**, bumble-bee leaving flower-head of Red Clover, showing length of proboscis.

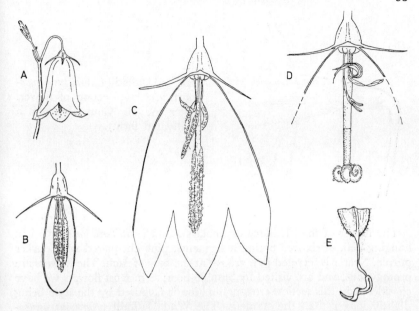

FIG. 48. Harebell (*Campanula rotundifolia*). A, sketch of part of inflorescence. B, bud at time of dehiscence of the stamens. C, newly opened flower; the apical part of the style is covered with pollen. D, style and stamens of an older flower; the style-branches have reflexed and the stigmas are now receptive. E, a single stamen from an open flower, showing the expanded base and the shrivelled filament and anther. A, Iwerne Courtney, Dorset. B-E, Flatford, Essex.

so far that they touch the remaining pollen, so self-pollination may take place if insect visitors fail.

Little need be said about the mechanism of the large bell-shaped flowers among the Monocotyledons. Meadow Saffron (*Colchicum autumnale*) functions in much the same way as the Bindweeds, though the stamens and stigmas are much farther apart. Nectar is secreted at the bases of the filaments, in the narrow space between these and the perianth segments where it accumulates in short grooves; it is further protected by woolly hairs. The flowers are usually protogynous, and are visited by honey-bees, bumble-bees and (much less effectively) by numerous flies. The flowers of *Crocus* species (*Iridaceae*) generally open rather wider than those of *Colchicum*, so that bees often alight on the stamens and stigmas projecting in the centre of the flower, in a way reminiscent of their behaviour on the Meadow Cranesbill (p. 55).

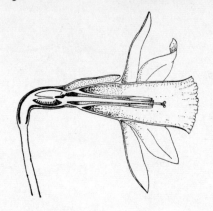

FIG. 49. Daffodil (*Narcissus pseudonarcissus*). Half-section of flower. After A. H. Church (from a cultivated form).

The flower of the Fritillary or Snake's Head (*Fritillaria meleagris*) is a hanging bell, of six free perianth segments, pink chequered with darker purple. Nectar is secreted in a groove at the base of each. The flowers are protogynous, and are visited by bumble-bees; as in most flowers we have considered in this section cross-pollination is favoured by the style being slightly longer than the stamens. The Wild Daffodil (*Narcissus pseudonarcissus*) (Fig. 49) is another flower pollinated by early flying bumblebees, though in this case the tubular part of the nodding flower is made up partly of the perianth tube, about 18 mm. long, at the base, and partly by a tubular outgrowth, the *corona*, extending for some 30 mm. above the level of the spreading corolla lobes. The large queens of *Bombus terrestris* just fill the bell, forcing the anther column and the style to one side, while the rather short tongue of the bee can just reach the nectar around the base of the style through the narrow spaces between the broad bases of the filaments. However, various other bees (for example the long-tongued *Anthophora acervorum*) and drone-flies (*Eristalis*) can also reach the nectar and can pollinate the flowers. The flowers are fertile to their own pollen, but apparently few seeds are set in the absence of insect visits, so the amount of seed depends greatly on the weather in March and early April when the plants are flowering (see Caldwell and Wallace, 1955).

Several of the flowers described in this section are what Kerner called 'revolver flowers'. An insect entering the flower is faced with a ring of narrow tubes – like the chambers of a revolver – through which it must probe to reach the nectar. As it sucks the nectar it may take up any one of the corresponding positions within the flower, and will often move round to feed at several positions in succession. The classic examples are such flowers as the Bindweeds and trumpet Gentians, but the same principle can be seen in the Columbines (*Aquilegia*), and in many tubular flowers, where the individual 'barrels' of the 'revolver' are separated by the stamens or by ridges on the corolla, as in Herb Robert (*Geranium robertianum*) or Wallflower (*Cheiranthus cheiri*) (Fig. 35). Possibly such a construction serves to detain the insect longer, and tends towards more effective pollination. At all events, such flowers are becoming, in a sense, multiple pollination units.

There are some more extreme developments of multiple pollination units from single flowers. In the Turk's-cap Lily (*Lilium martagon*), there is a tube on each of the perianth segments formed by a furrow covered in by flanges (Fig. 50). These tubes are very narrow, and suit the slender tongues of *Lepidoptera*, such as the hummingbird hawk-moth. The Yellow Flag (*Iris pseudacorus*) resembles the Martagon Lily in having multiple tubes giving access to the nectar. Here, however, there are three tubes

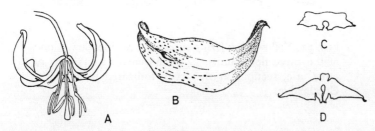

FIG. 50. Turk's-cap Lily (*Lilium martagon*). A, side view of young flower with one perianth segment removed. B, single perianth segment, showing entry to nectar-groove. C, cross-section of perianth segment, one-third of the way up the groove. D, the same, three-quarters of the way up the groove. The stamens, and later the style, bend up towards the perianth segments.

instead of six, three of the perianth segments (the 'standards') serving for display only (Fig. 51). The lower side of each tube is formed by the narrow stem, or haft, of one of the other perianth segments (the 'falls'), each of which has a large free blade on which insects can alight. Lying over the haft of each fall is a greatly expanded and flattened style, looking like a petal and forming the upper side of the tube, and arching over a single anther. The tube is large enough for the bumble-bees or long-tongued flies which pollinate the flowers to crawl right in (Plate 23). At the level of the section in Fig. 51d the tube is divided by the filament into two narrow channels which contain nectar. It is clear that the functional unit is not the whole *Iris* flower but a third of it; what is particularly interesting is the remarkable analogy between each individual tube and the whole of a Labiate flower (p. 206).

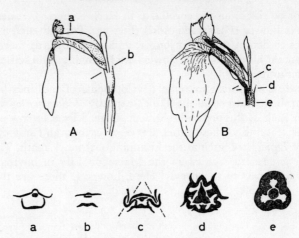

FIG. 51. Yellow Flag (*Iris pseudacorus*). A, side view of one of the three floral units. B, three-quarter front view of same. a-e, sections at the levels indicated in A and B, Cambridge.

ZYGOMORPHY

The flowers which have achieved the closest adaptation in form to their pollinators are those that have become, like their pollinators, bilaterally symmetrical. These *zygomorphic* flowers, as they are called, have evidently evolved quite independently in many families of flowering plants. Perhaps zygomorphy has always arisen, in the first place, as an adaptation to pollination by bees. However, as we shall see, by no means all zygomorphic flowers are bee-pollinated.

In a radially symmetrical flower, an insect can take up any one of a number of positions in relation to the axis of the flower. In a zygomorphic flower, the insect tends always to take up a single position, and it is this which allows much more precise adaptation of the flower to particular pollinators. By contrast with radially symmetrical flowers, zygomorphic flowers are usually placed more-or-less horizontally, and their orientation on the plant generally varies little.

In a zygomorphic flower the stamens and style may be so placed that they come into contact with the underside or with the upperside of the visitor. The first arrangement, called *sternotribic*, is found in the British members of the *Leguminosae* – the Vetches, peas and their allies – and in a number of smaller groups of zygomorphic flowers. Most of these flowers come into Faegri and van der Pijl's category of 'flag blossoms'. The second arrangement, with the pollen normally transferred on the upper side of the insect, is called *nototribic*. It is particularly characteristic of the Dead-nettle family (*Labiatae*) and several other large and important families with gamopetalous corollas, for example *Scrophulariaceae* and the largely tropical families *Acanthaceae* and *Gesneriaceae*, and of the Orchids. Most of these flowers fall into Faegri and Van der Pijl's category of 'gullet blossoms'.

VETCHES AND OTHER STERNOTRIBIC FLOWERS

The *Leguminosae* are a large and varied family. Almost all the European species belong to one sub-family, the *Lotoideae* (*Papilionateae*), and it is these which have the familiar pea-flower corolla. Usually there is a more-or-less tubular calyx, divided at the tip into five longer or shorter lobes.

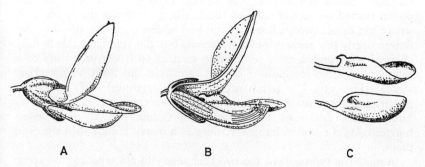

A   B   C

FIG. 52. Common Melilot (*Melilotus officinalis*). A, single flower from the side. B, flower with half of calyx and corolla removed to show stamens and stigma. C, wing and keel petal, seen from outside. Alderney, Channel Is.

The corolla consists of five free petals, of which the uppermost, known as the *standard*, is usually large and conspicuous. The two petals below this are called the *wings*, and the two remaining petals below and between these are pressed together and folded over one another to form a boat-shaped structure enclosing the stamens and the ovary, known as the *keel* (Figs. 52–56). The petals are more or less firmly interlocked at the bases of their blades by folds, projections, or zones where the cells interlock or adhere, the details varying from one species to another. The lower nine stamens are fused to form a tube, within which the nectar is accumulated; the tenth stamen is usually free, so allowing access to the nectar from the upper side. A visiting insect usually clings to the wings, and inserts its proboscis between the standard and the upper edges of the keel, sliding it down to reach the nectar in the base of the stamen tube.

The simplest type of mechanism is found in flowers like those of Sainfoin (*Onobrychis viciifolia*) and the Melilots (*Melilotus*) (Fig. 52). The upper edge of the keel is open, so that when a visitor forces its way into the flower the wings and keel are pressed down, uncovering the relatively rigid stamens and style which come into contact with the underside of the body of the insect. As the insect leaves the flower, the wings and keel spring back into place again, once more covering the stamens and stigma. The flowers of the Clovers (Plate 24b–d) work similarly, but the lower parts of the petals and the stamen tube adhere strongly, leaving only the upper parts of the wings and keel free to move.

The Vetches (*Vicia* and *Lathyrus*) are similar, but have a secondary pollen presentation mechanism reminiscent of that of *Campanula*. The style is bent up sharply from the tip of the ovary and carries a dense brush of fine hairs below the stigma. The anthers dehisce in the bud, and the pollen is shed on to the hairs of the brush, or into the tip of the keel, where the brush sweeps it out as the keel is depressed. By the time the flower opens the anthers have retracted, but the stigma brush is fully charged with pollen as it comes into contact with the underside of a visiting bee. The mechanism is well shown in the Bush Vetch (*Vicia sepium*) (Fig. 53). The petals are relatively large and stiff; it takes a powerful insect to reach the nectar, and the flowers are mainly visited by bumble-bees. Although the flowers are normally held more-or-less horizontally, they often hang vertically as a heavy bumble-bee clings to them.

In some common genera the two keel petals do not separate to expose the style and stamens when the keel is depressed by a visitor. Instead, the pollen is shed into the conical end of the keel, whose edges are fused except for a small hole or slit at the tip. When the wings and keel are pressed down by a visiting insect, the stamens beneath act as a piston, forcing out a string or ribbon of pollen like tooth-paste from a tube onto the underside of the visitor. In due course, the stigma protrudes through the slit, and pollination can take place. The details of the mechanism in Bird's-foot Trefoil (*Lotus corniculatus*) are shown in Fig. 54. A similar arrangement is found in Kidney Vetch (*Anthyllis vulneraria*), Horseshoe Vetch (*Hippocrepis comosa*), the Restharrows (*Ononis*) and the Lupins (*Lupinus*).

Finally, in Lucerne (*Medicago sativa*) (Fig. 55), Gorse (*Ulex*) (Plate 26a–c), Broom (*Sarothamnus scoparius*) (Fig. 56) and some other plants, there is an explosive mechanism. Lucerne is the most similar to the simple type of the Melilots and Clovers. The stamen tube is held under tension between the keel petals by a pair of hollow projections on their upper edges, and to some extent by projections from the upper edges of the

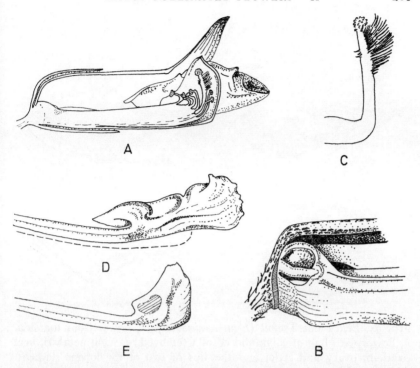

FIG. 53. Bush Vetch (*Vicia sepium*). A, flower with half of calyx and corolla removed. B, detail of base of stamen-tube, showing the openings to the nectary on either side of the uppermost filament. C, tip of ovary and style, showing stylar brush and stigma. D, wing petal from outside; the dotted line indicates the outline of the lower part of the keel. E, keel petal from outside; the region of adhesion to the wing is shaded. Sidbury, Devon.

wings. The pressure exerted by a visiting bee dislodges these projections, releasing the stamen tube and the style, which spring up, striking the under side of the visitor. In a newly opened Gorse flower the two keel petals adhere lightly together by their upper edges; in this case it is the keel that is held straight by the stamen-tube and style, rather than vice versa. The stamens dehisce just before the flower opens. The flowers are nectarless, but they are often abundantly visited by bumble-bees and honey-bees which forcibly enter the flower as though seeking nectar (Plate 26). This action causes the keel petals to break apart, uncovering the stamens and style and bringing them sharply into contact with the insect, so dusting it with pollen on the underside of the abdomen. Once

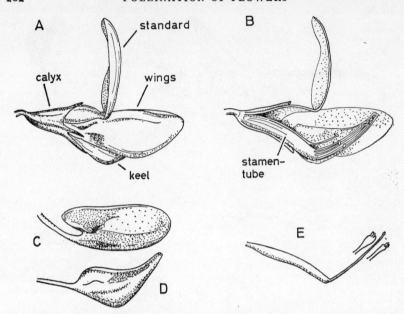

FIG. 54. Bird's-foot Trefoil (*Lotus corniculatus*). A, flower from the side.
B, flower with half of calyx and corolla removed. C, wing petal. D, keel
petal. E, ovary and style, and the tips of two of the longer stamens.
Alderney, Channel Is.

'exploded', the spent flower hangs limply open and is seldom visited again
by bees. The mechanism of the Broom flower is similar, but a little more
complicated and much more vigorous. Of the ten stamens five are shorter,
and strike the bee on the underside, while five are longer and, with the
long curved style, commonly strike the bee on the back of the abdomen.

It is inherent in most of these pollination mechanisms in papilionate
flowers that the stigma lies close to its own pollen; in some, the stigma is
actually embedded in pollen when the flower opens. In some cases the
stigma is not receptive until it has been abraded. Bird's-foot Trefoil and a
number of the larger-flowered clovers are known to be self-incompatible,
and Sainfoin is largely so. There are strong indications of self-incompatibility in many other species (though some are known to be self-
compatible and regularly selfed), and it is probably the main factor
controlling the breeding system throughout the family.

Species of the Fumitory family (*Fumariaceae*) have – apparently quite
independently – evolved floral mechanisms remarkably like some of those
found in the *Leguminosae*. The flower of *Corydalis cava*, a central and

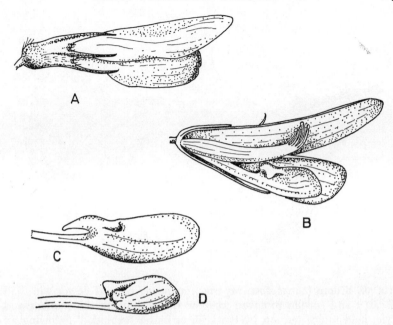

FIG. 55. Lucerne (*Medicago sativa*). A, young flower from the side. B, an 'exploded' flower with half of calyx and corolla removed to show stamens and stigma. C, wing petal. D, keel petal. Alderney, Channel Is.

southern European species occasionally found in this country as a garden escape, is shown in Fig. 57. There are four petals, of which the uppermost is the largest and has a long spur at its base. The spur receives the nectar secreted by a long backward-pointing process from the upper filaments. The two lateral petals are curved inwards at their margins and fused at the tip, so that they form a sheath or hood enclosing the rigid style. The stigma is large and lobed, and is covered with pollen by the stamens which dehisce and wither before the flower opens. The bumble-bees which visit the flowers depress the hood as they probe for nectar in the spurred upper petal. In young flowers they dust themselves with pollen in the process; in older flowers they may leave pollen on the now receptive stigma. Not surprisingly, the flowers appear to be entirely self-incompatible The whole mechanism of *Corydalis cava* bears a striking and curious similarity to that of the Vetches. *Corydalis lutea* (Fig. 58), which is a familiar garden plant and sometimes naturalised, has an explosive mechanism reminiscent of that of Lucerne or Gorse. It is visited by various bees, but is probably self-fertile. The Fumitories (*Fumaria*) have

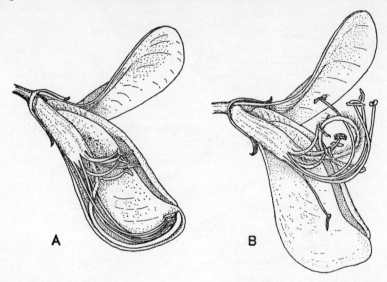

FIG. 56. Broom (*Sarothamnus scoparius*). A, newly opened flower with half of calyx and corolla removed to show the position of the stamens and style, held under tension by the keel. B, flower 'exploded' following an insect visit. Copplestone, Devon.

similar flowers to *Corydalis*, but are not much visited, and probably almost always self-fertilised.

Apart from the *Leguminosae*, by far the most impressive sternotribic flowers are the Monkshoods (*Aconitum*, *Ranunculaceae*) (Fig. 59). The flowers are beautifully adapted in size and form to pollination by bumble-bees, and in fact are a classic case of complete dependence on bumble-bees for pollination. The Monkshoods are related to the Hellebores described in Chapter 3, and although superficially so unlike them, the real differences between the flowers are few. The five sepals are strongly coloured, the uppermost forming a large erect helmet-shaped hood, covering the two upper nectaries, which are greatly enlarged. The remaining nectaries are small, or absent altogether. The flowers are protandrous. The young stamens are bent downwards, but stand erect as they mature and dehisce, and then bend back out of the way, so that the underside of a bee landing on the lower sepals and clambering up to reach the nectaries in the helmet is well dusted with pollen. After the stamens have all dehisced, the maturing stigmas are exposed, and pollination can take place. The mechanism of the Monkshoods is seen at its most highly developed in the beautiful yellow *Aconitum vulparia* (*A. lycoctonum*) of the Alps and northern

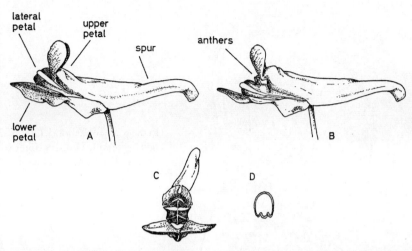

FIG. 57. *Corydalis cava.* A, side view. B, flower opened as if being probed by an insect, revealing anthers. C, front view. D, cross-section of spur showing median nectar-groove. The lateral petals enclose the stamens and style like the keel petals of Leguminosae; their claws are fused to the upper edges of the lower petal. The sepals drop when the flower opens.

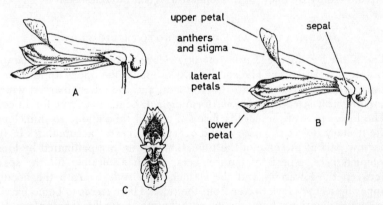

FIG. 58. Yellow Corydalis (*Corydalis lutea*). A, side view of newly opened flower. B, flower 'exploded' as after an insect visit; notice the anthers lying against the upper petal. C, front view.

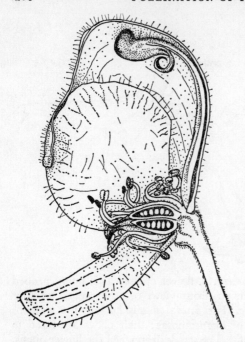

FIG. 59. Monkshood (*Aconitum anglicum*). Half-section of flower. Dartington, Devon.

Europe. The helmet is tall and narrow, and the nectaries are some 20 mm. long, with spirally coiled spurs secreting copious nectar which can be exploited only by a few of the longest-tongued bumble-bees.

DEADNETTLES AND OTHER NOTOTRIBIC FLOWERS

The White Deadnettle (*Lamium album, Labiatae*) (Fig. 60) is a good example of a nototribic flower. The corolla is two-lipped, the upper lip forming a hood over the style and the four stamens, and the lower, marked with a few greenish streaks and dots, forming a landing platform for insects. The lower part of the corolla forms a curved tube about 10 mm. long; the nectary lies at the base of the ovary, and nectar accumulates in the narrow part at the base of the tube. The flowers are pollinated by long-tongued bees, especially bumble-bees, which accurately fill the space between the lower lip and the stamens and style beneath the hooded upper lip as they suck nectar from the tube. There seems to be no barrier to self-pollination; how much outbreeding actually takes place has apparently not been established. The Yellow Archangel (*Lamiastrum* (*Galeobdolon*) *luteum*) has bright yellow flowers very similar in shape, though with a rather shorter tube, so they can be exploited by a greater variety of bees. The speed and precision with which bumble-bees visit

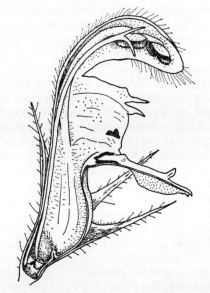

FIG. 60. White Deadnettle (*Lamium album*). Flower with half of calyx and corolla removed. Iwerne Courtney, Dorset.

these flowers and transfer the pollen is impressive to watch (Plate 27).

Many other members of the *Labiatae* have similar long-tubed bumble-bee flowers,though differing a good deal among themselves in colour and structural details. Some, especially those with smaller flowers and shorter tubes, are commonly visited also by hive-bees and the smaller wild bees. There is very noticeable variation from one species to another in the part of the visitor with which the anthers and stigma come into contact. This is easily seen when Yellow Archangel and Bugle (*Ajuga reptans*) (Plate 28b) are growing together; *Bombus agrorum* transfers the pollen of Yellow Archangel on its thorax and that of Bugle on its head. The stamens of Bugle, like those of Wood-sage (*Teucrium scorodonia*) (Plate 28c), project beyond the very short upper lip, though in the case of Bugle they obviously derive some protection from the bracts of the flower above. The functional protandry which is very common in the *Labiatae* is readily seen in these two plants. In a newly opened flower the stigma stands above the stamens and is not readily touched by a visiting insect. After the stamens have shed their pollen, the style bends down so that the stigma is exposed below the stamens at the entrance to the flower.

An elegant variation on the mechanism of the Deadnettle flower is found in the Sages (*Salvia*), and is seen at its most highly developed in such species as the bright blue Meadow Sage (*S. pratensis*) (Plate 29a) and the dingy yellow Continental *S. glutinosa* (Plate 29b). Only two stamens are functional, the other two being reduced to small vestiges. In the two functional stamens the connective (the tissue between the two anther-lobes) is greatly elongated. One lobe of each stamen is normally developed and lies beneath the hood of the upper lip; the other is abortive, and the connective at its lower end forms an expanded plate partly blocking the entrance to the flower. The filaments are reduced to short flattened strips, joining the connective to the corolla, and providing an elastic torsion joint about which the connective can hinge. When a bee pushes its head into the flower, the connective plate is pushed upwards and backwards, and the fertile anther lobes swing downwards bringing their pollen into contact with the abdomen of the bee. In older flowers the mature stigmas project in front of the flower, so that visiting bees rub against them as they enter. Not all *Salvia* species have the see-saw mechanism developed to this degree of perfection. In the garden Sage (*Salvia officinalis*), for instance, the connective is much shorter, and both anther-lobes produce pollen, though the lower lobes produce much less than the upper. The mechanism works in the same way as that of *S. pratensis*, but is evidently more primitive. On the other hand, in the scarlet-flowered bird-pollinated *S. splendens* from Brazil the lever mechanism has evidently been lost. The fertile anther-lobe is still borne on the end of a long connective, but the other end of the connective is not broadened or curved down to block the entrance; the pollen is simply rubbed off on to the head of the hovering visitor as it probes with its bill into the unobstructed mouth of the flower. The lever mechanism is also degenerate in the small-flowered species of *Salvia* that are habitually selfed, like our own dingy purple Clary (*S. horminoides*).

Many flowers with comparable 'Labiate' corollas are to be found in the *Scrophulariaceae*. We have in the British Isles two species of Lousewort (*Pedicularis*), a scanty representation of a genus which is very rich in species and varied in flower form in the Alps, Scandinavia and North America. *P. sylvatica* (Fig. 61) is a common plant of damp heath and moorland, flowering in spring and early summer. The pink flowers are two-lipped. The upper lip forms a narrow, laterally flattened hood, enclosing the four stamens, with the stigma just protruding from the underside near the tip. The lower lip is three lobed, forming a flat landing platform slightly oblique to the plane of symmetry of the rest of the flower. The two pairs of stamens face one another; the pressure of the sides of the hood keeps them pressed together, preventing the escape of pollen. The entrance to the upper lip forms a narrow slit some 8–10 mm. long, usually somewhat widened for about 3 mm. at its upper end (where it is separated from the pore through which the style protrudes by two narrow teeth), and with its margins roughened and rolled below. A strong rib on each side of the corolla runs downward and backward from the junction of these two parts of the margin. A visiting bee grasps the base of the slanting lip with its forelegs and the corolla tube just below the lip with its middle pair of legs, and inserts its head obliquely into the

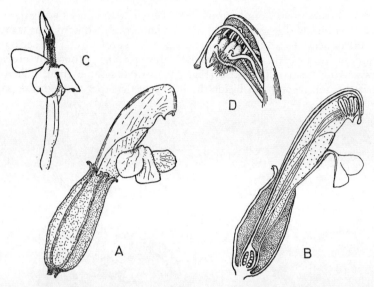

FIG. 61. Lousewort (*Pedicularis sylvatica*). A, side view of flower. B, half-section of flower. C, front view of flower. D, detail of anthers and upper part of style (semi-diagrammatic). A, B, Colaton Raleigh, Devon. C, D, after Kerner.

wider part of the entrance to the hood, touching the stigma in the process. As it probes for nectar it prises apart the sides of the hood, at the same time drawing forward the upper part of the hood and releasing the pressure on the stamens, allowing pollen to fall from between them on to its head. The pollination of the second British species, *P. palustris* (Red Rattle), seems to be generally similar.

The pollination of the Louseworts has been studied in Norway by Nordhagen (Lagerberg *et al.*, 1957) and recently in North America by Sprague (1962) and Macior (1968a and b). It is quite clear that the variation between the flowers of different species is closely connected with their adaptation to particular pollinators, and the development of particular pollination mechanisms. Thus some of the North American species are pollinated in essentially the same way as the British species described above, but in *P. racemosa*, which has a long down-curved beak to the upper lip and an extremely oblique lower lip, the transfer of pollen takes place on the lower surface of the abdomen of visiting bumble-bees.

Two examples studied by Macior using high-speed cinematography are particularly interesting. *P. canadensis* has yellow flowers, similar in shape to those of *P. sylvatica* but with a longer style which projects some distance below the tip of the upper lip. It is visited mainly by the queens of various bumble-bee species. Macior's photographs showed that, as the bee introduces its head and tongue into the corolla tube to suck the abundant nectar, the stigma sweeps through the pronotum crevice between the head and thorax (Fig. 62A). At the same time, pollen is released from the stamens inside the hood, and escapes through the pore

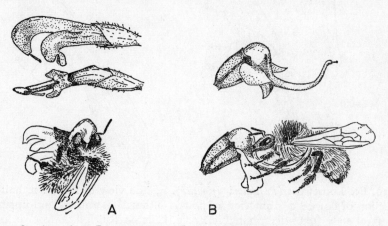

A   B

FIG. 62. American Louseworts. A, flower of *Pedicularis canadensis* seen from the side and below, and sketch showing foraging position of visiting bumble-bee. Iowa (after Macior 1968b). B, Side view of flower of *P. groenlandica*, and sketch showing position taken up by bumble-bee foraging for pollen. Colorado (after Macior 1968a).

surrounding the style, lodging in the pronotum crevice and on neighbour-ing parts of the insect's body. Most of this pollen is swept away as the bee grooms the hairy surface of its thorax in flying from flower to flower, but the pronotum crevice is inaccessible to the sweeping movements of the middle pair of legs, and the pollen there remains to be swept out by the stigma of another flower. *P. groenlandica* is a pink-flowered species, lacking nectar, and visited for pollen by worker bumble-bees of several species. The onset of flowering coincides with emergence of the worker bees; and queens are too large to operate the mechanism of the flower. The flower resembles the head of an elephant waving its trunk in the air – the basal part of the hood containing the anthers forming the 'head', the long upturned beak of the upper lip which ensheaths the style forming the 'trunk', and the three-lobed lower lip forming the 'ears' and 'jaw'. A visiting bee takes up a position astride the beak, which passes under the thorax and then curves up between the thorax and abdomen so that the stigma is in contact with the front surface of the latter (Fig. 62B). The bee

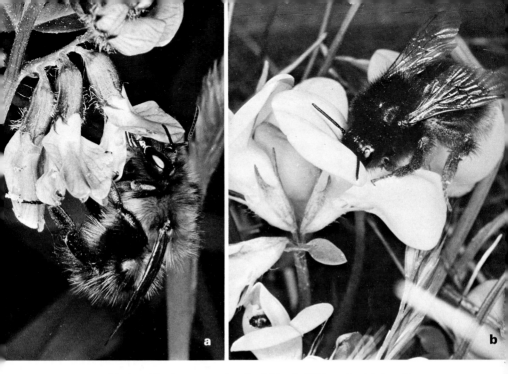

PLATE 25. **a**, bumble-bee *(Bombus agrorum)* on Bush Vetch *(Vicia sepium)* **b**, bumble-bee *(Bombus lapidarius* worker*)* on Bird's-foot Trefoil *(Lotus corniculatus)*. **c**, honey-bee on Bird's-foot Trefoil; the bee has approached the flower slightly from one side, and the keel of the flower, with the projecting tip of the style, can be seen in front of the bee's abdomen.

PLATE 26. **a-c**, honey-bees visiting flowers of Common Gorse *(Ulex europaeus)*; the bee in **a** is forcing an entry into a fresh flower; in **c** it is leaving the flower following the 'explosion' of the stamens and style from the keel.

grasps the stout central ridge of the hooded basal part of the upper lip with its mandibles, and by rapid movements of its wings shakes out pollen which falls on to the lower petal and is scattered as a yellow cloud enveloping the insect's body. Most of this pollen is groomed from the body and transferred to the corbiculae on the bee's hind legs, but much of that on the front face of the abdomen remains and may pollinate another flower. The behaviour of visiting bees, and possibly the precise structure of the flowers, may vary to some extent as Sprague observed bees inserting their heads between the sides of the upper lip in the manner more usual in *Pedicularis*. Nevertheless, these two North American examples suggest that further study of the European Louseworts (and indeed many other of our nototribic flowers) might disclose a good deal of interest, and possibly some surprises.

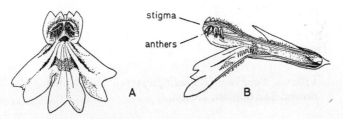

FIG. 63. Eyebright (*Euphrasia pseudokerneri*). A, front view of flower. B, flower with half of calyx and corolla removed, from the side.

The Louseworts are a particularly varied genus, but their flowers have one feature in common that is found also in all the related semi-parasitic members of the *Scrophulariaceae* (the *Rhinanthoideae*) and in the related family *Orobanchaceae* (Broomrapes and Toothworts) as well. This is the arrangement of the stamens in two inward-facing pairs, which are normally pressed together preventing release of the dry powdery pollen until an insect visits the flower. The Eyebrights (*Euphrasia*) (Fig. 63, Plate If), have a typically 'labiate' corolla, but in the other members of the *Rhinanthoideae* the lower lip tends to be reduced and the flower more-or-less tubular. In the Yellow Rattles (*Rhinanthus*) the flower is strongly flattened, and the stamens function in much the same way as in the Louseworts, releasing pollen as the sides of the flower and the lower parts of the filaments are prised apart by the tongue of a visiting bumble-bee. In other genera the anthers have spine-like projections beneath, and pollen is shed when these are touched by a visiting insect.

The remaining zygomorphic flowers in the *Scrophulariaceae* are a diverse collection. One of the most familiar, the Foxglove (*Digitalis purpurea*) (Fig. 64), is in effect a zygomorphic variation of the bell and trumpet flowers already described (p. 192). The Monkey Flower (*Mimulus guttatus*)

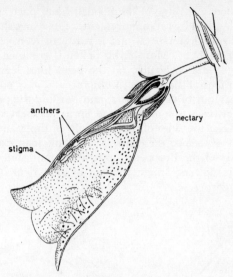

FIG. 64. Foxglove (*Digitalis purpurea*). Half-section of
flower. Dartington, Devon.

is rather similar, but shows an interesting peculiarity in the irritability of
its two-lobed stigma. A bee entering the flower brushes first against the
stigma, whose two lobes then fold quickly together, and lie tightly pressed
against the upper surface of the corolla. The insect then comes into
contact with the anthers, which lie immediately behind the lower stigma
lobe (Plate 31c). A very similar mechanism is found in the large-flowered
Bladderworts (*Utricularia, Lentibulariaceae*) (Fig. 65). The related Butterworts (*Pinguicula*) again have a similar mechanism but here there is no
active movement; the movement of the visiting insect suffices to draw
back the lower stigma-lobe and uncover the anthers as it leaves the
flower. In the Alps, the white-flowered *P. alpina* is visited principally by
flies, while the blue-flowered Common Butterwort (*P. vulgaris*) is visited
by bees. However, in north-west Europe the Common Butterwort is
seldom visited by insects and is apparently regularly self-pollinated
(Willis and Burkill, 1903b; Hagerup, 1951).

The Bladderworts and Butterworts have spurred flowers; in the
Bladderworts the visits of small insects are further impeded by the greatly
inflated lower lip, which is pressed against the upper lip forming a spring-
loaded 'door' at the entrance to the flower. The flowers of *Utricularia
vulgaris* are said to be visited by bees; visits by long-tongued hoverflies,
especially *Helophilus lineatus*, were observed by Heinsius (Knuth) and
Silén (1906b).

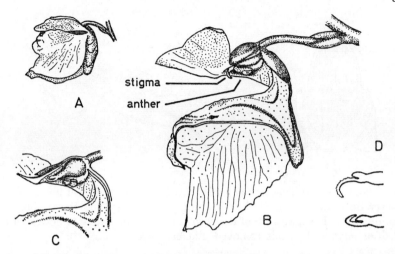

stigma

anther

A

D

B

C

FIG. 65. Greater Bladderwort (*Utricularia vulgaris*). A, side view of flower. B, side view of a flower with half of the lower lip removed. C, gynoecium and neighbouring parts of a flower with half of the calyx and corolla removed, showing the stigma-lobes closed after being touched. D, diagram showing stigma-lobes before and after stimulation. Shapwick, Somerset.

Returning to the *Scrophulariaceae*, similarly constructed *personate* flowers are found in the Snapdragons (*Antirrhinum*) and Toadflaxes (*Linaria* and related genera). The Common Toadflax (*Linaria vulgaris*) (Fig. 66) is a common example. Nectar is secreted by the base of the ovary, and accumulates in the conical spur projecting from the underside of the base of the corolla tube. The personate corolla excludes almost all insects but the strong and 'intelligent' bees; the length of the spur debars short-tongued bees from reaching the nectar. Normally, a hive-bee or bumble-bee visiting the flower lands on the lower lip, which bears a darker yellow or orange guide mark, and inserts its head between the upper and lower lips, prising open the corolla. The tongue of the bee is guided to the nectar by a smooth channel about 1 mm. wide between two orange hairy ridges on the floor of the corolla tube. As the insect sucks the nectar its back comes into contact with the stamens and stigma which lie against the upper side of the corolla tube. Bees visiting the flowers for pollen may work the flowers upside down, collecting the pollen directly from the anthers; more than a quarter of a sample of 269 bumble-bees foraging on a population of Yellow Toadflax which Macior (1967) studied in Wisconsin had evidently been working in this way. The flowers of the Common Toadflax are homogamous, but self-incompatible. The mechanism of the

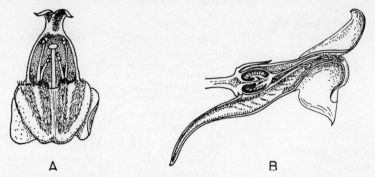

A   B

FIG. 66. Common Toadflax (*Linaria vulgaris*). A, front view of flower with lower lip pulled down to show the stigma, stamens, and the proboscis-guide on the lower surface of the corolla-tube. B, side view of flower with half of calyx and corolla removed. Exeter, Devon.

introduced Purple Toadflax (*Linaria purpurea*) (Plate 32a) is similar. On the other hand, in the Ivy-leaved Toadflax (*Cymbalaria muralis*) and in several of the annual weedy species (for example *Kickxia spuria*) the self-incompatibility has been lost, and the flowers are regularly self-fertilised.

We cannot leave spurred zygomorphic flowers without mentioning three other, quite unrelated groups – the Violets and Pansies (*Viola*, *Violaceae*), the Balsams (*Impatiens*, *Balsaminaceae*), and the Orchids

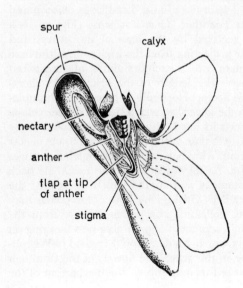

FIG. 67. Sweet Violet (*Viola odorata*). Half-section of flower. After A. H. Church.

(*Orchidaceae*) which form the subject of the next chapter. Our native Balsam, *Impatiens noli-tangere* (Touch-me-not), is a rare plant, but the introduced Himalayan Balsam (*I. glandulifera*) now grows in profusion along the banks of many of our rivers and canals. The pink flowers are large enough to enclose a bumble-bee completely, bringing its back into contact with the stamens and stigma in the roof of the flower, and the busy comings and goings of dozens of bees to a patch of this plant are often a fine sight. Violet flowers (like those of the Balsams) have five unequal free petals (Fig. 67). The lowest of these is prolonged at the base into a spur, which accumulates the nectar secreted by the flat green tail-like append-ages of the two lowermost anthers. The anthers form a cone surrounding the short hooked style, and release pollen on to the tongue of a visitor when the style or the apex of the cone is lightly touched. The style itself is hollow, and flexible at the base. Any pressure on the style causes a small drop of liquid to be exuded from the pore at the tip; it is drawn back again into the cavity of the style, together with any pollen it may have come into contact with, when the pressure is released. This drop of liquid may serve a secondary function in moistening a small area on the tongue of a visiting insect, so causing the powdery pollen to adhere (Beattie, 1969). Pollen grains will germinate only inside the stigmatic cavity, and although the flowers are self-compatible, no seed is set in the absence of insect visits. The main visitors to the flowers of violets are bumble-bees, other early-flying bees including *Anthophora acervorum* and species of *Os-mia*, *Andrena* and *Halictus*, and hoverflies (Beattie, 1972). It is interesting that many Violets produce cleistogamous flowers, self-pollinated in the bud, after the season of the conspicuous chasmogamous flowers is over (p. 286). There are also self-pollinated species of *Viola* with normal flowers, such as *V. arvensis*, in which the pollen can readily fall into the open stig-matic cavity.

There remain a number of zygomorphic flowers which do not fit readily into the categories so far considered. Some may have developed zygomorphy independently along lines of their own; others are the products of 'adaptive radiation' from the more 'central' bee-pollinated flowers already described. Honeysuckle (*Lonicera periclymenum*, *Caprifolia-ceae*) (Fig. 68) probably belongs to the first category. The long-tubed, pale, sweet-scented flowers, with long projecting stigmas and stamens, illustrate admirably the syndrome of pollination by night-flying moths. They are visited by hawk-moths and by various other moths including the silver-Y (*Plusia gamma*), which is probably one of the more important visitors in Britain (Plate 33).* The buds open between about 7 and 8 p.m. On the first night of opening the flowers are clear creamy white in colour, and exhale a powerful scent until after nightfall. The large

* The bumble-bee *Bombus hortorum* also visits the flowers, during the day, but often fails to touch the anthers and stigma.

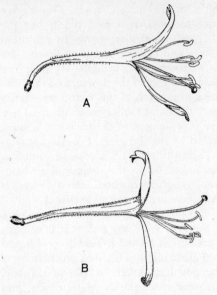

FIG. 68. Honeysuckle (*Lonicera periclymenum*). A, newly opened flower. B, older flower. Exeter, Devon.

versatile anthers on their long filaments stand directly in front of the straight corolla tube, but the style is bent down towards the lower lip. In the course of the following day, the flower takes on a yellowish colour, and the corolla tube a slight downward curve. At the same time the style bends upwards, so that by the second evening the stigma dominates the entrance to the flower, which is now in a functionally female phase. The fragrance, and the secretion of nectar, continue for several days, though

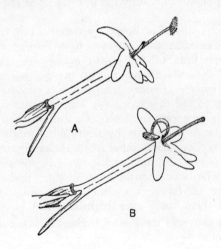

FIG. 69. Red Valerian (*Kentranthus ruber*). A, young flower with dehiscing stamen. B, older flower, stamen reflexed and style elongated. Exeter, Devon.

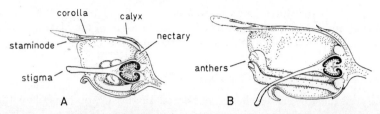

FIG. 70. Water Figwort (*Scrophularia auriculata* [*S. aquatica*]). A, half-section of newly-opened flower, in functionally female stage. B, half-section of older flower with dehiscing anthers. Alderney, Channel Is.

gradually diminishing, and during this time cross-pollination may take place. Zygomorphy seems to be rather incidental to the mechanism of the Honeysuckle flower. The same may be said of Red Valerian (*Kentranthus ruber, Valerianaceae*); with its very slender spur and bright pink or red corolla this is a good example of a butterfly flower (Fig. 69). Its colourful heads are favoured by butterflies of many species, and by the humming-bird hawk-moth.

The Figworts (*Scrophularia, Scrophulariaceae*) (Fig. 70, Plate 32b) are among the few plants that appear to be adapted particularly to pollination by wasps. The flowers (like those of a number of other wasp-pollinated species) are tinged with a dingy purplish brown, and the nectar is readily accessible on the upper side of the ovary at the base of the short, broad corolla tube. Rather curiously, by contrast with the bee-pollinated flowers in the same family, the stamens and stigma are held close to the lower lip in Figwort flowers. The flowers are protogynous, and the female phase is said to last for two days; the stamens then straighten, bringing the anthers to the mouth of the flower above the withered stigma. There seems to be no satisfactory explanation of the function of the staminode in the upper part of the flower (Trelease, 1881). Wasps, mainly *Vespula vulgaris* and *V. germanica*, seem generally to be by far the commonest visitors; this is certainly my own experience. However, hive-bees and various other bees also visit the flowers commonly, and there is no doubt that in particular places and seasons bees go to the flowers very freely, and may even be the dominant visitors.

Both the *Scrophulariaceae* and the *Labiatae* contain flowers with almost regular corollas. The Mulleins (*Verbascum, Scrophulariaceae*), which have the full complement of five stamens, may well be primitive, and illustrate an early stage in the evolution of the zygomorphy of other members of the family. On the other hand, the Mints (*Mentha, Labiatae*) have 4 stamens and Gipsywort (*Lycopus europaeus*) only 2 in the small, tightly clustered flowers. These are probably reduced and specialised from more 'normal' Labiate flowers; the heads of small two-lipped flowers with long stamens in the Thymes (*Thymus*) illustrate a possible stage in this process. At all events, the Mints have lost the characteristic floral mechanism of the *Labiatae*, and have in effect evolved 'brush-blossoms' of the kind that will be considered in a rather different connection later.

THEFT AND PROTECTION

Many of the flowers we have considered in this chapter regulate the behaviour of particular visitors with considerable precision, and exclude other insects with a fair degree of effectiveness. But with regulation comes the possibility of evasion. In some flowers there is little physical barrier to evasion. Hive-bees collecting nectar from *Brassica* crops may enter the flowers from behind and insert their tongues between the sepals and the claws of the petals, instead of entering the flower in the 'legitimate' way and bringing about pollination (Plate 34a–b). We have already seen how hive-bees may behave similarly on flowers of the Bluebell (*Endymion non-scriptus*). In flowers with the corolla fused into a tube it is not possible for an insect to get nectar in this manner, unless it can bite a hole through the tube. Many gamopetalous flowers are in fact robbed in this way. The usual culprits are the powerful but relatively short-tongued bumble-bees *Bombus terrestris* and *B. lucorum* (and other species), which use their mandibles to perforate the corolla tubes or spurs of the long-tubed Labiates, Comfrey, Toadflax, Daffodils, Columbines (Macior, 1966) and other flowers whose nectar would otherwise be inaccessible to them (Brian, 1957). The holes made by these short-tongued bumble-bees are often used subsequently by hive-bees and other insects; on the other hand, a flower with a perforated tube may still be visited legitimately and pollinated in the normal way.

A number of flowers show features which may be interpreted as adaptations preventing – or at least minimising – such theft of nectar. Thus the butterfly-flowers of the Pinks (*Dianthus*) have a firm leathery calyx, further protected at its base by stout overlapping bracts (Fig. 38). The inflated calyces of the Campions (*Silene*) may serve the same function in a different way. In general, the calyx obviously provides some degree of protection against perforation of the base of the corolla tube in many species, as well as protecting the developing bud and providing mechanical support for the mature corolla – which in many flowers will be roughly handled by strong and heavy insects. To separate these functions may be interesting – and a reasonable abstraction – to the floral biologist, but of little significance in the life of the plant where it is the end result – pollination and the setting of seed – which counts. The calyx evolved in relation to a variety of factors, and the form it takes must reflect a balance between them. It is, of course, possible to point to flowers which have tubular corollas apparently with no special protection. Some of these are often perforated, others are not. All that can be said is that different flowers have evolved different patterns of adaptation to the complex environments in which they exist – and the continued existence of all of these plants demonstrates that all of their patterns of adaptation are workable.

Isolated flowers of Charlock (*Sinapis arvensis*) are often robbed in the same way as *Brassica* flowers, but this seldom happens to the flowers in the denser inflorescences (Fogg, 1950). This suggests that protection against nectar theft may have been one factor in the development of the dense inflorescences seen, for instance, in Clovers (*Trifolium, Leguminosae*), *Buddleia* (*Loganiaceae*), Mints and Thyme (*Mentha* and *Thymus, Labiatae*), the Valerians (*Valerianaceae*), the Teasel and Scabious family (*Dipsacaceae*) and the Daisy family (*Compositae*). Grant (1950) and others have suggested that protection of the ovules from damage, especially by beetles, has been an important factor in the evolution of flowers, leading to the development of inferior ovaries,* and again favouring dense, head-like inflorescences – a trend whose culmination is seen in the Compositae. Dense inflorescences made up of numerous small flowers are so common, and the Compositae in particular are so manifestly successful, that there must be potent factors of floral ecology working in favour of this form of organisation.† Burtt (1961) has suggested that the particular advantage of the Composite head – made up of many small flowers opening over a period of a week or two, each with a single ovule – is that it allows a very wide range of *different* pollinations to take place. In this he contrasts it with the large zygomorphic flowers with their specialised pollination mechanisms and numerous ovules. As he says, 'In these elaborate single flowers a very large number of ovules are fertilised by pollen from one, or a few, male parents. This will lead in due course to *intensive* exploration of the possible recombinations between the plants that cross. In the *Compositae*, whose individual ovules are individually pollinated (neighbouring ones often receiving pollen from different plants) the exploration of possible recombinations is *extensive* through the population rather than *intensive* between individual plants.' The same argument applies to other plants in which the 'blossom' is a compact inflorescence rather than a single flower – not only to the long-tubed flowers listed already, but also to such examples of the 'brush-blossom' type as the heads of the Great Burnet (*Sanguisorba officinalis*) and the insect-pollinated catkins of the willows. The exposed stamens and stigmas of these 'brush-blossoms' are readily intelligible in terms of insect pollination. At the same time, brush blossoms have

---

* It is worth noticing that the specialised but predominantly bee-pollinated *Boraginaceae, Labiatae* and *Scrophulariaceae* all have superior ovaries.

† In some of the general features of their floral biology the Compositae are paralleled by another large and successful family, the Umbelliferae, though the inflorescences of these plants are less specialised and the structure of the individual flowers is quite different.

evidently provided a favourable starting-point for the development of wind-pollination when conditions have favoured it (see Chapter 9).

THE COMPOSITAE

With some 18,000 species, the *Compositae* are one of the largest families of flowering plants. They are related to the Bellflower family (*Campanulaceae*), and the individual flowers, or 'florets', may be compared with the *Campanula* flower described on p. 194. The florets making up the heads of *Compositae* fall rather sharply into two types, well shown in a Daisy or Ragwort flower. The small tubular 'disc-florets' of the middle of the head are actinomorphic, and conspicuous only in so far as many of them are massed together. The 'ray-florets' round the edge are zygomorphic, and their strap-shaped (*ligulate*) corollas are largely responsible for the conspicuousness of the head. In some *Compositae*, such as Hemp Agrimony (*Eupatorium cannabinum*) and the Knapweeds (*Centaurea*) and Thistles (*Carduus, Cirsium* and related genera), only tubular florets occur. On the other hand, in the tribe *Cichoriae* all the florets are ligulate, resulting in the familiar Dandelion type of flower.

Fig. 71 shows three disc-florets of the Oxford Ragwort (*Senecio squalidus*). The calyx is reduced to a ring of long silky hairs, the *pappus*, around the top of the inferior ovary. The corolla is tubular, very slender below, widening abruptly into a bell shape, shallowly divided like a *Campanula* flower into five lobes. The five stamens are inserted at the top of the narrow part of the corolla tube. Their filaments are free, but the anthers are united into a tube surrounding the style. The pollen is shed into the interior of the tube before the flower opens. At this stage, the style is relatively short, and the tufts of hair at the tips of the two style branches fit into the tube like the piston in a cylinder. As the flower opens, the style grows, gradually forcing pollen out through the top of the anther tube. In fact pollen presentation is not continuous. On stimulation, the filaments contract, measuring out a small amount of pollen to the visiting insect, while the rest of the pollen remains inaccessible inside the anther tube. Small (1915) remarks on this 'miserly' presentation of pollen; if we accept the importance of repeated insect visits to the biology of the Composite head we may see it as an essential and characteristic part of the pollination mechanism. When all the pollen has been swept from the anther tube the style emerges. The stigma branches begin to diverge, exposing the receptive upper surface which is now ready to receive pollen. The ray florets are purely female, and open before any of the flowers of the disc. To this extent the pollination mechanism of species with both disc and ray florets is more complicated than those with ligulate florets alone. In these plants, for example Lesser Hawkbit (*Leontodon taraxacoides*), the ligulate florets are hermaphrodite (Fig. 72), and

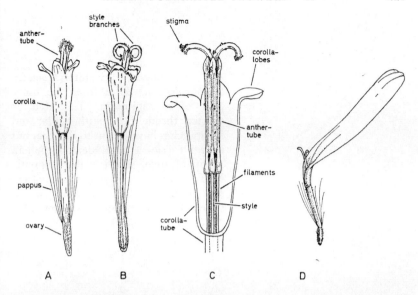

FIG. 71. Oxford Ragwort (*Senecio squalidus*). A, young disc-floret. Pollen is being forced out of the top of the anther-tube as the style grows up through it. B, older disc-floret with the divergent branches of the style projecting from the top of that anther-tube. C, upper part of disc-floret with half of corolla removed to show details of the anther-tube and filaments surrounding the style. D, ray-floret. Exeter, Devon.

function in the same way as the disc florets of the Daisy or Ragwort. Many *Compositae* are visited by a large variety of insects. Harper and Wood (1957) list 178 visitors to Ragwort (*Senecio jacobaea*), including thrips, *Hemiptera*, beetles, butterflies and moths, and many *Hymenoptera* and flies.

Sensitive stamens are found in a number of plant families. Thus in the Rock-roses (*Helianthemum* and related genera, *Cistaceae*) (Plate 49c) the stamens swing out towards the petals on stimulation, and in the Barberries (*Berberis, Mahonia, Berberidaceae*) they bend inwards towards the stigma. But the functional significance of the movement is seldom as obvious as it is in the *Compositae*. Not all *Compositae* show evident movement; Small found irritability of the stamens in 64 per cent. of the 149 species he observed. The movement is particularly striking in the Cornflower (*Centaurea cyanus*) and other members of the Knapweed genus (Fig. 73, Plate 36b). The conspicuous marginal florets are tubular but sterile and somewhat zygomorphic. The fertile florets have deeply

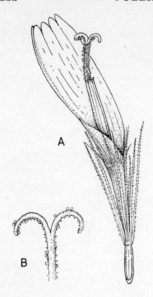

FIG. 72. Lesser Hawkbit (*Leontodon taraxacoides*). A, floret from the inner part of a capitulum. The style is exerted through the anther-tube and its branches have begun to diverge, but it still carries a good deal of pollen. B, tip of style, showing the stigmatic surface on the upper (inner) side of the style-branches, and the hairs which sweep the pollen from the anther-tube. Alderney, Channel Is.

divided blue corollas, and the long slender anther tubes stand prominently above the general surface of the head. The style carries a ring of sweeping hairs below the short stigmatic branches. On stimulation the filaments may contract several millimetres, ejecting a quantity of pollen on to the visitor. After an interval, the filaments recover their original length, and the process can be repeated. Percival (1965) found that in a freshly opened floret of *Centaurea montana* the amount of pollen delivered on stimulation was about a quarter of the whole amount in the anther tube, but the amount fell progressively on subsequent occasions. Stimulation of the filaments on one side of the floret will cause the anther tube to turn towards the insect causing the disturbance. In due course, the style grows up through the anther tube, and the stigma lobes diverge and can be pollinated.

The *Compositae* are generally thought of as a rather homogeneous family, but they vary a good deal in the details of their florets and capitula (see Small 1917, 1918), and consequently in their floral biology. The purely female ray-florets of many species have been mentioned already; these bring an element of protogyny to the head as a whole, although individual hermaphrodite florets are protandrous. In Coltsfoot (*Tussilago farfara*) the ray-florets are female, and the disc-florets are effectively purely male. In the Dwarf Thistle (*Cirsium acaule*) the plants are gynodioecious, and the smaller female heads can be picked out at a distance from the larger hermaphrodite heads. Both of these arrangements, like the gynodioecism of many *Labiatae*, are evidently adaptations tending to

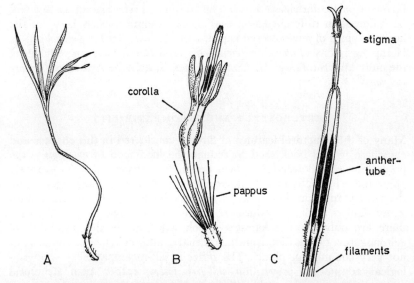

FIG. 73. Greater Knapweed (*Centaurea scabiosa*). A, sterile marginal ('ray') floret. B, fertile floret. C, detail of anther-tube and style. Alderney, Channel Is.

promote a greater degree of outbreeding. As in other families, self-incompatibility is often found in *Compositae*. This would be expected from the nature of the pollen-presentation mechanism – as in many *Leguminosae* (p. 202). Thus the self-pollination brought about by the style branches curving back to touch residual pollen at the top of the anther tube, or by the styles picking up pollen from other florets when the capitula close at night, will in many cases not result in fertilisation. However, there may be a good deal of outbreeding even in self-compatible species; thus Watts (1958) found up to 11.5 per cent. outcrossing in an experiment with cultivated lettuce (*Lactuca sativa*), probably brought about mainly by hoverflies.

There are perhaps three main factors limiting the size of the individual heads of *Compositae*. A very large capitulum is mechanically vulnerable; a corymb of smaller heads is less easily damaged, and if a few branches are broken this has little effect on the inflorescence as a whole. Large Composite heads provide a rich and compact store of food for insect larvae; the larger the head, the more seeds are put at risk by a single infestation. Here again, a corymb of separate smaller heads has an obvious advantage. Thirdly, if the marginal florets are female and the remainder herma-phrodite, selection in favour of a particular ratio of the two types will favour a particular size of head. The balance of selective advantage will be different in different cases, and every gradation exists between the large corymbs of few-flowered capitula of Yarrow (*Achillea millefolium*) or Hemp Agrimony (*Eupatorium cannabinum*) and the massive single heads of the cultivated Sunflower (*Helianthus annuus*) and Globe Artichoke (*Cynara scolymus*).

HETEROSTYLY AND INCOMPATIBILITY

Many of the structural features of flowers considered in this chapter and elsewhere in this book tend to reduce the likelihood of *autogamy* – the pollination of a flower by its own pollen. But structural features rarely affect the probability of a flower receiving pollen from another flower on the same plant, and this *geitonogamy* is genetically no different from autogamy. Unless a plant produces no more than a single flower at a time there are only two mechanisms which will exclude the possibility of geitonogamy. One is dioecism, the normal condition in higher animals but not very common in plants. The other is self-incompatibility – physiological adaptation preventing self-*fertilisation*, rather than structural adaptation preventing self-*pollination*. However, there is one type of structural adaptation which favours *xenogamy* – pollination by pollen from another individual. This is the phenomenon of *heterostyly*, of which the 'pin-eyed' and 'thrum-eyed' forms of the Primrose (*Primula vulgaris*) (Fig. 74) are the classic example. In the pin-eyed flowers the stamens are inserted about half-way up the corolla tube, and the long style reaches to the mouth of the tube. In thrum-eyed flowers the stamens are inserted close to the top of the tube, and the style is short, with its stigma at the level of the anthers in the long-styled flowers. The short-styled flowers have larger pollen grains and smaller stigmatic papillae than the long-styled flowers. Any one plant bears either pin-eyed or thrum-eyed flowers, never both. If the numbers of plants bearing the two kinds of flowers are counted they will be found to be approximately equal. Haldane (1938) examined 2302 plants from a number of localities and found no significant departure from equal numbers of the two types. Darwin (1877) found that 'legitimate' pollination – of a 'pin' flower by 'thrum' pollen or vice versa – results in a much higher degree of fertility than 'illegitimate' pollination of a stigma by pollen from a flower of the same type.

The system normally behaves as though it were controlled by a single gene. Long-styled plants are homozygous recessives; short-styled plants are heterozygous. Because of the incompatibility, the only crosses that can normally take place are those with the recessive homozygote as one parent and the heterozygote as the other – a simple Mendelian 'backcross' in which equal numbers of the same two types are to be expected among the progeny.

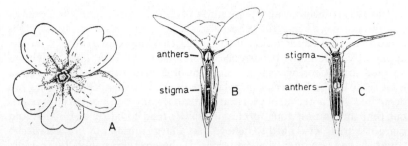

FIG. 74. Primrose (*Primula vulgaris*). A, face view of corolla-limb of thrum-eyed flower. B, thrum-eyed flower with half of calyx and corolla removed. C, pin-eyed flower with half of calyx and corolla removed. Cambridge.

As Darwin remarked, it is surprising how rarely insects can be seen visiting the flowers of primroses during the day. Knuth recorded visits by bumble-bees and butterflies, and he supposed that these insects are the main pollinators. Christy (1922) observed many visits by bees, a number by butterflies and some by bee-flies (*Bombylius major*). Crosby (1960) has suggested that the bee-fly may be an important visitor, little observed because of its fast erratic flight and speed of working. It seems unlikely that this is generally true everywhere, and Darwin's suggestion that the principal pollinators may be night-flying moths remains as a possibility (but see p. 377). A considerable diversity of pollinators – small beetles, flies, *Lepidoptera* and bees – may be important at particular times and in particular places. Much more observation in the field is needed.

The Cowslip (*Primula veris*) (Plate 37a), Oxlip (*P. elatior*), Bird's-eye Primrose (*P. farinosa*) and many other *Primula* species show similar heterostyly. *Primula scotica*, like some other species, is homostylous. Homostylous flowers, with the stigma and anthers at the same level, occur widely but usually in small numbers in populations of the Primrose. In two areas in the south of England – around Sparkford in Somerset and around Chesham and High Wycombe in the Chilterns – homostylous flowers are common, and locally predominant in the Primrose populations. The homostylous form is apparently determined by a mutation of the dominant gene determining the short-styled or 'thrum' condition. The homostyle plants are self-fertile, and under natural conditions probably out-cross to the extent of 5–10 per cent. (Crosby, 1959). The self-fertility of the homostyle flowers puts pollen from 'thrum' flowers at a disadvantage, so homostyles tend to spread in the population at the expense of thrums. Pin-eyed flowers continue to exist alongside homostyles, probably because homozygous homostyles have a lower viability than the heterozygotes (Crosby, 1949). Crosby concluded that the gene for homostyly should sweep through any population of Primroses in which it becomes established, converting the whole from a heterostylous to a homostylous condition. This seems unlikely to be the case, especially in view of the apparently widespread occurrence of homostyles; and evidence which has accrued since Crosby's original analysis of the situation shows that in at least some of the Somerset populations the proportion of homostyles has decreased over a period of years rather than the reverse. Ford (1964) points out that the homostylous species of *Primula* tend to occur in isolated and marginal areas. It is hard to believe that a large number of *Primula* species could have remained heterostylous if heterostyly could be rapidly destroyed by the chance appearance of homostyle individuals. Ford is probably right in concluding that the breeding system – with its attendant heterostyly or homostyly – is likely to be powerfully controlled by selection, and that the reason for the local abundance of homostyle Primroses is to be sought in the special ecological conditions of the areas where they occur.

Many other flowers show dimorphic heterostyly similar to that of the Primrose, including Perennial Flax (*Linum perenne* (agg.), *Linaceae*), Buckwheat (*Fagopyrum esculentum*, *Polygonaceae*), Water Violet (*Hottonia palustris*, *Primulaceae*), Bogbean and Fringed Water-Lily (*Menyanthes trifoliata*, Plate 37b, and *Nymphoides peltata*, *Menyanthaceae*), and the Lungworts (*Pulmonaria* spp., *Boraginaceae*). *Linum perenne* and the Lungworts set little or no seed with 'illegitimate' pollinations; probably at least some degree of incompatibility accompanies heterostyly in every case.

The flowers of Purple Loosestrife (*Lythrum salicaria*) are trimorphic, with styles and stamens of three lengths (Plate 38a). Pollen grains are of three sizes, corresponding to the three stamen-lengths – the longest stamens producing the largest pollen. The three flower-types occur on average in roughly equal numbers though, as the figures summarised by Haldane (1936) show, there is a good deal of variation in the proportions of the three types between one population and another. As in *Primula*, the 'legitimate' pollinations – by pollen from stamens of the same length as the style – give a much higher seed-set than other pollinations. Tristyly is controlled by two pairs of alleles which assort independently. One of these determines whether or not the style shall be short. The other has no effect upon short styles, but determines whether the remainder shall be mid or long (Fisher and Mather, 1943).

In all these cases the incompatibility mechanism could ensure outbreeding without the development of heterostyly. It is interesting that Thrift (*Armeria maritima*) and Common Sea Lavender (*Limonium vulgare*, *Plumbaginaceae*) exhibit dimorphism very similar to that of *Primula*, except that the style and stamens do not differ in length between the two forms in Thrift, while the Sea Lavender shows only slight heterostyly. Probably heterostyly is always secondary and incidental to the incompatibility that

PLATE 27. **a-b**, bumble-bees *(Bombus agrorum ♀)* on flowers of Yellow Archangel *(Galeobdolon luteum)*.

a

PLATE 28. **a**, bumble-bee *(Bombus agrorum ♀)* on Hedge Woundwort *(Stachys sylvatica)*. **b**, hover-fly *(Rhingia campestris)* sucking nectar from Bugle *(Ajuga reptans)*. **c**, bumble-bee *(Bombus lapidarius)* on Wood-sage *(Teucrium scorodonia)*; the flower that the bee is visiting is in the functionally male stage, the two flowers below are older with anthers recurved and stigma projecting over the entrance to the flower.

c

b

accompanies it, developing in response to selection pressures tending to maximise the proportion of compatible pollen reaching a stigma (Baker, 1949 1966).*

On average, any particular individual of a dimorphic species can be pollinated successfully by 50 per cent. of the individuals in the population. An individual of a trimorphic species can be pollinated by two-thirds of the other individuals. The commonest forms of incompatibility among flowering plants are determined by a large number of allelomorphs acting at a single locus, or occasionally two loci. Pollen from a particular flower is prevented from fertilising ovules carrying the same allelomorphs by an antigen-antibody reaction between the pollen tube and the tissue of the style, but can fertilise all other flowers. If the number of allelomorphs is large, this means that while every flower is incompatible with its own pollen, it will be compatible with the majority of other flowers in the population.† The reaction of the pollen is most commonly determined by the (single) allelomorph carried by the grain itself. Incompatibility results when this is the same as one of the (two) allelomorphs carried by the tissues of the style. This 'gametophytically determined' incompatibility is found in many families, including *Cistaceae*, *Convolvulaceae*, *Ericaceae*, *Onagraceae*, *Plantaginaceae*, *Leguminosae*, *Ranunculaceae*, *Rosaceae*, *Saxifragaceae*, *Scrophulariaceae*, *Solanaceae*, *Orchidaceae*, *Iridaceae*, *Amaryllidaceae*, *Gramineae*, *Campanulaceae* and *Araceae* (see Pandey, 1960, for list). Less often, the reaction of the pollen is determined by the genetic constitution of the parent plant, and incompatibility results when this is the same as that of the style. This 'sporophytically determined' incompatibility is known from a number of families, some large and important, including *Caprifoliaceae*, *Chenopodiaceae*, *Compositae*, *Cruciferae* and *Geraniaceae*.

THE WILD ARUM

Cuckoo-Pint or Lords-and-Ladies (*Arum maculatum*) possesses one of the most remarkable pollination mechanisms to be found in the British flora. It produces its curious inflorescence (Plate 39) on a short stout stalk a few inches above the ground in late April or early May. The two main organs of the inflorescence are the spadix, which is an inflorescence axis, and the leafy spathe. The female flowers form a zone at the base of the spadix and consist merely of ovaries topped by stigmas. Above them are a few sterile flowers, and above these is a zone of male flowers consisting only of

* See Vuilleumier (1967) for a list of genera exhibiting heterostyly, and a general discussion of the possible origins of the condition.

† This *multiple allelomorph incompatibility* is probably primitive or at least of very long standing in the Angiosperms, while heteromorphic incompatibility of the kind discussed above has probably originated independently and (in evolutionary terms) relatively recently in the various groups in which it occurs.

short-stalked stamens. The uppermost flowers are again sterile with hair-like appendages, while above this the spadix is prolonged into a club-shaped structure, which is usually purplish in colour. The spathe forms a chamber round the lower part of the spadix, opening above the upper sterile flowers to reveal the club of the spadix. Entry to the chamber is by a space round the spadix, between it and the spathe. Pollination is brought about by small insects becoming trapped in the flower chamber during its first, female stage of development, followed by their release after the male stage, in which they become dusted with pollen.

The mechanisms involved in this process were studied by Fritz Knoll (1926), who carried out most of his work on *Arum nigrum*, a Mediterranean species which is larger than *A. maculatum*, is visited by larger insects, and is therefore easier to study. In this species the spathe opens overnight. During the first day of opening a strong faecal smell is produced by the spadix. In the morning insects are attracted; these are mostly dung-frequenting flies or beetles. If they alight on the club of the spadix or the inner surface of the spathe they lose their grip and fall. This results from the special nature of the surface of these organs. The cells of the surface are smooth, and produced into downwardly directed conical papillae which provide no foothold for the insects. The effect is enhanced by an oily secretion which covers the surface. On the spathe this type of surface is found on the inside from the top down to the upper part of the chamber. As they drop to the base of the spathe, the insects encounter the ring of bristles formed by the sterile flowers, which have a smooth greasy surface. If the insects are small enough they fall through into the chamber; larger insects are arrested by the bristles and can fly off. If the insects that are trapped have come from another *Arum* inflorescence they may pollinate the female flowers (which have receptive stigmas during the first day), probably by climbing upon them in their attempts to escape. The pollen tubes grow rapidly, and the stigmas soon wither, so that by the time the inflorescence sheds its own pollen, which is during the second night, self-pollination is impossible. The pollen is shed in great quantity and thoroughly dusts the trapped insects. By the morning of the second day a change has taken place in the cells of the appendages of the sterile flowers, so that their surfaces have become wrinkled, while the papillae on the rest of the spadix shrink; these changes enable the insects to climb up the spadix and escape. If they are then trapped by a plant in its first day they can cause cross-pollination. On the second day the spadix no longer produces a smell.

Knoll carried out experiments using imitation spathes of coloured glass. The inner surfaces of these were dusted with talcum powder, which made it impossible for insects to cling to them. The models, when provided with real *Arum nigrum* spadices, caught just the same kinds of insects as the real plants, though fewer of them, apparently because the smell of the detached spadices was weaker. These models also demonstrated the principle of capture by falling: it had earlier been thought that the insects entered the chamber voluntarily to seek shelter and warmth. The models were also used to show that the attraction of the pollinating insects from a distance is purely by scent. Light coloured and dark coloured spathes gave identical results, and Knoll concluded that the colouring of very dark spathes (as in *Arum nigrum*) or very light ones (as in *A. italicum*) was only significant in so far as they made the spathes stand out from their surroundings and so induced insects to alight. Scentless models captured some insects if placed touching, and facing the same way as, scented models. Models containing a vessel of stale blood in place of a spadix also caught insects.

A remarkable feature of the spadix of *Arum*, which has long been known, is that it generates heat. This led to the theory that it was the warmth which attracted insects to enter the chamber. Knoll performed experiments with models having an artificial spadix which was electrically heated; this showed that the heat was no attraction. The rapid respiration which gives rise to the heating is in fact probably connected with the metabolism necessary to produce the smell. The amount – several grams – of starch consumed in the course of a few hours is out of all proportion to the few milligrams of malodorous compounds (ammonia, amines, amino-acids, skatole, indole) that are produced. Probably the main function of the heating is to help vaporise these compounds and intensify their dissemination (Meeuse, 1966). The smell itself (like that of other 'fly-trap' flowers described in Chapter 10) is a purely deceptive attraction. The insects receive no food from the *Arum* apart from drops of a sweet secretion from the withered stigmas.

Our native British *Arum maculatum* has essentially the same mechanism as *A. nigrum*. The spathes open about midday, and heating of the spadix and scent production are at their height during the afternoon and evening of the same day – though the odour of *A. maculatum* is not very noticeable to most people. The inflorescences are visited chiefly by small flies of the genus *Psychoda*, mainly females of *P. phalaenoides* (though Grensted (1947) found that in exposed places *P. grisescens* was at least as common), of which large numbers may be found in the chamber (see Prime, 1960). Müller and Knuth record finding hundreds of specimens of this species, and in one inflorescence from the castle garden at Plön Knuth estimated that there were some 4,000 of these insects. *Arum italicum* (a south European species found locally near the coast in the south of England) also attracts mainly *Psychoda* species, but *A. conophalloides* from Asia Minor and Persia has a scent which attracts blood-sucking midges of the families *Ceratopogonidae* and *Simuliidae*. One spathe of this species which Knoll examined contained 600 Diptera of which 461 were identified; these were females of three species, one of which was represented by 427 insects. Evidently, the attraction of these plants is both effective and highly specific.

CHAPTER 7

THE BRITISH WILD ORCHIDS

In the variety and perfection of their adaptations to insect pollination the Orchids stand alone. They include not only some of the most perfect examples of adaptation to insect visitors, but also a surprising range of intriguing and beautiful mechanisms.

The Orchids are an attractive and interesting group of plants in many ways, and it is not surprising that their pollination has received a great deal of attention, quite apart from its own intrinsic interest. But although earlier botanists had described the structure of the Orchid flower, the pollination mechanisms of the common Orchid species was first fully understood by Charles Darwin. His fascinating book, *The various contrivances by which Orchids are fertilised by Insects*, published in 1862, is the record of a vast amount of accurate and perceptive observation. From 1842 until his death forty years later Darwin lived at Down, close to the crest of the North Downs in Kent. Many of the British Orchids are plants of chalky soils, and it was around Down that many of Darwin's observations were made. The British Orchids show well the characteristic pollination mechanism of the family,* even if they cannot parallel some of the more bizarre variations found among the tropical genera (see Chapter 10, pp. 295-299).

At first sight the Orchid flower has little in common with the flowers of any other family. However, when it is examined in detail, it appears that it is in fact an exceedingly specialised version of the kind of flower seen in the Lilies and their relatives (*Liliaceae*). A Lily or Tulip flower has six perianth segments, three outer and three inner, six stamens, again in two whorls of three, and an ovary made up of three fused carpels – though the ovary is superior in the Tulip or Lily but inferior in the orchids. The Orchid flower also has six perianth segments, though one of the inner whorls is much larger than the others, and is called the lip, or *labellum*. The labellum is strictly the uppermost petal, but in most Orchids the ovary is twisted through 180° so that the flower is, in fact, upside down. The labellum then appears at the bottom, where it forms a convenient alighting platform for insects: it often has a spur or nectary at its base. It is in the stamens that the greatest modifications of the Orchid flower are

* Summerhayes (1951) gives information on the pollination of all the British species.

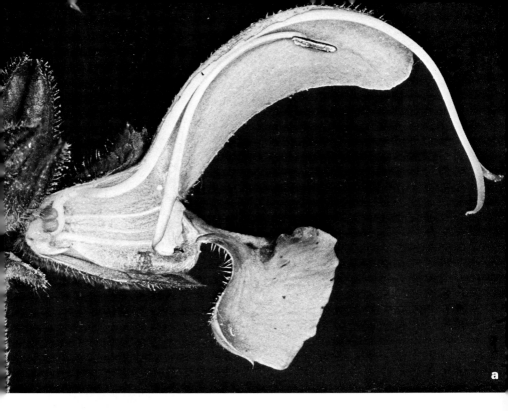

PLATE 29. **a**, half-section of flower of Meadow Sage *(Salvia pratensis)*. **b**, bumble-bee *(Bombus agrorum)* on flower of *Salvia glutinosa*; the fertile anther-cells have swung down into contact with the abdomen of the bee.

PLATE 30. **a**, bumble-bee *(Bombus agrorum ♀)* making a 'legitimate' visit to Lousewort *(Pedicularis sylvatica)* ; notice the distension of the upper lip of the flower. **b**, *Bombus lucorum* stealing nectar through a hole in the corolla-tube of Lousewort. **c**, bumble-bee *(Bombus agrorum ♀)* on Yellow Rattle *(Rhinanthus minor)*.

a

b

c

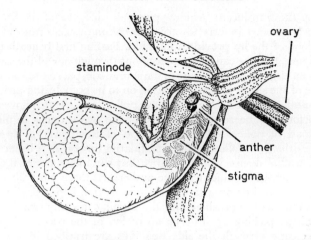

FIG. 75. Lady's Slipper (*Cypripedium calceolus*). Flower with half of lip removed to show details of the column and the path taken by a visiting insect.

found. Most of the stamens have been either completely lost or reduced to sterile vestiges. Two of the outer whorls are missing: Darwin believed that they had become fused with the sides of the labellum, where he found two slender vascular bundles, but it is generally thought now that they have vanished without trace.* The remaining stamens, together with the stigmas, have become fused into the stout column which projects in the centre of the flower, above the lip. But of these four stamens never more than two are fertile, and in most Orchids the only fertile stamen is the remaining one in the outer whorl. Only two stigmas are functional: the third forms the *rostellum*, which generally projects from the top of the column.

The Orchids fall into two main groups. Least specialised are the Lady's Slipper (*Cypripedium calceolus*) and its relatives; they differ so greatly from the other Orchids that some systematists place them in a separate family of their own. In the Lady's Slipper (Fig. 75, Plate 40a) two stamens of the inner whorl are fertile, the third forming the front of the column. The one remaining stamen of the outer whorl is represented by a thick petal-like staminode, overarching the stigma. The lip forms a deep pouch, and the nectar secreted inside it attracts small bees. These enter the lip through the obvious large opening, but their escape by the same route is barred by the smooth slippery sides of the lip and its inrolled edges. Eventually,

* Hagerup (1952b) considered that vestiges of all six stamens can be detected in *Herminium*, and perhaps other genera.

guided by the translucent 'window-panes' in the sides of the lip near its base, the bee makes its way towards the column, where numerous hairs on the floor of the lip provide a foothold. Passing first beneath the long rough stigma, the bee can then squeeze out through one of the two small openings on either side of the base of the staminode, past one of the anthers, and some of the sticky pollen is smeared on to its body in the process. On visiting another flower, the bee will leave pollen from the first flower on the stigma before it squeezes out past one of the two stamens to pick up a further load of pollen. The construction of the flower of *Cypripedium* determines the path followed by a pollinating insect; it is in fact a neat 'one-way-traffic' device which ensures that the insect passes the stigma before it comes into contact with the anthers. As there is usually only one flower open at a time on a stem, cross-pollination is practically assured. An insect too small to brush against the stigma can easily crawl out of the side openings, picking up little or no pollen in the process. Insects too large to escape through the side openings are trapped in the lip and usually die.

In the great majority of Orchids there is only a single, much-modified anther. Like most anthers, this is two-lobed, but by contrast with the loose powdery pollen of most plants, the pollen grains are bound together by slender elastic threads into a pair of pollen masses or *pollinia*. The two anthers which are fertile in *Cypripedium* are reduced to projections on the column, often forming part of the *clinandrum* – the little hood protecting the fertile anther. These Orchids in their turn fall into two main groups. In the greater number of Orchids the anther is borne by a comparatively slender stalk on the back of the column. Its apex is close to the rostellum, and it is by their apices that the pollinia become attached to a visiting insect: the point of attachment gives the group their name *Acrotonae*. The *Acrotonae** are often woodland plants, and they include almost all the epiphytic Orchids of the tropics. By contrast, in the more specialised *Basitonae* the single anther is perched on top of the column, and the two pollinia are furnished with minute stalks or *caudicles* at thei bases, attached to viscid discs formed of rostellum tissue. The *Basitonae* are often grassland plants, and although much fewer in both species and individuals than the *Acrotonae* over the world as a whole, they are well represented in temperate regions and include many of the best-known European Orchids.

Most of the British *Acrotonae* belong to the tribe *Neottieae*. A good example is the Marsh Helleborine (*Epipactis palustris*), which is locally frequent in fens and dune slacks. The whitish flowers (Fig. 76) are borne in a loose raceme on a stem from a few inches to a foot or more high, and

---

* Many modern systematists do not consider the division of the Orchids into *Acrotonae* and *Basitonae* to be of fundamental classificatory importance. However, the terms are useful in considering the structural basis of the pollination mechanisms of the plants.

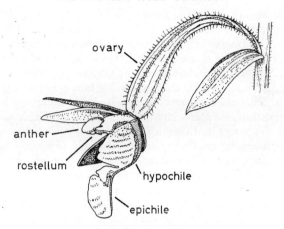

ovary

anther

rostellum

hypochile

epichile

FIG. 76. Marsh Helleborine (*Epipactis palustris*). Side view of flower with half of perianth removed. Braunton, Devon.

are mainly pollinated by bees. The summit of the column in a newly opened flower is occupied by the large, projecting, almost globular rostellum, with the broad squarish stigma below it. The anther overhangs the rostellum: even before the bud opens the anther cells dehisce, releasing the pollinia which come to lie with their tips touching the rostellum. The pollinia are rather friable; the individual pollen-grains cohere together in groups of four, and these are rather loosely bound together by fine elastic threads. As the rostellum matures, its outer surface develops into a soft elastic membrane, so tender that it can be penetrated by a hair. At a slight touch it becomes viscid, so that the pollinia stick to it. The tissue within the rostellum develops into a lining of sticky matter which, on exposure to air, hardens in five or ten minutes. An object brushing upwards and backwards against the rostellum easily removes the whole of the elastic skin of the rostellum as a little cap, which sticks firmly to it by its adhesive lining. The lip of the Marsh Helleborine is of peculiar structure. The base forms a cup (the *hypochile*), containing nectar. The broad flat tip (*epichile*) is attached to the base by a slender elastic 'waist', so that the weight of an insect alighting on it bends it sharply downwards. Thus a bee visiting the flower for nectar and alighting on the lip depresses it, and as long as it is feeding is well clear of the rostellum and the pollinia. But as soon as it makes to leave the flower and takes its weight from the lip, this springs up, compelling the bee to fly slightly upwards. In so doing, its head brushes upwards and backwards against the rostellum and the projecting end of the anter, neatly removing the pollinia. Owing to the depression of the elastic lip the pollinia are in the right position to strike the stigma when the insect feeds at the next flower. The flowers are visited by honey-bees which remove the pollinia in the manner described, and by various smaller *Hymenoptera* and flies which remove the pollinia on their backs. If the flowers are not visited by insects, the friable pollinia sooner or later break up, and the loose pollen falls down over the rostellum and stigma. This may happen even before the flower opens, and the relative importance of cross- and self-pollination probably varies greatly at different times and in different localities.

The Broad-leaved or Common Helleborine (*Epipactis helleborine*) is certainly the most generally common member of the genus in the British Isles. It is a shade-loving plant, but is perhaps more common along shady roadsides and wood margins than in extensive woods. In structure, its flower is very like that of the Marsh Helleborine, though the lip is not hinged in the middle, and the rostellum is more bulbous and projects farther over the stigma, and the flowers open rather more widely. The flowers vary greatly in colour from dull purple to greenish, and seem to be pollinated almost entirely by common wasps (*Vespula vulgaris* and *V. germanica*) (see pp. 141–144), which are attracted to the flowers in large numbers. The bees and other insects which visit the Marsh Helleborine ignore this species. Whatever other features may be attractive to wasps, even to the human eye the colour of the flowers has something in common with the dingy brownish purple of the wasp-pollinated Figworts (*Scrophularia* spp.) (Chapter 6, p. 217). The mechanism of pollination is evidently much the same as in the Marsh Helleborine, but removal of the pollinia depends entirely on the insect's head striking upwards against the rostellum as it backs out of the flower, and the more protuberant rostellum no doubt aids in bringing this about. This species like the last can also be self-pollinated. According to the Danish botanist Hagerup (1952a), pollen grains which fall on to the rostellum are trapped by its viscid secretion, which later spreads out over the stigmas, where the pollen grains germinate and bring about fertilisation. In common with Orchids known to be regularly self-pollinated, the Broad-leaved Helleborine shows remarkably regular production of well-developed capsules. In a long flower-spike the capsules are beautifully graded in size as they mature in succession from the bottom of the inflorescence to the top. In this species, self-pollination may act only as an insurance against the failure of insect visitors; the species of *Epipactis* which are regularly self-pollinated will be discussed later.

The Helleborines of the related genus *Cephalanthera* illustrate the way in which the *Epipactis* type of pollination mechanism may have originated. In size and structure the flowers are not unlike those of *Epipactis*, but there is no rostellum and the pollen grains are only weakly bound together with a few elastic threads. The Sword-leaved or Narrow Helleborine,

C. longifolia, is apparently mainly pollinated by small bees. Its white flowers open rather widely, but the narrow tubular space between the lip and the column forces an insect penetrating the flower close against the stigma, which is covered with a copious sticky secretion. Leaving the flower, the insect brushes past the anther, which arches over the front of the stigma, and the friable pollen adheres to the stigmatic secretion smeared on the insect's back. The anther has an elastic hinge at its base, and springs back to its original position as soon as the insect has gone. It is arguable whether the absence of a rostellum, and the other characters of the *Cephalanthera* flower, are primitive or degenerate; but at least the flower illustrates one way in which the *Epipactis* condition may have evolved from a state of affairs much like that seen in an ordinary flowering plant with friable pollen and a sticky stigma. It is particularly interesting that the sticky stigmatic secretion serves to stick the pollen to the visiting insect, since it is the sticky secretion of the rostellum – generally thought to be derived from a vestigial stigma – which takes over this function in all the more advanced Orchids. Whether the flowers of the Sword-leaved Helleborine are self-compatible with their own pollen does not seem to be recorded, but there is little in the structure of the flower to prevent self-pollination. In Britain it is a rare plant, and is seldom pollinated very effectively, perhaps because suitable pollinating insects are few and far between in its woodland habitats. The much commoner Large White or Broad Helleborine (*C. damasonium*), which is regularly self-pollinated, will be described later in this chapter.

The Common Twayblade (*Listera ovata*) (Plates 41, 42) shows a rather similar mechanism to that of *Epipactis*, but adapted to the visits of smaller insects. The greenish flowers, borne in a long slender raceme above the two leaves which give the plant its name, are particularly attractive to ichneumons, which visit the flowers in considerable numbers. Beetles and various other insects also visit the flowers commonly. The lip is broadly strap-shaped, bent sharply downwards from a point near its insertion, and deeply notched at the tip. It forms a landing platform leading up to the column; a groove, secreting much nectar, runs up the centre of the lip from the notch at its tip. The anther lies behind the rostellum, protected by a broad expansion of the back of the column. As in *Epipactis*, the anther-cells dehisce before the bud opens, and the pollinia are left quite free, supported in front by the concave back of the rostellum. A visiting insect crawls slowly up the narrowing lip, feeding on the copious nectar, which leads it to a point just below the rostellum. On the gentlest touch the rostellum exudes, almost explosively, a drop of viscid liquid which, coming into contact with the tips of the pollinia and with the insect, cements them firmly to its head and sets in a matter of seconds. After the pollinia have been removed, the rostellum slowly straightens from its arched position close to the lip, leaving clear the way to the stigma. Now an insect crawling up the nectar groove can pollinate the flower with pollinia brought from another younger flower. It has been suggested that the almost explosive ejection of the viscid matter from the rostellum may startle the pollinating insects sufficiently to make them fly to another plant before they start feeding again. However, many visiting insects seem little disturbed by the explosion of the rostellum and, as insects generally work up the inflorescence from the bottom, this probably makes little difference to the effectiveness with which cross-pollination is brought about. The most interesting feature of the pollination of the Twayblade is the way in which a precision mechanism has evolved depending on relatively undiscriminating pollinators. It is striking and curious that an ichneumon or a skipjack beetle will operate the mechanism neatly and accurately, but the flower evidently does not provide the appropriate cues for orientation of the bees which visit this species casually for nectar, and in general there are not effective pollinators.

An essentially similar mechanism is seen in the Bird's Nest Orchid (*Neottia nidus-avis*), though in this case droplets of nectar are secreted diffusely from the broad, slightly concave lower half of the lip of the buff or parchment-coloured flowers. The Bird's Nest Orchid is a saprophytic plant growing in the deep humus of shady woods, and like the Hellebor-ines found in similar situations it seems to be little visited by insects. The abundant well-filled capsules are probably mainly the result of self-pollination. Such visitors as there are seem principally to be two-winged flies (*Diptera*).

The Autumn Lady's Tresses (*Spiranthes spiralis*) is an attractive though inconspicuous little Orchid which is often rather common in short turf in the south of England in late summer. The flower-spikes appear before the leaves, and bear the small tubular sweetly scented whitish flowers in a spiral on the upper part of the stem. This gives the flowering spikes a rather plait-like appearance, hence the English name of the plant. The flowers project almost horizontally from the stem and never open widely (Fig. 77). The rostellum is a slender flattened structure, projecting forwards above the stigma. The central part of its upper surface consists of an elongated mass of thickened cells (which Darwin called the 'boat-formed disc'). The tips of the pollinia become attached to the top of this; its underside consists of a mass of viscid matter, protected by the delicate membrane of the underside of the rostellum. At a touch, the membrane of the rostellum splits down the middle of the underside, and round the edges of the 'boat-formed disc', exposing the viscid matter and leaving the disc free but supported between the prongs of a fork formed by the sides of the rostellum. In a newly opened flower, the column lies close to the lip, leaving only a narrow passage for the tongue of a bee visiting the flower for the nectar in the cup-shaped base of the lip. The flower cannot be pollinated, but the bee inevitably touches the lower surface of the

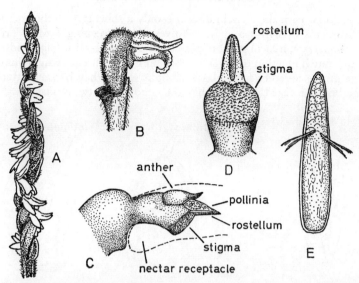

FIG. 77. Autumn Lady's Tresses (*Spiranthes spiralis*). A, inflorescence.
B, single flower, with lower sepal removed. C, detail of column; the
dotted lines indicate the outline of the perianth. D, front view of column.
E, disc with pollinia attached. B-E, after Darwin.

rostellum, which splits, allowing the 'boat-formed disc' with its attached
pollinia to become cemented to the upper side of its proboscis. Following
removal of the pollinia the remains of the rostellum wither, and, as in the
Twayblade, the column slowly bends up away from the lip leaving the
stigma freely exposed to pollen brought by a visiting bee from another
flower. The flowers at the bottom of a spike always open first, and those at
the top last, so that while the most recently opened flowers at the top of
the spike have pollinia waiting to be removed, those at the bottom of the
spike are ready for pollination. This fact, combined with the invariable
behaviour of the visiting bumble-bees,* starting at the bottom of a spike
and crawling up, visiting the flowers in succession until they reach the top,
means that if insects visit the flowers cross-pollination is practically
assured. *Spiranthes* is striking in its beautiful adaptation to cross-pollination, but according to Hagerup (1952a) the flowers are readily self-
pollinated if insect visitors are scarce. A curious feature is that the rostellum is provided with stigmatic papillae to its apex, and these may well
increase the chance of self-pollination. In the structure of its rostellum

* Darwin observed visits by bumble-bees at Torquay, but there seem to be no
more recent records of visitors to this species, and more observation is needed.

and pollinia, with their viscid disc, it is only a short step to the elaborate arrangements of *Orchis* and its relatives. The Creeping Lady's Tresses (*Goodyera repens*), a local plant of pinewoods in the Scottish Highlands, has a flower much like that of *Spiranthes*, but the elastic threads which bind the pollen grains together form elastic ribbons where they become attached to the broader viscid disc, very like the caudicles of the *Basitonae*.

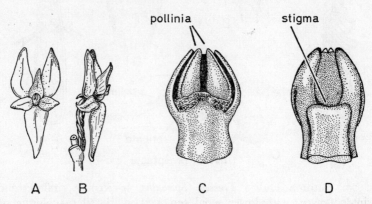

FIG. 78. Bog Orchid (*Hammarbya paludosa*). A, front view of flower. B, side view of flower. C, front view of column. D, back view of column. B-D, after Darwin.

The Bog Orchid (*Hammarbya paludosa*) is the smallest British Orchid; it and the rare Fen Orchid (*Liparis loeselii*) are the only British representatives of the tribe *Epidendreae*, which includes many tropical epiphytes. The flower of the Bog Orchid is peculiar in that the lip is uppermost – that is, in its morphologically 'correct' position. But this comes about not, as might be expected, by the ovary being untwisted (as it is in the Alpine *Nigritella*); instead the ovary is twisted through a complete 360° to bring the lip back to the same position. This orientation of the flower means that in the Bog Orchid, unlike the Orchids we have so far considered, the anther is in front of the rostellum; it is protected by two membranous expansions of the column, enclosing it on either side (Fig. 78). The anther-cells dehisce before the buds open, and shrivel until they form no more than small cups holding the bases of the pollinia. At the same time a group of cells at the tip of the rostellum break down to form a drop of viscid liquid which catches the tips of the pollinia. If a visiting insect touches this drop, the pollinia become stuck to the underside of the insect's head and are withdrawn when it leaves the flower. The stigma lies in a pocket-shaped cavity between the rostellum and the base of the column, at the front of the narrow passage between the rostellum and the lip. When an

PLATE 31. **a**, bumble-bee *(Bombus agrorum)* at mouth of corolla-tube of Foxglove *(Digitalis purpurea)*. **b**, *Bombus agrorum* in flower of Foxglove. **c**, flowers of *Mimulus glutinosus*; stigma of the left-hand flower has been touched and the lobes have closed together, exposing the anthers behind it.

PLATE 32. **a**, honey-bee visiting flower of Purple Toadflax *(Linaria purpurea)*. **b**, common wasp *(Vespula* sp.*)* visiting flower of Figwort *(Scrophularia nodosa)*.

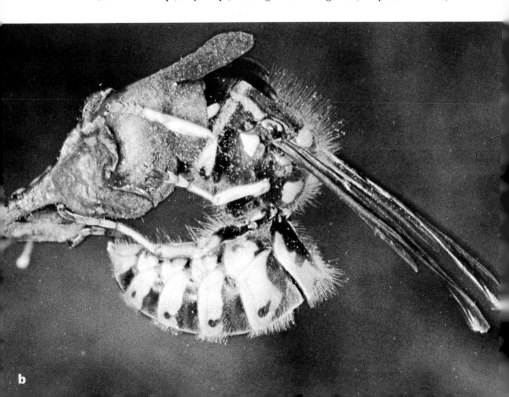

insect bearing pollinia visits another flower, the pollinia are forced into the stigmatic cavity. Little is known of the species pollinating the Bog Orchid, and parts of the account given here are conjecture. But the flowers are undoubtedly attractive to insects; the pollinia are almost all quickly removed, and the amount of seed set shows that pollination is manifestly effective.

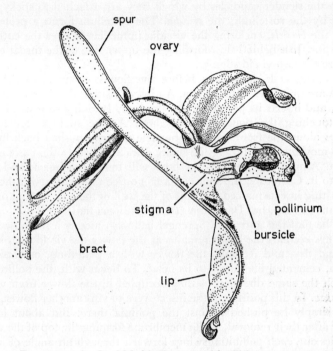

FIG. 79. Early purple Orchid (*Orchis mascula*). Side view of flower with half of perianth cut away to show details of the column. Sidford, Devon.

The *Basitonae* (tribe *Ophrydeae*) include the most highly organised of Orchid flowers, and some of the most specialised for insect pollination. They also include some of the best-known Orchids, and they show the characteristic pollination mechanism of the family in its most elegant and highly developed form. The basic type is well illustrated by Darwin's example, the Early Purple Orchid (*Orchis mascula*) (Fig. 79). This is often common in woods and pastures and along roadsides in spring and early summer, with its bright purple flowers borne above the rosette of dark-spotted leaves in a rather loose spike on a stem which may be anything from a few inches to a foot or more high. The lip is broad, flat or some-what reflexed at the sides, slightly lobed, and with a long stout spur at its base. The two lateral sepals spread widely, while the upper sepal and the two upper petals form a hood over the single anther, which lies just above the wide entrance to the spur. The pollen-grains are aggregated into small compact masses, which are bound together by slender elastic threads into a pair of club-shaped pollinia; the elastic threads run together at the base to form the slender caudicles by which they are attached to sticky discs formed by the rostellum, the *viscidia*. The rostellum forms a protective pouch, the *bursicle*, enclosing the viscidia immediately over the entrance to the spur. Just behind this, forming the upper side of the throat of the spur, are the two viscid stigmas.

The anther-cells open even before the flower expands, so that the pollinia are quite free within them. An insect visiting the flower lands on the lip, and inserts its proboscis into the spur. In doing so it can hardly avoid touching the pouch-like rostellum. At the slightest touch this ruptures along the front, and the bursicle is easily pushed back by the insect's movements, exposing the sticky discs attached to the bases of the pollinia. Almost infallibly one or both will touch the insect, and stick firmly to it. Curiously, the spur contains no free nectar. Darwin thought that visiting insects pierced the cells of the wall of the spur to feed on the abundant cell-sap, but Daumann (1941) considers that this is not so, and it may be that the spur is, as Sprengel believed, merely a sham nectary. In the few seconds the insect remains at the flower the viscid matter sets hard and dry, and it leaves the flower with a pollinium, or a pair of pollinia, cemented like horns to its head. To begin with, the pollinia lie in much the same direction as they occupied in the flower from which they came. In this position, if the insect were to visit another flower, they would simply be pushed against the pollinia there. But about half a minute after their removal, as the membrane forming the top of the viscid disc dries out, each pollinium swings forward through an angle of about 90°. This movement, completed in a time which would allow the insect to fly to another flower, brings the pollinia into exactly the right position to strike the sticky stigmas. The whole process is easily reproduced if a well-sharpened pencil is substituted for the tongue of an insect (compare Plate 43). Hermann Müller and a friend of Darwin's observed visits by bumble-bees of various species, but such visits are certainly very sporadic, and perhaps take place only early in the season. Hobby (1933) saw visits by *Empis tesselata* in Yorkshire. There is room for much more observation and experiment on the pollination of this common Orchid.

A mechanism identical in all its essentials is found in other members of the genus *Orchis*, and in the Marsh and Spotted Orchids (*Dactylorhiza* spp.). Of these plants, the Green-winged Orchid (*Orchis morio*) seems to be pollinated mainly by social and solitary bees, the others variously by bees and *Diptera* (largely hoverflies and empids). According to Hagerup, *Dactylorhiza maculata* depends almost entirely on the drone-fly (*Eristalis intricarius*) near the northern edge of its range in the Faroes and Iceland.

The other nearly related species show variations in detail, some of them very interesting. The Fragrant or Scented Orchid (*Gymnadenia conopsea*), the Pyramidal Orchid (*Anacamptis pyramidalis*) and the two Butterfly Orchids (*Platanthera*) are all specialised butterfly and moth flowers. This appears most obviously from their long slender spurs, but is evident, too, from other details of their structure.

The Pyramidal Orchid has rather small flowers in a dense pyramidal spike. The individual flowers are much like those of the Early Purple Orchid, but apart from their much more slender spurs there are other significant differences of detail. The lip bears two conspicuous projecting ridges, forming a guide like half a funnel leading into the mouth of the spur; Darwin compares them with the sides of a bird decoy. The pollinia are borne, not each on a separate viscidium as in the Early Purple Orchid, but on a single saddle-shaped viscidium. The pollinia are placed very low on the column over the mouth of the slender spur, so that the two stigmas, which are confluent in the Early Purple Orchid, are widely separated. The flowers have a rather strong sweet smell. Butterflies and moths, especially burnet moths (*Zygaena* spp.), visit the flowers in large numbers (Plate 44). The proboscis of a visiting insect is guided straight into the mouth of the spur by the converging ridges on the lip. As it is inserted into the spur it brushes past the bursicle, which moves back at a slight touch and exposes the viscidium. As soon as the saddle-shaped viscidium is exposed to the air it begins to curl inwards, clasping the insect's proboscis around which it fits like a collar. Indeed, if the proboscis of the insect is slender, the two ends of the viscidium may overlap, encircling it completely. Within a few seconds the viscid matter has set, and the pollinia are firmly cemented in place, though owing to the curling of the viscidium they now diverge more widely than they did in the anther. After a short interval they begin to swing forward, and soon come to project one on either side of the insect's proboscis. In this position they are exactly placed to come into contact with the two stigmas when the insect visits another flower.

Several points about this mechanism are noteworthy. The size and disposition of the parts of the flower and the form of the collar-like viscidium are perfectly adapted to the slender flexible tongues of butterflies and moths. For the pollinia to come into contact with the stigmas they must be symmetrically placed on the insect's proboscis, and this is ensured by the guiding ridges on the lip. Both the number of pollinia removed and the number of capsules produced by the Pyramidal Orchid suggest that this adaptation is highly effective. Darwin records a noctuid moth with eleven pairs of pollinia of this species on its proboscis: 'The proboscis of this latter moth presented an extraordinary arborescent appearance'! A pencil, which will do duty for the tongue of a bee in the flower of an Early Purple or Spotted Orchid, is ordinarily too coarse to remove the pollinia of *Anacamptis*, but the tongue of a butterfly is easily simulated by a needle.

The only other British Orchid with the two pollinia attached to a single viscidium is the very rare Lizard Orchid (*Himantoglossum hircinum*).* The Man Orchid (*Aceras anthropophorum*) illustrates a stage linking this with the usual condition of two quite separate viscidia. The two viscid discs are placed close together, so that each is somewhat D-shaped, and although it is possible to remove one alone, an insect almost invariably removes the pair together.

The Fragrant Orchid (*Gymnadenia conopsea*) with its heavily scented pink flowers is also pollinated by butterflies and moths. The lip lacks the guiding ridges of the Pyramidal Orchid, and the rostellum does not form a bursicle, so the two viscidia are freely exposed to the air. The long narrow viscidia become fixed lengthwise to the proboscis of a visiting insect, and stick sufficiently firmly even though the viscid matter does not set hard as it does in *Orchis* and its allies. The spur is copiously filled with free nectar. Darwin thought this was related to the instant adhesion of the viscidia; there is no advantage to be gained by delaying the insect at the flower as there is in *Orchis*, where the viscid matter needs a second or two to harden enough to attach the pollinium firmly to the insect.

The Butterfly Orchids (*Platanthera bifolia* and *P. chlorantha*) resemble the Fragrant Orchid in their naked viscidia. In fact the pollination mechanism of the Lesser Butterfly Orchid (*P. bifolia*) (Plate 45a) is very much like that of the Fragrant Orchid, though the rather spidery fragrant white flowers are more obviously adapted to attracting the night-flying moths which visit them in considerable numbers. The small round viscidia are placed facing each other close together over the mouth of the spur and the stigmas, and become attached to the tongues of visiting moths. After a short interval the pollinia swing forward to come into contact with the stigmas of another flower as in *Orchis*. The Greater Butterfly Orchid (*P. chlorantha*) (Plate 45b), though closely related to *P. bifolia* and very like it in the superficial form of its flowers, is strikingly different in the form of the column and the disposition of the pollinia and viscidia. The viscidia are placed wide apart, at either side of the entrance to the spur, with the pollinia forming an arch over the two large confluent stigmas. As in the Lesser Butterfly Orchid, the flowers are largely visited by night-flying noctuid moths, but the pollinia become attached to the sides of the insect's head, often to its compound eyes. Judging from the amount of seed set, the two species both have an efficient means of pollination, and

* Schmid (1912) observed visits by the solitary bee *Andrena carbonaria*: the flowers appeared to have little or no attraction for most other insects.

it is evident that although they rely on much the same pollinating insects they have developed an effective barrier to hybridisation between them.

The Frog Orchid (*Coeloglossum viride*) (Plate 45d, p. 69) and the Musk Orchid (*Herminium monorchis*) are pollinated by various small crawling and flying insects. They represent a further variation on the *Orchis* theme, and provide interesting parallels with the Twayblade, which has a similar range of pollinators. The flowers of the Frog Orchid are very like those of *Orchis* in structure, but are quite inconspicuous and greenish or brownish in colour. The lip is broadly strap-shaped, with a short spur at its base; the remaining perianth segments form a helmet over the column. The two viscidia are rather widely spaced; but the stigma is small and lies in the centre of the flower between them. Nectar is secreted in the spur but, in addition, there are two small nectaries on either side of the lip close to its base, and almost underneath the viscidia. The lip has a median ridge, which tends to make an insect landing on the labellum crawl up one side or the other, towards one of the drops of nectar beneath the viscidia, rather than up the middle. Feeding at one of these nectaries the insect can easily remove a single pollinium on its head. The forward movement of the pollinium, completed within a minute in *Orchis*, takes twenty minutes or half an hour in the Frog Orchid; time enough for its rather slow-moving pollinators to visit another spike. The pollen is then readily transferred to the central stigma as the insect explores the nectaries at the base of the lip. Beetles are probably among the commonest pollinators of the Frog Orchid. Silén (1906a) observed many visits by beetles of the genus *Cantharis* in northern Finland, and some visits by ichneumons and other insects. The Musk Orchid has a rather similar mechanism, though the green flowers are much smaller than those of the Frog Orchid. The flowers do not expand widely, and the lip, which has a shallow cup-shaped nectary at its base, does not differ greatly from the other petals. Its pollinating insects – a variety of minute *Hymenoptera*, beetles and flies – crawl into the flowers on either side of the lip to seek nectar in the cells of the cup at its base. So placed in the corner of the flower, the insect's fore-leg is immediately below one of the relatively large saddle-shaped viscidia, which becomes transversely attached to the femur, usually near its base. The two stigmas are transversely elongated with their broadest parts just below the viscidia, where they receive the pollen (after the pollinium has swung forward in the usual way) when the insect visits another flower. The flowers are readily self-pollinated by the pollinia falling out of the anther on to the stigmas if the flowers are not visited.

Quite the most remarkable pollination mechanisms among European Orchids – and indeed among the most remarkable to be found in any plants – are those of the 'insect orchids' of the genus *Ophrys*. These Orchids are well known for the striking resemblance of their flowers to various insects. What function, if any, this resemblance served was long a matter for speculation; Darwin was plainly puzzled by it. It was not until this century that it was discovered that pollination in most species is brought about by male insects performing part at least of their mating behaviour in response to the flower. This process, often called *pseudocopulation*, was first elucidated by M. Pouyanne (Pouyanne, 1917, Correvon and Pouyanne, 1923), who observed the Mirror Ophrys (*O. speculum*) for many years around Algiers, where he was Conseilleur de la Cour d'Appel. His observations on the *O. speculum* and other species were soon confirmed and amplified by those of Col. M. J. Godfery (1925 onwards) in the south of France, and not long afterwards pseudocopulation was described by Mrs E. Coleman (see p. 136) in the Australian Orchid (*Cryptostylis leptochila*), which is pollinated by the males of the ichneumon (*Lissopimpla semipunctata*).*

The genus *Ophrys* is represented by many species in southern Europe, north Africa and the Near East. The flowers are built on the same general pattern as those of *Orchis*; the disposition of the stigmas and parts of the column and anther are almost identical (and the pollinia show the same forward movement to bring them into the right position to strike the stigma, after they are withdrawn from the anther), but the lip is thick, dark brownish, and velvety-textured, often with metallic bluish patches and markings, there is no spur, and the two viscidia are covered by separate bursicles instead of the single one found in *Orchis*.

Ophrys speculum is a widespread plant in the western parts of the Mediterrean region. The lip is like an oval convex mirror, of a curious glistening metallic violet-blue colour, with a narrow yellow border thickly fringed with long red hairs. The short side-lobes of the lip overlap the mid-lobe, recalling the wings of an insect at rest. The thread-like dark red upper petals simulate antennae. Pouyanne found, as a result of twenty years' observation, that *O. speculum* is regularly visited by one insect only, the scoliid wasp (*Campsoscolia* (*Dielis*) *ciliata*), and of that species only by the males. The females ignore the flowers, although both sexes visit other flowers for nectar.

Campsoscolia ciliata is a wasp rather larger than a honey-bee; each segment of its abdomen is fringed with long red hairs. The males appear about a month before the females, and can often be seen during March skimming with a swift zig-zag flight over the dry sunny banks where the wasps make their burrows. The females spend much of their lives underground, hunting for the worms with which they provision the burrows for their larvae, and scarcely leave the soil except to mate and to feed.

The flowers of *Ophrys speculum* are eagerly sought out and visited by the

* Dodson (1962) has described pseudocopulation in the tropical American Orchid *Trichoceros antennifera*, pollinated by flies of the genus *Paragymnomma*; some further possible cases of pseudocopulation are discussed by Van der Pijl and Dodson (1966).

males,* though the insect does not use its proboscis and evidently neither seeks nor finds nectar or other food. Although the flowers have no appreciable scent to us, the male wasps can detect their presence at some distance; a few spikes of *O. speculum* held in the hand where *Campsoscolia* was flying soon attracted the insects, sometimes several hustling one another on the same flower, and apparently quite oblivious of the observer. The attraction of the flowers resides in the lip; flowers with the lip cut off were completely ignored. Detached flowers laid face upwards on the ground were quite as attractive as they were on the flower spike; but if laid face downwards, with the 'mirror' hidden, although the insects were still attracted they had difficulty in finding the flowers. The wasps were clearly conscious of the presence of *O. speculum* flowers even when they were hidden from sight. A dozen spikes were wrapped in newspaper; five or six wasps alighted on the paper, agitatedly coming and going, and two tried to crawl inside.

Alighting on the flower of *O. speculum*, the male *Campsoscolia* sits lengthwise on the lip, with his head just beneath the rostellum (Fig. 80B), and plunges the tip of his abdomen into the fringe of long red hairs at the end of the lip, with brisk, tremulous, almost convulsive movements, in the course of which the insect rarely fails to carry off the two pollinia on its head. Pouyanne was struck by the resemblance of the behaviour of the insect to copulatory movements; later, when he was able to observe the males pursuing the females, he described them alighting on their backs and performing exactly the same movements as they did on the flowers.

The lip of *O. speculum* does bear a certain curious resemblance to the female of *Campsoscolia ciliata*, with her broad abdomen, likewise fringed with red hairs. At first sight the mirror is puzzling, but as Pouyanne realised when he was able to observe the females of *Campsoscolia* at close quarters, it corresponds exactly with the position of the bluish shimmering reflection on the wings of the wasp when she is resting or crawling over the ground. To our eyes the resemblance of the flower to *Campsoscolia* is crude and unconvincing, but combined with scent it attracts the male bees effectively enough for some 40 per cent. of the flowers to produce capsules.

Our own Fly Orchid (*O. insectifera*) is related to *O. speculum*, and its pollination – first observed by Godfery (1929), and more recently studied in detail by Wolff (1950) in Denmark and Kullenberg (1950, 1961) in Sweden – bears many points of resemblance to the pollination of that species. The flower-spike is slender, seldom much over a foot high, and bears up to about ten rather widely spaced flowers. The lip is rather long and narrow, dark reddish brown in colour, with a metallic bluish patch in

* Kullenberg (1961) observed 19 visits in the course of about 11 hours watching spread over a period of 8 days from the end of February into early March in Morocco in 1948.

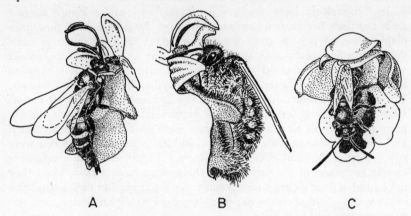

A   B   C

FIG. 80. Insect visitors to *Ophrys* flowers. A, male of the solitary wasp *Gorytes mystaceus* visiting a flower of the Fly Orchid (*Ophrys insectifera*). Sweden. B, male of the solitary wasp *Campsoscolia ciliata* on a flower of *O. speculum*. Morocco. C, male of the bee *Andrena maculipes* visiting a flower of *O. lutea*. Lebanon. After photographs by Kullenberg (1956, 1961).

the centre, and shallowly lobed at the tip (Plate IIa). The remaining two petals are small and narrow, and blackish, forming the 'antennae' of the 'fly'. The three sepals are rather small and blunt, and pale greenish in colour. The Fly Orchid is widespread in Britain, but very local and often inconspicuous and difficult to see. Perhaps its most characteristic habitat is about wood margins on calcareous soils, but it occurs too in woods, in chalk and limestone grassland, and (as it often does in Scandinavia) in calcareous fens.

Only two species of insects are known as regular visitors to the Fly Orchid, the solitary wasps *Gorytes mystaceus* and *G. campestris*. Here again, the wasps are attracted to the flowers in the first place by scent. Upon settling, *Gorytes* sits lengthwise on the lip – like *Campsoscolia* on *O. speculum* – with its head close to the column (Fig. 80A). Often it remains on the flower for many minutes, every now and then restlessly vibrating its wings and changing its position before settling down again and performing movements which look like an abnormally vigorous and prolonged attempt at copulation. While it is on the flower the wasp seems quite oblivious of the observer's presence. Very similar accounts of visits by *G. mystaceus* are given by Godfery from the south of France and by Wolff and Kullenberg from Scandinavia. Visits by *Gorytes* to flowers of the Fly Orchid have apparently never been observed in the British Isles, but they presumably take place, because Darwin found that pollinia had been

PLATE 33. **a-b**, silver-Y moths *(Plusia gamma)* visiting flowers of Honeysuckle *(Lonicera periclymenum)*.

PLATE 34. Nectar theft: **a**, honey-bee sucking nectar 'legitimately' from *Brassica* flower. **b**, honey-bee stealing nectar from back of *Brassica* flower. **c**, bumble-bee *(Bombus agrorum)* sucking nectar 'legitimately' from flower of Comfrey *(Symphytum officinale)*. **d**, bumble-bee *(Bombus* cf. *terrestris)* stealing nectar from Comfrey through hole bitten in corolla-tube. See also Pl. 18**b**.

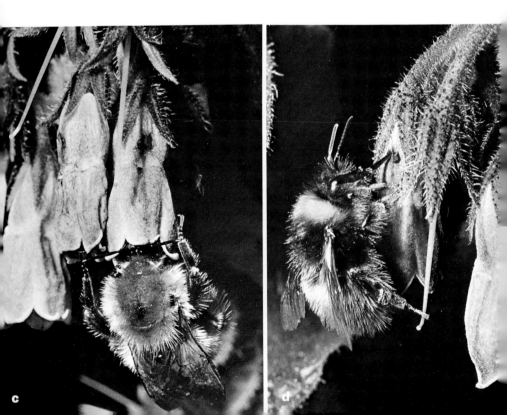

removed from 88 out of 207 flowers he examined; in a small Wiltshire colony I visited in early June, 1969, of 13 flowers (on six plants) all except three of the oldest had had at least one pollinium removed, and eight had been pollinated. Godfery remarked that with only one – apparently accidental – exception he never saw *Gorytes* visit any orchid but *O. insectifera*; and apart from the two species of *Gorytes*, *O. insectifera* never receives more than rare casual visits from other insects. The flowers are self-compatible with their own pollen, but most of the flowers which are pollinated at all must receive pollen from other plants. The relationship between the Fly Orchid and *Gorytes* is so highly specific that hybridisation with other species of *Ophrys* cannot be more than the rarest accident. Obviously the Fly Orchid is closely adapted to the visits of *Gorytes*. The upper surface of the lip bears a remarkable general resemblance to the back of the female in contour and the nature of its hair covering. The Fly Orchid has a long flowering season, from early May to the latter part of June, perhaps as an adaptation to ensure that at least some flowers are open at the time when the male wasps emerge. Kullenberg noticed the interesting fact that in a Swedish locality the flowers were visited by *G. mystaceus* in the early part of the flowering season, but a fortnight later they were being visited by *G. campestris*. But rarely more than a quarter of the flowers are pollinated, and sometimes the proportion may be very much lower than this, especially in large colonies and in seasons when the orchids are particularly numerous.

There is in fact a good deal of diversity of detail in the pollination of *Ophrys*. One of the species studied by Pouyanne was the widespread Mediterranean orchid *O. lutea*. In this plant the lip is brilliant yellow, with a dark raised area in the centre, and a pair of narrow metallic bluish patches on either side of the dark marking near the base. *O. lutea* flowers in March around Algiers, when even in north Africa calm sunny days are few and far between, and its visitors are much more difficult to observe than those of *O. speculum*. The number of capsules produced varies enormously from place to place; in the localities Pouyanne examined it ranged from as few as 3 per cent. to as many as 70–80 per cent. of the flowers produced. It was in this last favourable locality that he was able to witness repeated visits to the flowers by small bees, the males of *Andrena nigro-olivacea* and *A. senecionis*. The bees made the same kind of movements as *Campsoscolia ciliata* on *O. speculum*, but in contrast to the insects visiting that species and the Fly Orchid, the visitors to *O. lutea* always took up a position with their heads outwards on the flower, so that the pollinia were borne away on the tip of the abdomen (Fig. 8oc). Evidently to the male bees the 'decoy' represents a female bee sitting head downwards on a large yellow flower. Kullenberg (who observed visits by a number of further *Andrena* species in Morocco and the Lebanon, and visiting cultivated specimens of *O. lutea* in Sweden) found that by cutting off and reversing the lip of *O. lutea* he could induce the males to visit the flowers the 'normal' way round, with their heads next to the column!

Another species Godfery observed in the south of France was the Late Spider Orchid (*O. fuciflora*), which just reaches south-east England where it grows in a few colonies on the chalk in Kent. The flowers were visited by the large grey males of the bee *Eucera tuberculata*. The bees became aware of the flowers remarkably promptly and pounced on them, staying only momentarily but quickly and neatly removing the pollinia as they flew away. Kullenberg (1961) observed many visits by *Eucera longicornis* to plants in experimental cultivation in Sweden, and this bee also effectively pollinated the flowers. Of the pollination of the other British 'Spider Orchid' (*O. sphegodes*) very little is known. It is a very common plant in some parts of southern Europe, but the few available records suggest that very little seed is set. Few insect visitors have been observed, but visits and pollination by the beetle *Trapinota hirta* have been observed in north Italy (Ferlan in Kullenberg, 1961). Godfery (1933) found that of 27 flowers he examined near Swanage, in Dorset, four had both pollinia removed and six had pollen on the stigma. Here is an interesting field for observation by anyone with access to colonies of this plant.

The pollination of these Orchids shows some very interesting biological features. Some points which were puzzling when the pollination of *Ophrys* was first observed are more intelligible in the light of what is now known about animal behaviour. Instinctive actions are often released by quite simple stimuli – or combinations of stimuli – and although these stimuli may in effect be perfectly specific to the normal releasing situation in nature, instinctive patterns of behaviour may be released by superficially very different situations in laboratory experiments. Elaborate and prolonged behaviour patterns are often performed as a chain of simple reactions released successively; the original stimulus releases the first link in the chain, which produces a situation which releases the second link and so on.

It seems that *Ophrys* has evolved flowers which parasitise the copulatory patterns of the pollinating insects; they have, so to speak, 'discovered' the releasing stimuli normally provided by the females. In every case the male insects are attracted to the neighbourhood of the flowers by scent. Scent plays an important part in the mating behaviour of many solitary bees and wasps, and the males generally seem to be attracted to the females by scent. The males of *Halictus* have scent glands, with which they mark flowers that they visit in the course of their daily routine. They show evident excitement on revisiting the perfumed spots, which they do quite regularly. The perfumed spots are visited also by the females; this undoubtedly helps to bring the sexes together, and probably also helps to maintain the coherence of the colony. The scents produced by the lips of *Ophrys* species appear to resemble the scents produced by the corresponding female bees and wasps. Thus Kullenberg considers that both the Fly Orchid and the female of *Gorytes mystaceus* produce scents reminiscent of the organic compounds farnesol and hydroxycitronellal. These two compounds produce no effect on males of *Gorytes*, but farnesol is strongly attractive to the males of the solitary bee *Macropis labiata* (whose female produces a similar scent), and the reaction of the wasp *Crabro cribrarius* to citronellal suggests sexual excitement. But although scent attracts and excites the males, contact stimuli are necessary before they will attempt copulation.

The male insects, attracted by scent, locate the *Ophrys* flowers by sight, but once they have alighted on the flower their behaviour is guided almost entirely by touch stimuli, and these are responsible for the insect taking up the position which brings about removal of the pollinia. But as Kullenberg points out, the behaviour of the insect does not always resemble an introduction to copulation at all closely. Often its movements are quite abnormally prolonged and vigorous, or it may suddenly whirr its wings or bite fiercely into the labellum. Kullenberg regards these acts as 'displacement activities'. The males of the Hymenoptera which visit *Ophrys* flowers often emerge some weeks before the females, and it has often been said that visits to *Ophrys* flowers stop once the females have emerged. This is certainly not strictly true. Faded flowers of *O. insectifera* were visited quite late in the season, and Kullenberg even found that, given the choice between *O. lutea* and their females, the pollinating Andrenas always landed on the flowers.

How this remarkable relationship between *Ophrys* and its bee or wasp pollinators arose is an interesting question. Probably a fairly specific attractive scent evolved first,* and the origin of a flower pattern close enough to the releasing pattern of the female to provoke a reaction from the male may have been quite accidental. Once such a relation was established, the way would have been open for the flower to evolve closer and closer adaptation to the insect partner. Among the solitary bees and wasps not only do the males emerge first, but also there is commonly an excess of males, so there is brisk competition for the females as they emerge – a state of affairs peculiarly favourable for the evolution of pseudocopulation by *Ophrys*.

* Scents with a specific attraction for particular groups of insects are probably widespread among the Orchids. The Twayblade is probably a case in point, but there are some very impressive examples among tropical Orchids, for instance in *Stanhopea* and related genera (Chapter 10, p. 297). Dodson and Hills (1966) using gas chromatography, have demonstrated the existence of about fifty compounds in the fragrances of this group of Orchids, the number in the scent of any one species ranging from 3 to 18. Many of the compounds are widespread in occurrence but each Orchid has its characteristic blend to which only one or a few species of bee respond.

The very close adaptation of the flower to a particular species (or a small group of species) of insect provides a very specific pollination mechanism, but it also ties the fortunes of the plant to those of the insect. Many of the *Ophrys* species set only a small proportion of capsules, and except in occasional places where the pollinating insects are unusually abundant the number of flowers pollinated generally bears more relation to the abundance of insects than to the number of flowers produced. It is a pollination mechanism for a rare plant, and it is probably no accident that *Ophrys* spikes bear fewer flowers than most other Orchids. The *Ophrys* species cannot transgress far outside the geographical range of their pollinators, but with their light seeds the detailed distribution of the pollinator probably has rather little effect on the detailed distribution of the Orchid, provided the two overlap to some extent. While *Ophrys speculum* produces a fair proportion of capsules everywhere (being pollinated by the widespread *Campsoscolia*), *O. lutea* fruits heavily only in the neighbourhood of dense colonies of the pollinating andrenas. Pouyanne found one large colony of *O. lutea* in rather damp flat clayey meadows, which must have provided ideal conditions for germination and vegetative growth, for *O. lutea* was so abundant that it could almost have been mown; yet the flowers produced barely 5–10 per cent. of seed capsules. In this case there must have been a great many capsules in aggregate, but it may be that some *Ophrys* species reproduce almost entirely from small parts of their populations – a state of affairs appropriate enough in a light-seeded Orchid, but unthinkable for the majority of plants with their quite limited powers of seed dispersal.

With a large number of flowers remaining unfertilised, the *Ophrys* species provide a situation in which alternative pollination mechanisms should evolve quite readily. Probably a good many of the closely related species have diverged in this way, by parts of populations becoming adapted to the visits of different insects This appears to be happening in *O. lutea*, and in another Mediterranean species, *O. fusca*. In both, races of strikingly different flower size exist, and it is difficult to escape the conclusion that they have become adapted to different groups of *Andrena* species. One line of evolution which has apparently appeared only rarely in *Ophrys* is that leading to self-pollination, but we have an excellent example in the British flora in the Bee Orchid (*O. apifera*) (Plate IIc).

Structurally the Bee Orchid is very like other members of the genus. Its two significant differences are that the anther-cells open a little more widely, and that the caudicles are a little longer and more flexible. Apparently insects occasionally visit the flowers; such visits are probably commoner in the Mediterranean region than in the British Isles. In Morocco, Kullenberg observed visits by *Eucera* and *Tetralonia* males (the main genera of bees visiting the allied species *O. scolopax*, *O. bombyliflora*, *O. tenthredinifera* and *O. fuciflora*). However, these insects seem often to fail to come into contact with the pollinia and it is doubtful whether they are effective pollinators. It is noteworthy that some of the *Eucera* males observed by Kullenberg visiting the allied *O. tenthredinifera* attempted copulation with the labellum but here again failed to reach the pollinia. In Britain also pollinia are sometimes removed from flowers of the Bee Orchid, hybrids with the two Spider Orchids have been reported, and a correspondent of Darwin's saw a bee 'attacking' a Bee Orchid flower. However, these are rare occurrences, and normally, when the flowers have been open for a day or two, the pollinia fall out of the anther and hang down in front of the stigma.* With the spike shaking in the wind, the pollinia are caught against the sticky stigma and held fast. Kullenberg, who made most of his observations on the Bee Orchid in Morocco, clearly doubted whether self-pollination would take place regularly without the disturbances caused by insect visits, and in north Africa this may be so. My own experience of this species in the south of England confirms the conclusion of other observers that in this country the Bee Orchid is automatically self-pollinated with a very high degree of regularity. Often a spike of *O. apifera* can be found with the uppermost flower freshly opened and the pollinia still in their cells, the next flower with the pollinia dangling freely from the column, and the lowest flower faded, with the pollinia caught against the stigma and a plump capsule developing beneath the flower. In contrast to the other species almost every flower produces a capsule.

The orchids are an insect-pollinated group *par excellence*, and at first sight it seems surprising to encounter self-pollinated members among them. But Orchids have been subject to the same conflicting selection-pressures as other plants, especially when insect visitors are scarce, or changing conditions have begun to make a delicately adjusted mechanism progressively less effective. Self-pollination has an obvious advantage in maintaining a high and reliable seed-production; and its genetic effects may also be beneficial in the short run, even though ultimately, with changing conditions, they may be crippling to the point of bringing the plant to extinction. Self-pollination is dealt with in Chapter 9 (see also Chapter 12, p. 382), but since the self-pollinated Orchids often differ little from their insect-pollinated relatives in structure, and the differences are often interesting, they will be discussed here.

The Sword-leaved or Narrow Helleborine (*Cephalanthera longifolia*) is an insect-pollinated flower and has already been described. A very much commoner plant in Britain is the Large White or Broad Helleborine (*C. damasonium*) (Fig. 81), one of the most characteristic plants of the beech 'hangers' of the chalk escarpments. Unlike those of the rarer species,

* A small proportion of the pollinia fall forward from the anther-cells in this way in old flowers of the Fly Orchid, but as the caudicles are much shorter and less flexible this does not usually bring about self-pollination.

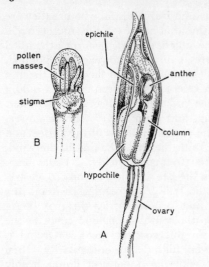

FIG. 81. Broad Helleborine (*Cephalanthera damasonium*). A, side view of flower with half of perianth removed. B, front view of upper part of column. Micheldever, Hants.

the flowers of *C. damasonium* open for only a short time, and even then never open widely, but almost every flower produces a well-developed capsule. The flowers are certainly occasionally visited by bees and other insects, which probably bring about cross-pollination in the same way as in *C. longifolia*; there is no nectar, but Darwin observed that insects bite small pieces from the orange ridges on the lip, which are said to taste of vanilla. The pollinia are more friable than those of *C. longifolia*, and even in the bud pollen-tubes grow out from the grains close to the top of the column and penetrate the upper part of the stigma. Thus if insect visitors fail some seed production at least is assured. However, Darwin found that although plants covered with a net during the flowering period to exclude insects produced well-developed capsules, much of their seed was inviable, and their effective seed-production was only about a quarter of that of plants left freely exposed to insect visitors.

Self-pollination in the Broad-leaved or Common Helleborine (*Epipactis helleborine*) has already been mentioned; in that species it probably supplements cross-pollination by wasps, though the relative importance of selfing and crossing probably varies from place to place and from time to time. There is an interesting series of species, more or less closely related to *E. helleborine*, which are regularly self-pollinated. They are all very local plants, and much less variable than *E. helleborine*, from which some or all of them probably originated. The Narrow-lipped Helleborine (*E. leptochila*) occurs in the same sort of shady habitats as *E. helleborine*, and indeed may be found growing in mixed colonies with it; but it is a much rarer plant and is confined to the south of England. The flowers differ from those of *E. helleborine* in various small structural particulars. They are rather large in size, and generally yellowish-green in colour with a whitish lip. All the perianth segments are relatively longer and more pointed than in *E. helleborine*, and this is true particularly of the narrow pointed epichile of the lip which gives the plant its name. The column and the attachment of the anther are slender, and the anther projects far forward over the column. The rostellum is present in the bud, but dries up about the time the flower opens. Although cross-pollination by insects probably takes place on rare occasions, the flowers are usually selfed by the pollinia swelling up and the loose pollen falling on to the stigma. The Dune Helleborine (*E. dunensis*) is confined to the slacks of a few large sand-dune systems in Lancashire and Anglesey. Structurally it is very like *E. helleborine*, but the flowers are smaller, dingy greenish in colour, and never open as widely as they do in that species. The rostellum is rarely developed, or at least dries up very soon after the flower opens, so cross-pollination cannot occur in more than rare cases. Usually the flowers are self-pollinated in the same way as in *E. leptochila*. Though the flowers project more or less horizontally when they first open, they soon droop, and hang down following fertilisation.

A further species, the Green-flowered Helleborine (*E. phyllanthes*), shows even more marked adaptation to self-pollination. It occurs in a number of widely scattered small colonies in dune slacks and woods from Lancashire and Londonderry southwards. *E. phyllanthes* differs most obviously from *E. helleborine* in its pendulous flowers. The rostellum and column wither almost as soon as the flower opens, or even earlier, and the lip not long afterwards. The flowers are self-pollinated by the pollinia swelling up, so bringing the pollen into contact with the stigma; this takes place in the bud (Fig. 82) or soon after the flower opens. No insect has ever been observed visiting the flowers.

Within a single colony of *E. phyllanthes* the plants are very uniform, but every colony tends to differ a little from every other (see Chapter 9, p. 290). It is likely that self-pollinated forms like these species of *Epipactis* have arisen repeatedly from insect-pollinated species, and flourished for a time before changing conditions and the competition of more versatile species have sooner or later fragmented their ranges as a prelude to their ultimate extinction.

Other examples of self-pollination are known in the *Basitonae*, apart from the Bee Orchid. In several of the smaller-flowered species the separate small pollen-masses (*massulae*) of which the pollinia are made up are only very loosely bound together, so that the pollinia readily break up. This seems to be the common manner of pollination of the Dense-flowered Orchid (*Neotinea intacta*), a small Mediterranean species which reappears locally on limestone soils in the west of Ireland (where it is best known in the Burren district of Co. Clare). The flowers contain nectar, which

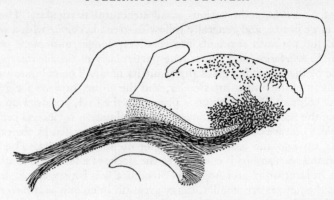

FIG. 82. Green-flowered Helleborine (*Epipactis phyllanthes*). Longitudinal section of column from unopened bud, showing germination of pollen. Drawing by O. Hagerup, reproduced by permission from Young (1962).

probably attracts insects in the plant's Mediterranean habitats at least, but in Ireland the massulae are generally found scattered around the inside of the flowers as soon as they open (Summerhayes, 1951). Hagerup (1952a) found that in Scandinavia the same appears to happen in the flowers of the Small White Orchid (*Pseudorchis albida*). In this case the process is aided by the fact that the stigmas project forward from the column, where they are well placed to catch the massulae falling from the pollinia. In the Alps, *Pseudorchis* is said to be butterfly-pollinated. It may be that evolution of a means of self-pollination has enabled it to penetrate into areas where insects are scarcer. Hagerup considered that self-pollination of the flowers in the bud is common among European orchids, especially in the species with small and inconspicuous flowers.

As has already been indicated, a good many Orchids are capable of self-pollination if the flowers are not visited by insects. In these cases a balance must exist between cross- and self-pollination, and the change from preponderant cross-pollination to preponderant self-pollination is only a matter of degree. In the Helleborines and some of the other *Acrotonae* there is little in the structure of the flowers to hinder self-pollination, and the case of *Cephalanthera damasonium* suggests that the balance between cross- and self-pollination may be held mainly by the degree of self-incompatibility of the flowers to their own pollen. In the *Basitonae*, where there are generally highly developed and effective mechanisms favouring cross-pollination, there seems to have been less pressure towards the evolution of incompatibility (species of *Orchis* and *Ophrys* are self-compatible), and habitual self-pollination is brought about by modification in the structural arrangements of the flower.

PLATE 35. **a**, solitary bee *(Andrena* sp.*)* on male catkin of Common Sallow *(Salix cinerea)*. **b**, hover-fly *(Syrphus* sp.*)* feeding on pollen of Common Sallow. **c**, hebrew character moth *(Orthosia gothica)*, a common nocturnal visitor to Sallow, sucking nectar from female catkin.

a

b

c

a

PLATE 36. *Compositae:*
a, drone-fly *(Eristalis tenax)*
on Ox-eye Daisy
*(Chrysanthemum
leucanthemum)*; notice the
orchid pollinia (probably of
the Heath Spotted Orchid
(Dactylorhiza maculata)) on
the proboscis of the insect.
b, hover-flies *(Syrphus* cf.
luniger and *S. ribesii)* on
Dandelion *(Taraxacum
officinale* agg.*)*. **c**, flower-
head of Greater Knapweed
(Centaurea scabiosa).

b

c

Finally a few general comments about the family may be made. The Orchids are typically long-lived plants, and many of the species grow in stable and rich plant communities in which competition is intense. We know all too little of the factors controlling the numbers of the rarer species in such communities, but we may conjecture that the copious production of minute seeds so characteristic of the Orchids has evolved largely in response to these conditions. Linked with this, the family has become committed – by and large – to a breeding system which depends on frequent intercrossing to maintain a high degree of heterozygosity and a great measure of genetic diversity among individuals and their offspring. A system of this kind is most effective where large numbers of offspring are produced – as in the orchids – and selection-pressure is consequently high (see Chapter 12, p. 381). The variability which is so obvious in the flower-colour and -pattern in a colony of Spotted Orchids undoubtedly extends into the subtler but more important characters of adaptation to habitat. With their outcrossing breeding systems, but large numbers of seeds produced from highly specialised large single flowers, the Orchids provide at the same time a parallel and an antithesis to their rivals as the largest family of flowering plants, the *Compositae* (see Chapter 6, p. 220). A simple explanation of any part of the biology of a species or family is always hazardous, because any one feature, such as seed-production or pollination-mechanism, influences and is influenced by so many other interacting factors. None the less, it seems fair to say that the Orchids have taken one particular solution to the problems of living in a competitive world very near to its logical conclusion.

CHAPTER 8

POLLINATION BY WIND
AND WATER

ALMOST all of the plants we have considered so far depend for pollination on insects. However, many plants rely for pollination on other agencies, of which by far the most important is wind. The clouds of pollen blowing like yellow smoke from pines and other conifers are a familiar sight in early summer. Other wind-pollinated plants include many of our commonest native trees, almost all the grasses, sedges and rushes (*Gramineae*, *Cyperaceae* and *Juncaceae*), many seashore plants and weeds belonging to the Goosefoot and Dock families (*Chenopodiaceae* and *Polygonaceae*) and a diverse assortment of other species.

Wind-pollination (or *anemophily*) has the obvious advantage to the plant of being independent of the possibly erratic occurrence and capricious behaviour of insects. It is effective when insects are scarce or absent. On the other hand, effective wind-pollination requires the production and dissemination of very large amounts of pollen. If effective pollination requires no more than one pollen grain to reach a stigma with an area of one square millimetre (about the area of an oak stigma), every square metre of the plant's habitat must receive around a million pollen grains to make pollination reasonably certain. In fact, pollen production is ample to achieve this sort of density. It has been estimated that a single birch catkin produces about five-and-a-half million pollen grains and a hazel catkin nearly four million; a single floret of rye produces over fifty-thousand grains (Pohl, 1937).* The 'pollen rain' falling from the air during 1943 at eight sites widely scattered over Great Britain was studied by H. A. Hyde of the National Museum of Wales, by identifying and counting the pollen grains caught on gelatine-coated glass slides. He found that the total annual catch of grass pollen averaged about 2100

---

* The data quoted by Pohl show a general tendency for the wind-pollinated plants among the species he studied to produce more pollen than the insect-pollinated plants, whether pollen-production is expressed in terms of stamens, flowers, inflorescences or whole plants. However, the relationship is by no means clear-cut, and many insect-pollinated plants produce very large amounts of pollen. Indeed, the single flower producing the greatest amount of pollen was the insect-pollinated Corn Poppy (*Papaver rhoeas*) – though, as a particularly large and striking instance of a pollen-flower (see p. 54 and Plate 5a) this might be taken as the exception that proves the rule.

grains per sq. cm.; the total tree-pollen count averaged just over half that number. Taking the eight sites together the most abundant tree-pollen types were ash, oak and elm. Some of Hyde's results are summarised in Table 9.1. For various reasons these must be nearly minimum figures. Horizontal 'gravity slides' are now known to be inefficient in trapping pollen under ordinary windy conditions, especially for the smaller grains.* The slides had to be protected from the rain, and although the apparatus was carefully designed to expose the slides as freely to the air as possible, this protection may have decreased the pollen-catch still further. At every site the slides were exposed well above the surrounding vegetation, usually on a building, and most of the sites were in built-up areas. In any case, the countryside of Britain is only thinly and sporadically wooded. Taking these factors together, there is no doubt that much higher figures for particular tree-pollen types would be found using more efficient trapping surfaces in and around woods; similar considerations apply to herbaceous plants though to a lesser degree. In the much more heavily forested landscape of Sweden, according to Erdtman, the total annual pollen-rain may amount to 30,000 grains per sq. cm.

Table 5.   The pollen rain over Great Britain, 1943. Data recalculated from Hyde (1950). This table includes all the pollen types of which the average catch at eight sites during the year was more than 5 grains per sq. cm. The sites were: Llandough Hospital, Penarth, Glamorgan; National Museum of Wales, Cardiff; University College of Wales, Aberystwyth (old buildings on sea front); St Mary's Hospital, Paddington, London; Botany School, Cambridge; Derbyshire Sanatorium, Chesterfield; King's Buildings, Edinburgh; City Hospital, Aberdeen. No station is quoted for minimum catch if this was zero at more than one station

Pollen type	Average catch (grains/ cm²/yr)	Least and greatest catch (grains/cm²/yr)	
Gramineae (grasses)	2106.4	724.6 (Paddington)	4454.8 (Chesterfield)
Fraxinus (ash)	271.2	88.6 (Aberdeen)	505.4 (Llandough)
Quercus (oak)	177.6	9.8 (Aberdeen)	504.2 (Cambridge)

* More satisfactory results are obtained by using an apparatus which will trap the pollen present in a known volume of air, such as the Hirst automatic volumetric spore trap used by Hyde and his collaborators from 1954 onwards: a trap of this kind was used to obtain the results shown in Fig. 83. However, for the purpose of the present discussion, the results from volumetric trapping are less easy to interpret than those from gravity slides.

Pollen type	Average catch (grains/ cm³/yr)	Least and greatest catch (grains/cm³/yr)	
Ulmus (elm)	145.6	24.2 (Chesterfield)	773.0 (Cardiff)
Plantago (plantain)	133.6	52.8 (Paddington)	238.2 (Llandough)
Platanus (plane)	96.3	0.0 (Aberdeen)	427.6 (Paddington)
Urtica (nettle)	88.3	21.6 (Aberdeen)	156.4 (Llandough)
Rumex (dock)	61.9	34.2 (Paddington)	117.0 (Chesterfield)
Pinus (pine)	59.9	13.8 (Chesterfield)	105.2 (Aberdeen)
Corylus (hazel)	46.8	17.6 (Edinburgh)	94.2 (Llandough)
Betula (birch)	41.3	21.6 (Chesterfield)	71.2 (Cardiff)
Taxus (yew)	36.1	9.2 (Aberdeen)	137.6 (Cambridge)
Populus (poplar)	30.2	0.2 (Aberystwyth)	132.6 (Cambridge)
Salix (willow)	27.8	6.4 (Aberdeen)	66.6 (Chesterfield)
Chenopodiaceae	27.3	7.6 (Edinburgh)	92.8 (Cambridge)
Sambucus (elder)	16.8	2.8 (Aberystwyth)	59.4 (Chesterfield)
Tilia (lime)	16.6	0.0	83.6 (Cambridge
Alnus (alder)	16.2	6.0 (Aberdeen)	34.8 (Cardiff)
Ericaceae	14.4	2.0 (Cambridge)	43.2 (Edinburgh)
Aesculus (horse chestnut)	13.9	0.0 (Edinburgh)	72.8 (Cambridge)
Juncaceae:		5.4 (Paddington)	33.4 (Aberystwyth)
Luzula (woodrush)	9.9		
Juncus (rush)	5.3		
Fagus (beech)	11.9	0.6 (Chesterfield)	46.6 (Edinburgh)
Umbelliferae	11.5	2.6 (Aberystwyth)	36.0 (Chesterfield)
Artemisia (mugwort)	11.3	5.0 (Chesterfield)	26.6 (Cardiff)
Ranunculaceae	11.2	0.6 (Aberdeen)	19.4 (Cambridge)
Compositae (misc.)	11.1	8.2 (Paddington)	21.8 (Cardiff)
Acer (sycamore and maples)	8.0	0.8 (Aberystwyth)	18.6 (Edinburgh)
Rosaceous trees	8.0	2.4 (Aberystwyth)	16.6 (Cardiff)
Cruciferae	6.9	2.2 (Aberystwyth)	31.0 (Llandough)
Compositae: Cichorieae	6.8	2.2 (Edinburgh)	15.2 (Llandough)
Castanea (sweet chestnut)	6.4	0.0	27.8 (Cardiff)
Carpinus (hornbeam)	6.4	0.0	16.8 (Paddington)
Cyperaceae (sedges)	5.3	1.8 (Chesterfield)	11.6 (Aberystwyth)

Wind-pollinated plants usually have pollen grains with a smooth dry surface, in contrast with the sticky and often highly ornamented grains that are often common in entomophilous species. Consequently their pollen grains are dispersed singly or in twos and threes, rather than sticking together in larger groups. The pollen grains of entomophilous species are of very varied sizes; those of anemophilous species vary much less, and are commonly between 10 and 25 μm in diameter in angiosperms and between 30 and 60 μm in diameter in conifers. This is probably because, on

POLLEN GRAINS IN THE AIR AT CARDIFF 1955 - 59

MEAN 24 HOUR COUNTS PER CUBIC METRE

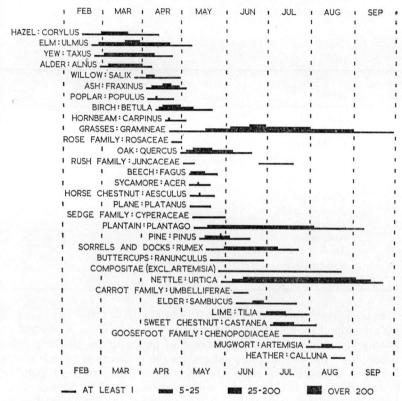

FIG. 83. Pollen calendar for Cardiff based on counts obtained with a Hirst automatic volumetric spore trap situated on the roof of the National Museum of Wales, Cardiff, from 1955 to 1959. The concentrations indicated were attained on average during the periods shown except that certain trees (Poplar, Willow, Hornbeam, Beech, *Acer* and Horse Chestnut) reached a mean 24 hour count of over 5 grains per cubic metre in certain years only. Figure drawn by Mrs K. F. Adams, reproduced by permission from Hyde & Williams (1961).

the one hand, larger grains are trapped more efficiently from a moving air-stream by a solid object than smaller grains while, on the other hand, the high rate of fall of very large grains will limit their range of dispersal. Rates of fall of wind-borne pollen grains in calm air range from between

1 and 2 cm./sec. for the smallest and lightest to about 40 cm./sec. for the largest and heaviest, though accurate measurements are difficult to make. As can be seen from Table 9.2, the rates of fall for many common anemophilous species are around 2–6 cm./sec.

Table 6.   Rates of fall of pollen grains in still air (for authors see Gregory, 1961)

Species		Rate of fall (cm./sec.)
Abies alba	Silver fir	38.7
Betula pendula	Birch	2.4
Corylus avellana	Hazel	2.5
Dactylis glomerata	Cocksfoot	3.1
Fagus sylvatica	Beech	5.5
Larix decidua	European larch	9.9–22.0
Picea abies	Norway spruce	8.7
Pinus sylvestris	Scots pine	2.5
Quercus robur	Oak	2.9
Salix caprea	Great sallow	2.16
Secale cereale	Rye	6.0–8.8
Tilia cordata	Small-leaved lime	3.24
Ulmus glabra	Wych elm	3.24

In fact, under normal conditions, the rates of fall of pollen grains in still air are only of secondary importance. Most anemophilous plants possess adaptations which prevent the release of pollen under perfectly calm conditions – for instance, in most catkins the pollen shed by one flower lodges on the horizontal surface of the bract of the flower below, and in grasses the pollen is held by the curved, spoon-shaped lower ends of the anther loculi (Plate 47a). If there is even slight wind to dislodge the pollen grains, they will be kept in suspension by the turbulence of the air. Turbulence arises in two ways. Obstacles in the path of the wind produce eddies, which break up the smooth flow of the air behind them. Also, on clear days much turbulence is produced by convection currents, as the ground and the air above it are warmed by the sun. In scale, turbulence ranges from the eddying around the twigs and branches of a tree on a windy day to the great global wind-systems whose effect is so dramatically shown in photographs of the earth taken from space-craft.

The pollen grains released from a flower are dispersed by atmospheric turbulence in the same way as a puff of smoke; the variation in density across the cloud through its centre approximates to the well-known bell-shaped 'Normal distribution' (Fig. 84). If the pollen is being released continuously over a period it is dispersed in the same way as a plume of smoke from a chimney, trailing down-wind – with density showing a Normal distribution across the plume, but falling off downwind at a rate rather less than the inverse square law which it approaches under extremely turbulent conditions.

Pollen grains are deposited on the stigmas mainly by impact as the air streams past. Gravity is of little significance except in conditions of complete calm (rarely found except on clear, still nights) and in the thin 'boundary layer' close to the ground and other solid objects.

The mechanism of both dispersal and deposition are very complex, and cannot be considered in detail here; they are discussed more fully by Gregory (1961). However, it remains to say something about the distance to which pollen grains are dispersed under natural conditions; a measure of this is the *standard deviation* from the source. Investigations using sticky microscope slides as pollen traps to study the deposition of pollen at different distances from isolated trees show that, on the average, pollen deposition is greater very close to the source and at substantial distances from it than would be expected from a Normal distribution – the distribution is of the kind known as *leptokurtic* (Fig. 84) (Bateman, 1947c, Wright, 1953, Griffiths, 1950, Wang *et al.*, 1960). Estimates of the standard deviation for some American trees are given in Table 7. For herbaceous plants the standard deviations are probably equally diverse, but in general much smaller – perhaps of the order of a tenth of these figures.

Table 7. Standard deviation of pollen-dispersal distance for some wind-pollinated trees

Species		Standard deviation (*feet*)	Author
Fraxinus americana and F. *pennsylvanica*	Ash	55–150	Wright, 1953
Pseudotsuga menziesii	Douglas fir	60	,,
Populus deltoides and P. *nigra* var. *italica*	Poplar	1000 or more	,,
Ulmus americana	Elm	1000 or more	,,
Picea abies	Norway spruce	130	,,
Cedrus atlantica	Atlas cedar	240	,,
C. libani	Cedar-of-Lebanon	140	,,
Pinus cembroides var. *edulis*	Pinyon	55	,,
Pinus elliottii	Slash pine	220	Wang, Perry and Johnson, 1960

There is no doubt that the great majority of wind-borne pollen grains of all species are deposited quite close to their source. Equally, it is clear that

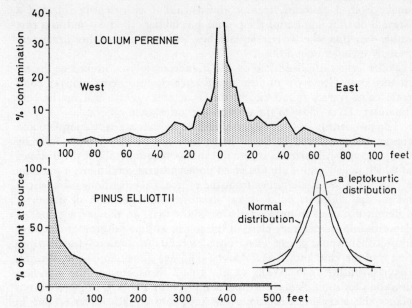

FIG. 84. (Top) Percentage pollination of test plants of Perennial Rye Grass (*Lolium perenne*) at various distances to the east and west by pollen from a 'contaminating' plot of a variety of the same species with red-based shoots (Griffiths 1950). (Bottom) Relative number of pollen grains caught on sticky slides placed in the open at various distances from an isolated tree of Slash Pine (*Pinus elliottii*) in Florida (Wang, Perry & Johnson 1960). (Inset) A leptokurtic distribution compared with a Normal distribution.

some pollen is carried for very long distances. Thus Hesselman (1919) recorded a total pollen catch of 1620 grains/sq. cm. between 16th May and 25th June on the Västra Banken lightship, 30 km. from the Swedish coast in the Gulf of Bothnia; from data collected over a shorter period he estimated that the pollen rain was a little over half as great at the Finngrundet lightship, 55 km. from the coast. In 1937 Erdtman, using a vacuum-cleaner apparatus, demonstrated the presence of pollen grains in the air even in mid-Atlantic, though the number was less by a factor of ten or twenty thousand than in the air over southern Sweden in April and May. Nevertheless, long-distance transport on this scale would account for the presence of *Nothofagus* pollen in peat on Tristan da Cunha, some 3000 miles distant from the nearest source in South America.

Spectacular long-distance transport of this kind is of little significance for pollination. The relatively few grains carried to such distances would normally be completely swamped by vast quantities of locally produced pollen. Particular meteorological conditions may sometimes lead to local deposition of large amounts of pollen at a considerable distance from its source. Rempe (1937) investigated the distribution of pollen in the course of a series of flights over Göttingen. He found that under conditions of strong convection pollen concentration was almost unchanged up to an altitude of 1000 m., and that considerable amounts of pollen were still present at 2000 m.; he found larger amounts of pollen in cumulus clouds (up-currents) than outside. By contrast, under stratus cloud the pollen concentration fell off sharply with altitude. In the course of one flight, Rempe encountered a heavy fall of spruce pollen associated with the base of a dissolving cumulus cloud over Göttingen, at a time when spruce was past flowering in the lowlands. The wind was north-easterly, and the source of the pollen was evidently in the Harz Mountains, where spruce was still flowering and cumulus cloud building up, some 34 km. to the north-east. In this instance, with a wind-speed of 14 km./h., the pollen would have taken about 2½ hours to reach Göttingen. As Rempe pointed out, transport of pollen for distances up to some 300 km./day is readily explained. It is probably for reasons of this kind that 89 per cent. of the spruce pollen counted by Hesselman on the Västra Banken lightship was trapped during 2 of the 40 days of observation. Cumulus clouds break up and dissolve at the end of the day, so pollen carried high into the atmosphere during the day largely settles to the ground during the night. Some pollen may be caught up and carried again to high levels on subsequent days, but probably little remains viable beyond the first day or two, because at least some pollen grains, and probably all, are quickly damaged by more than a few hours' exposure to the ultra-violet radiation in sunlight (Werfft 1951).

These various considerations account for many of the features commonly found in wind-pollinated flowers. The stamens are usually large and well exposed, often hanging freely on long filaments or in catkins. The stigmas too are large and well exposed, and often finely divided and feathery – a narrow surface is more efficient in trapping pollen from moving air than a broad surface of the same area. The functions served by the perianth in entomophilous flowers are irrelevant to wind pollination; indeed a well developed perianth would be a hindrance to free transfer of pollen, and accordingly the perianth is generally much reduced or absent altogether. The concentration of pollen is very high in the immediate neighbourhood of the dehiscing stamens; stigmas close to them would be so thickly covered with their own pollen that there would be little chance of fertilisation by pollen from other individuals. It is not surprising, therefore, that many wind-pollinated plants have the sexes in separate flowers, or if the stamens and stigmas are borne close together in space they are separated in time by strongly marked dichogamy.

Anemophilous plants tend to produce small numbers of ovules. In many cases each flower produces only a single seed, as in oak, hazel, or the grasses and sedges. The rushes (*Juncaceae*) produce many-seeded capsules, but this is unusual among wind-pollinated plants, and probably reflects the occurrence of a good deal of self-pollination.

Two other points underline the ways in which insect and wind pollination are complementary. Most of the wind-pollinated deciduous forest trees flower very early in the year, at a time when few insects are active. But this is also the time when the trees are bare of leaves; the flowers are then most freely exposed to disperse and receive pollen, and the surrounding area 'competing' with the stigmas for pollen is least. Secondly, it is noteworthy that, in a temperate climate such as our own, most of the species that are dominant over wide expanses of the landscape are wind-pollinated. On the one hand these are the plants for which wind-pollination is most efficient (and for ecological reasons forest trees tend to produce large single seeds). On the other hand, it may well be that effective pollination of these dominant plants would require a greater population of insects than a temperate climate can support.

THE CONIFERS

The conifers are among the few groups of plants which are consistently wind-pollinated, and they are perhaps the only group in which wind-pollination is certainly primitive. They are probably not closely related to the true flowering plants, and the structure and development of their 'flowers' are very different from those of the true flowering plants (*Angiospermae*) outlined in Chapter 3. The flowers are typically aggregated into *strobili* or 'cones', as in the Scots Pine (*Pinus sylvestris*) (Fig. 131). The male cones are grouped around the bases of the elongating new shoots, and mature about the end of May or early June. Each is about 5–8 mm. long, and made up of numerous spirally arranged yellow 'stamens', each consisting of a scale (*microsporophyll*) with a narrow upturned crest, bearing two pollen sacs on its under side. The pollen grains each have two 'wings' or bladder-like expansions of the wall which serve to reduce the density and rate of fall of the grain – though their main significance may be rather in orienting the grain relative to the nucellus as it is drawn into the micropyle by the pollination fluid (see Doyle, 1945). The young female cones are formed near the tip of the current year's shoots. At the time of pollination, in June, the cone is around 8 mm. long. The axis of the cone bears small bract scales and much larger and thicker ovuliferous scales, each with two ovules near the axis on its upper surface. Each ovule is covered by a single integument, with a rather wide micropyle facing the axis. The tips of the ovuliferous scales, which make up the outside of the young cone, gape slightly apart, so that pollen can reach the ovules.

Pollen grains which lodge around the mouth of the micropyle are drawn down into contact with the nucellus by a drop of liquid exuded into the micropylar canal and then reabsorbed. After pollination, the scales thicken and seal the interior of the cone. Fertilisation does not take place until the following summer. The cones ripen in the second year after pollination, when the scales of the now dry, woody 'fir-cone' gape apart once more to release the winged seeds.

Most of the other common conifers differ only in minor particulars from the Scots Pine, but the seeds usually ripen in the same year as the cones are pollinated. In some, such as the Douglas Fir, the bract scales are long and project from between the ovuliferous scales in the ripe cone. In the larches (*Larix*) the bract scales are large and brightly coloured in the young female cones, forming the attractive 'larch roses'. The cedars (*Cedrus*) are unique in flowering in the autumn rather than in spring.

Yew (*Taxus baccata*) is related to the conifers but its ovules are solitary. They are produced from buds borne in the leaf axils during the winter. The ovule itself is borne at the tip of a minute shoot in the axil of a scale just below the apex of the bud. The tip of the ovule emerges from between the bud scales very early in the year. By February or March it is ready for pollination. A sticky drop of liquid is exuded from the micropyle, in which pollen grains are trapped and find their way down the canal of the micropyle to the nucellus. The male cones of yew are also produced from buds formed in the leaf axils, but on separate trees. They are smaller than those of the Scots Pine and the scales are umbrella-shaped, with 5–9 pollen sacs on the lower surface. The pollen grains lack the characteristic 'wings' of the pine and its close relatives, and the large seed with its fleshy red aril is ripened within the year.

THE CATKIN-BEARING TREES, AND SOME OTHER TREES

The trees of the Birch and Hazel families (*Betulaceae* and *Corylaceae*) bear their flowers in catkins rather comparable in their general arrangement with the strobili of the conifers. In the birches, of which we have two common species in Britain (*Betula pendula* (Fig. 85) and *B. pubescens*), the male catkins are formed in the autumn, and as they mature in March and April they expand and hang freely dangling from the tips of the twigs. The flowers are borne in threes, with a few bracteoles, in the axil of each catkin scale. Each consists of a pair of deeply divided stamens, with a small bract-like perianth. The short carpellary catkins are borne on the same tree, and remain stiff and erect as they expand. The flowers lie in the axils of the catkin scales as in the male catkins. There is no perianth; each flower consists of an ovary bearing two styles, and containing two ovules. The fruit, ripened in late summer, is a small winged nutlet with

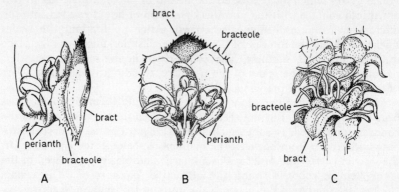

FIG. 85. Silver Birch (*Betula pendula*). A, side view of a group of male flowers in the axil of a single catkin scale. B, a similar group of flowers seen from the adaxial surface (i.e. the side towards the tip of the catkin) C, part of a female catkin. Exeter, Devon.

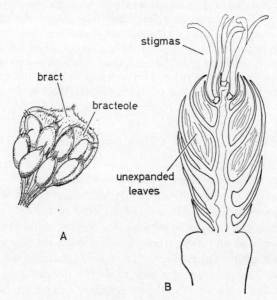

FIG. 86. Hazel (*Corylus avellana*). A, pair of male flowers in the axil of a single catkin scale. B, vertical section of female catkin. Dunsford, Devon.

a single seed. Alder (*Alnus glutinosa*) is similar in essentials to birch, but there are four stamens to each male flower, and the female catkins become woody and cone-like in fruit. Between late February and the end of March the long reddish male catkins expand and shed their yellow pollen, and the small dark female catkins, in clusters a few inches back from the tips of the shoots, are enlivened by the red of their stigmas.

The long yellow 'lamb's tail' catkins of Hazel (*Corylus avellana*) (Fig. 86) are among the first signs of approaching spring as they expand in the first mild weather of February and March. If the air is perfectly calm little pollen escapes from the catkin, but pollen blows out in clouds as the catkins bob and dangle in the wind. The male flowers have no perianth, and are borne in pairs in the axils of the catkin scales. The female flowers are also borne in pairs, and possess a small perianth. The catkin is reduced to a plump bud containing only a few flowers, which at the time of pollination are so undeveloped that they consist of little more than the crimson stigmas projecting between the scales at the tip. After fertilisation, the ovary develops into a one-seeded nut, with a leafy involucre developed from the bracteoles at the base of the flower. Hazel has a particular place in the history of pollination biology as the subject of Richard Bradley's experiments in the early eighteenth century (see p. 22).

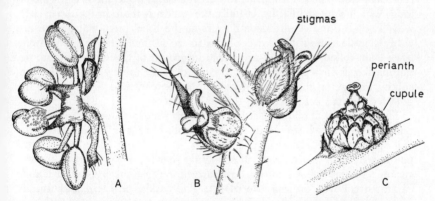

FIG. 87. Pedunculate Oak (*Quercus robur*). A, side view of a single male flower. B, two female flowers. C, young developing acorn, showing perianth and cupule. Exeter, Devon; A, and B, 22 May; C, 18 June. Compare Plate 46b.

The family which includes Oak (Fig. 87) and Beech (*Fagaceae*) is related to the two families we have just considered. The slender yellowish-green catkins of Oak appear with the opening leaf-buds in April and May. The catkin scales are very small, and the individual flowers have a rather larger perianth and more stamens than those of the catkin-bearing trees we have discussed so far. About six stamens is usual, but the number is variable. The female flowers are borne a few together in short spikes. Each is surrounded at the base by a scaly involucre which later develops to form the cupule or 'acorn-cup'; the minute green perianth forms a toothed border at the top of the ovary surrounding the three styles. The ovary contains six ovules, but only one develops. We have two native oaks in Britain. In the Pedunculate Oak (*Quercus robur*) the female spikes are relatively long, and the ripe acorns are typically borne on a stalk an inch or two long. In the Sessile Oak (*Q. petraea*) the female spikes are much shorter, and the acorns are only shortly stalked or almost sessile on the twigs.

Beech (*Fagus sylvatica*) has its male flowers in long-stalked tassel-like heads. The female flowers are borne in pairs, surrounded by the involucre which grows to form the four-valved cupule enclosing the beech nuts. The individual flowers are much like those of oak, and as in the oaks they appear as the foliage begins to expand.

Beech and the oaks are regularly wind-pollinated and are visited only casually by insects for their pollen. By contrast, the Sweet Chestnut (*Castanea sativa*) is largely insect-pollinated; so, too, are some tropical species of oak. The chestnut has stiff erect catkins up to six or eight inches long, which appear in July. Usually the catkin is male in its upper part, with a small number of female flowers at the base, but some catkins are entirely male. The male flowers have 10–20 stamens; the female flowers have a six-celled ovary and are borne in groups of three in each cupule. Pollination is brought about by bees and other insects which visit the catkins for nectar and pollen; wind-pollination can take place later as the pollen dries and is blown from the male flowers.

The same kind of relationship between wind-pollinated and insect-pollinated species is shown in another family of catkin-bearing trees, the Willow family (*Salicaceae*). The willows and sallows (*Salix*) are insect-pollinated; sallows are favoured collecting sites for entomologists in early spring. All the willows and sallows are dioecious; the male and female catkins are borne on separate plants. The catkins are stiff and usually erect. In the larger willows they are about 6 cm. long and a little less than 1 cm. thick, and appear with the leaves in April and May; in the sallows ('Palm', 'Pussy Willow') they are shorter and broader and appear on the bare twigs in March and April. The individual flowers, one in the axil of each catkin scale, are very simple. There is no perianth, but the flowers possess a nectary at the base; male flowers usually have two stamens (more in some species) while female flowers have a two-celled ovary with a short style and two stigmas (Fig. 88). The catkins are freely visited by insects in fine weather, but a great deal of pollen is dispersed by wind, and this must also be important for pollination. The poplars

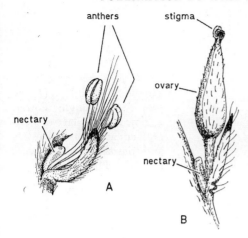

anthers   stigma

ovary

nectary

nectary

A

B

FIG. 88. Common Sallow (*Salix cinerea*). A, single male flower. B, single female flower. Uplyme, Devon. Compare Plate 35.

(*Populus*) are closely related to the willows, but are entirely wind-pollinated. The catkins are up to 10 cm. long, and are flexible and pendulous like the male catkins of oak and hazel; they appear in early spring before the leaves. The individual flowers lack the nectaries of the willows, but have a cup-like disc at the base, and the stamens of the male flowers are more numerous.

The remaining important wind-pollinated trees of our woods are the elms (*Ulmus*) and Ash (*Fraxinus excelsior*). The elm family are all wind-pollinated, and our native elms are among the earliest trees to flower in spring. Towards the end of January the tight clusters of dark reddish flowers begin to outline the crowns of the trees against the winter sky. The individual flowers are bisexual, unlike those of any of the trees we have considered so far. There is a small bell-shaped perianth divided into four or five lobes, as many stamens, and a one-celled ovary with two styles which in due course develops into the one-seeded winged fruit. The flowers are strongly protandrous, and at first sight may appear to be unisexual, especially in the male phase when the stamens project far out of the flowers on their long filaments (Fig. 89).

Ash also has bisexual flowers, which appear in coarse dark greenish masses on the naked twigs in March and April (Fig. 90, Plate 46d). Most members of the Olive family (*Oleaceae*), to which the Ash belongs, are entomophilous; familiar examples are Privet (*Ligustrum*), Lilac (*Syringa vulgaris*) and Jasmine (*Jasminum*). The flowers have no corolla; there are two stamens, as in other members of the family, and the long ovary which later develops into the flat one-seeded 'ash key' bears two rather large blackish stigmas at the tip. Some trees bear hermaphrodite flowers, others may bear purely male or purely female flowers or a mixture of herma-

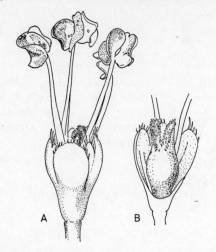

FIG. 89. English Elm (*Ulmus procera*). A, single flower. B, flower with part of perianth cut away to show ovary and stigmas. Exeter, Devon. Compare Plate 46c.

phrodite and unisexual flowers; the same tree may vary somewhat in behaviour from year to year. The Common Ash is an obvious example of a plant of entomophilous ancestry which has quite recently (in evolutionary terms) become adapted to wind-pollination. It is particularly interesting that the Mediterranean 'Manna Ash' (*Fraxinus ornus*) still possesses a white corolla and the fragrant flowers are pollinated by insects.

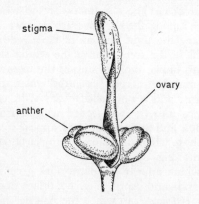

FIG. 90. Ash (*Fraxinus excelsior*). Single hermaphrodite flower. Sidford, Devon. Compare Plate 46d.

THE GRASSES AND SEDGES

The grasses (*Gramineae*) are by far the most important family of wind-pollinated herbs. The only exceptions to wind-pollination in the family are the relatively few species which are self-pollinated or apomictic (see Chapter 9). The small greenish flowers of grasses are grouped in *spikelets*

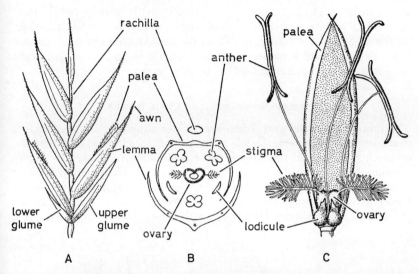

FIG. 91. A, spikelet of Red Fescue (*Festuca rubra*). B, diagram of a grass flower. C, flower of Meadow Fescue (*Festuca pratensis*) with lemma removed. A, after Hubbard (1954); C, after Rendle (1930). Compare Plate 47a.

(Fig. 91), each enclosed at its base by a pair of chaffy *glumes*. The individual flowers or 'florets' in the spikelet are borne alternately on either side of the slender axis or *rachilla*. Each flower is enclosed by a glume-like *lemma*, usually with a prominent mid-nerve and often with a slender awn at the back or tip, and *palea* usually with two prominent nerves. The flower itself consists of an ovary with two long feathery stigmas, three stamens with slender filaments and large versatile anthers, and a pair of small swollen scales called *lodicules* – perhaps the last vestiges of the perianth – which swell to open the floret when the anthers and stigmas are mature. The spikelets are variously arranged into slender wiry inflorescences. A glance at a few grasses in flower, or at the photograph on Plate 47a, will convey better than any description how effectively the stamens and stigmas are presented to the wind.

The sedges (*Cyperaceae*) are superficially similar to the grasses, and like them have much-reduced greenish or brownish flowers with prominent stigmas and large stamens. They differ from the grasses in the structure of the individual flowers and inflorescences. The flowers are borne singly in the axils of scales or 'glumes', forming catkin-like spikes. In most genera the flowers are hermaphrodite, with (usually) three stamens and an ovary, containing a single ovule, with a style divided into two or three long rough stigmas. The perianth may be represented by bristles (which form the 'cotton' of the cotton-grasses (*Eriophorum*)) or may be absent altogether. In the large and common genus *Carex* the flowers are unisexual, and are often grouped into separate male and female spikes, an arrangement recalling the wind-pollinated catkin-bearing trees.

OTHER WIND-POLLINATED HERBS

The flowers of grasses and sedges are so specialised for wind-pollination that they bear little resemblance to insect-pollinated flowers. Among the remaining wind-pollinated herbs an entomophilous ancestry is usually obvious. Thus the rushes (*Juncaceae*) have a small chaffy perianth, large stamens and large rough stigmas, but the arrangement of the parts of the flower leaves no doubt that they are closely related to the Lily family (*Liliaceae*). The plantains (*Plantago*, family *Plantaginaceae*) (Fig. 92) have

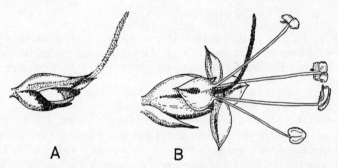

A   B

FIG. 92. Ribwort Plantain (*Plantago lanceolata*). A, young flower, with receptive stigma. B, older flower; stigma withered, anthers dehiscing. Alderney, Channel Is. Compare Plate 47d.

no close insect-pollinated relatives, but are comparable with the rushes in their adaptations to wind pollination – though in no way related to them. The corolla is small and membranous, and the strongly protogynous flowers have long rough stigmas and large versatile anthers borne on long filaments. Some species, such as the Hoary Plantain (*P. media*), are said to be at least partly entomophilous, but it must be remembered (see p. 268) that bees sometimes collect pollen from undoubtedly anemophilous plants such as grasses and beech and oak trees (Bogdan, 1962, Chambers, 1945).*

The pollen of the common Stinging Nettle (*Urtica dioica*) is among the

* Porsch (1956) lists a large number of beetles and flies (*Diptera*) that have been observed feeding on the pollen of anemophilous plants.

most abundant in the pollen rain of late summer. The Stinging Nettle has small greenish unisexual flowers, borne in catkin-like inflorescences hanging from the leaf axils; usually male and female flowers are found on separate plants. The female flowers have two smaller and two larger perianth segments, and a one-celled ovary with a sessile tufted stigma. The male flowers have four perianth segments and four stamens; the stamens are incurved and under tension in the bud, and spring back, scattering the pollen explosively, after the flowers open (Fig. 93).

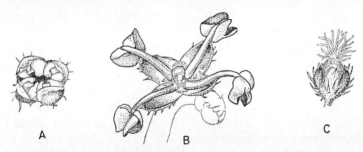

FIG. 93. Stinging Nettle (*Urtica dioica*). A, newly opened male flower. B, male flower after dehiscence of anthers. C, female flower. Exminster, Devon.

The docks (*Rumex*) are also abundantly represented in the pollen rain. Of the two main British genera in the Dock family (*Polygonaceae*) species of *Polygonum* are mainly pollinated by insects or self-pollinated. Some, like the Bistort (*Polygonum bistorta*) and the introduced *Polygonum cuspidatum*, have conspicuous inflorescences and well developed perianths; others, such as the Common Knotgrass (*Polygonum aviculare* agg.) have small inconspicuous flowers in the leaf axils. All have simple styles. The wind-pollinated docks and sorrels, on the other hand, have bulky lax inflorescences well exposed to the wind, rather small perianths and tufted stigmas (Fig. 94).

The members of the Goosefoot family (*Chenopodiaceae*) are largely plants of exposed maritime habitats or salt-steppes. Their insect-pollinated relatives are to be sought in the *Caryophyllaceae* and several other related families. The flowers of the *Chenopodiaceae* are small and greenish, with no corolla, and the short stamens and stigmas are usually quite freely exposed (Fig. 95). Pollen of *Chemopodiaceae* appears abundantly in the pollen rain; there is also a good deal of evidence that some species, at least, are commonly self-pollinated (for example *Atriplex patula*). Probably various degrees of balance between anemophily and self-pollination are to be found in the family.

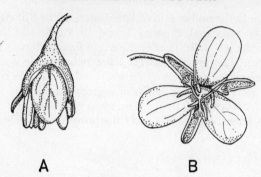

A　　　　　B

FIG. 94. Broad-leaved Dock (*Rumex obtusifolius*). A, newly opened flower, in functionally male phase. B, older flower with anthers shed and stigmas expanded. Alderney, Channel Is.

Many aquatic and waterside plants belonging to various families are wind-pollinated, for example Mare's Tail (*Hippuris*) (Plate 48d), Water Milfoil (*Myriophyllum*), many of the Water Starworts (*Callitriche*) (Fig. 100) and 'pondweeds' (*Potamogeton*) (Plate 50a) and the Bur-reeds and Bulrushes (*Sparganium* and *Typha*). Most of the remaining anemophilous herbs are scattered species belonging to predominantly insect-pollinated families, for example the wind-pollinated species of *Thalictrum* (Meadow-rues) in the *Ranunculaceae*, Salad Burnet (*Poterium sanguisorba*) in the *Rosaceae*, Dog's Mercury (*Mercurialis perennis*) (Plate 48b) in the *Euphorbiaceae* and the Mugwort and its relatives (*Artemisia*) in the *Compositae*. The interesting general point is the way in which the combination of characters associated

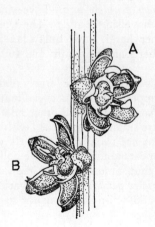

FIG. 95. Sea Beet (*Beta vulgaris* subsp. *maritima*). A, young flower about time of dehiscence of anthers. B, older flower; anthers mostly shed and stigmas mature. Alderney, Channel Is.

PLATE 37. Heterostyly: **a**, pin-eyed and thrum-eyed flowers of Cowslip (*Primula veris*). **b**, short-styled and long-styled flowers of Bogbean (*Menyanthes trifoliata*).

a

b

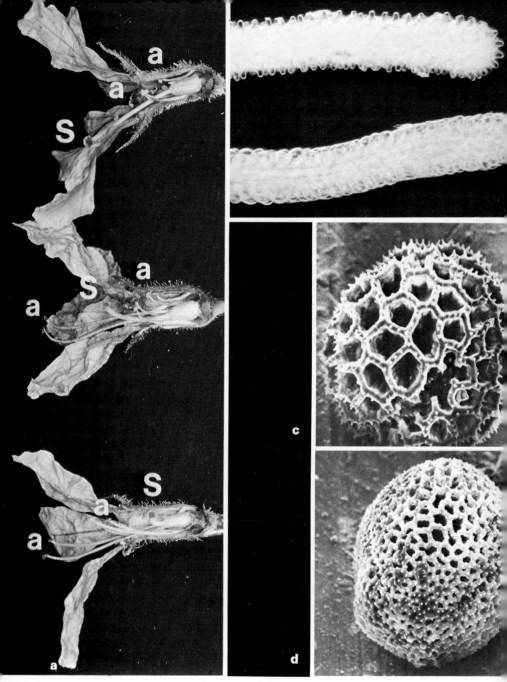

PLATE 38. Heteromorphic flowers: **a**, the three types of flower in Purple Loosestrife (*Lythrum salicaria*); **a**-anthers, **s**-stigma. **b**, 'cob' and 'papillate' stigmas of Thrift (*Armeria maritima*). **c-d**, scanning electron micrographs of Type A and Type B pollen grains of Thrift.

with wind-pollination – the 'syndrome of anemophily' – has appeared independently in flowers of widely varied basic structure.

A striking feature of Hyde's results was the amount of pollen he found of types that we ordinarily think of as purely entomophilous. This should be no cause for surprise. There are probably few entomophilous species of which no pollen is ever shed into the air, and this pollen may often bring about a small amount of local wind-pollination. Even a very small proportion of wind-pollination will provide a basis for natural selection to work on under conditions where wind-pollination may be advantageous and so lead to the evolution of anemophily. We tend to think of plants in terms of clear-cut categories. However, it is a common assumption that many insect-pollinated species may be self-pollinated if pollination is not brought about by insects, and a balance between two or more pollination mechanisms is probably common in other cases as well. It is not easy to demonstrate and evaluate a balance of this kind, but there can be little doubt that there is such a balance between entomophily and anemophily in, for example, Lime (*Tilia*), Ling (*Calluna vulgaris*), and the Rock-roses (*Helianthemum*), as well as in the Sweet Chestnut and willows mentioned already. All of these are visited by insects, often in large numbers, but pollen of all is released plentifully into the air, and all show at least some indication of the syndrome of anemophily. However, in two possible instances which have been analysed, Free (1964) concluded that wind-dispersed pollen is of negligible importance in the pollination of apple trees, while Hatton (1965) concluded that in Mistletoe (*Viscum album*) anemophily is predominant.

HAY FEVER

The profusion of pollen released into the air by wind-pollinated plants has some important incidental effects. We inhale a good deal of pollen with the air we breathe, and as sufferers from hay-fever will be only too keenly aware the pollen of particular species may set up allergies in some people. The pollen types causing hay-fever fall largely into two categories: pollen of the common grasses, and that of weeds and ruderal species. Timothy grass (*Phleum pratense*) is the commonest cause of hay-fever in Britain; in the United States the most troublesome species is Ragweed (*Ambrosia artemisifolia*). Pollen of the common wind-pollinated trees may cause hay-fever in early spring, but pollen of pine and related conifers seems to be almost inert. If the times of flowering of the anemophilous species in a district are known (Hyde, 1950, 1969, Wodehouse, 1945), the time of onset and cessation of the attacks will often indicate which species are likely to be responsible. However, a sure diagnosis can only be made by tests in which small amounts of extracts of suspected pollen types are scratched or injected into the patient's skin to determine those to which he is allergic. Anti-histamine drugs give some relief in mild cases of hay-fever. More severe cases may be treated by desensitisation – injection of gradually increasing doses of an extract of the pollen type responsible for the patient's symptoms – over a period of a few months before the start of the hay fever season.

POLLEN ANALYSIS

The outer layers (*exine*) of the walls of pollen grains and spores are very resistant to decay, and to chemical treatment. The pollen-rain falling on the surface of lakes and growing peat-bogs thus becomes incorporated in the accumulating lake-mud or peat; a cubic centimetre of peat may contain fifty- or a hundred-thousand pollen grains. By suitable chemical treatment, the pollen grains present in a sample can be recovered for identification; the proportions of the various pollen types present (the *pollen spectrum*) can then give a great deal of information about the surrounding vegetation at the time the sample was laid down.

Pollen grains vary greatly in the number, position and form of the apertures (germination pores and furrows), and in the thickening and sculpture of the exine. Most dicotyledons have three pores or three furrows arranged symmetrically round the equator of the grain, but some genera have larger numbers of pores or furrows variously disposed over the surface. Most monocotyledons have a single pore or furrow; some lack apertures altogether, for example most *Cyperaceae, Potamogeton*. It is commonly possible to identify a pollen grain to a genus; in a few cases the grains of particular species can be recognised, while in others it may only be possible to identify to the family. The pollen grains shown in Plate 3 will give some idea of the range of variation. Descriptions and many illustrations of pollen grains are given by Erdtman (1952, 1957) and by Hyde and Adams (1958).

The technique of pollen analysis was originated by the Swedish botanist Gustaf Lagerheim in the early years of this century, though others before him, such as C. A. Weber, had observed and studied the pollen grains in peat. Pollen analysis was first applied extensively by the geologist Lennart von Post, for the correlation of profiles in Swedish peat bogs. Von Post published his first results in 1916, and since that time pollen analysis has become the most important research method for studying the vegetational history of the last million years or so.

Usually a number of pollen spectra from successive levels in a deposit are combined into a *pollen diagram*, which shows graphically the variation over the course of time of the various pollen types recognised. In north-west Europe, pollen analysis was readily and fruitfully applied to studying the course of forest history since the last glaciation. The pollen diagrams from different sites show striking correspondences in the course of events, but also regular regional differences, as different trees have successively dominated the forests in the changing climates of the last ten or twelve thousand years. During the last few decades pollen analysis has yielded valuable information on the non-forest vegetation of the late-glacial period, on the course of forest-clearance by prehistoric man, and on the flora and vegetation of the interglacial periods. Many examples of pollen diagrams, and of the results of pollen analysis, are given by Godwin (1956), Pennington (1969) and West (1968).

Pollination by wind is common and important, but pollination by water is surprisingly rare. As Agnes Arber wrote in her classic book on *Water Plants* (1920), 'The most notable characteristic of the flowers of the majority of aquatic Angiosperms is that they make singularly little concession to the aquatic medium . . .' Insect-pollinated and wind-pollinated aquatics abound; water-pollinated species are relatively few, and most of them show obvious signs of a relatively recent entomophilous or anemophilous ancestry. The adaptations to pollination by water (*hydrophily*) are diverse, so that it is hardly possible to speak of a syndrome of hydrophily. Pollination may take place at the water surface, or completely under water.

One of the best known and most highly adapted examples of pollination at the water surface is provided by the Ribbon-weed (*Vallisneria spiralis*) (Fig. 96). This is a native of warmer climates than our own, but it is naturalised at a number of places in Britain where the water is warmed by industrial effluents. The plants are dioecious. The minute male flowers are borne many together in a tubular spathe near the base of the plant. Each has two stamens, tightly enclosed by the three sepals. At maturity, the flowers break free and float to the surface of the water, where they open and the stamens dehisce, exposing the pollen which adheres in a globular mass to the tip of each stamen. The female flower is borne to the surface of the water on a slender flexible peduncle, where it lies more or less horizontally in a shallow dimple in the surface film. When the sepals open, exposing the three large fleshy stigmas, the flower is some 3–4 mm. across; the tubular spathe enclosing it at the base is a little over a centimetre long. The stigmas are unwettable, like the leaves of a number of floating aquatics, owing to the dense velvety pile of water-repellent hairs with which they are covered. The male flowers are carried about by water currents and the wind. If one chances to encounter a female flower, it slides down the depression in the surface-film and comes to rest with the projecting stamens in contact with the stigmas. It seems that the mechanism may not work in this simple way in *V. americana*, in which the stamens are often erect and do not project beyond the sepals of

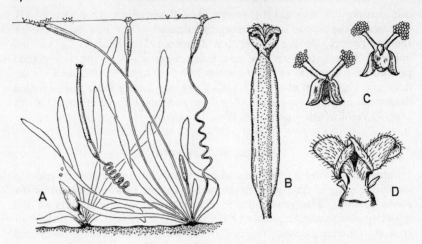

FIG. 96. Tape-grass (*Vallisneria spiralis*). A, semi-diagrammatic sketch to illustrate habit of plant (The width and length of the leaves and the depth of the water are more than proportionately reduced). B, female flower. C, male flowers. D, stigma-lobes of female flower. A, based on Kausik (1939) and Sculthorpe (1967); B-D, after Kausik.

the male flower. In this case, pollination probably depends on the male flowers being toppled into the female flower when the latter is momentarily submerged.

In *Vallisneria*, as in the cleistogamous aquatic plants, the actual transfer of pollen takes place in the air, so that in effect this is not truly hydrophilous pollination at all. However, in many species of the related genus *Elodea*, including the familiar Canadian Water-weed (*E. canadensis*), it is the pollen-grains themselves which are dispersed across the water-film to reach the stigmas. As in *Vallisneria*, the plants are dioecious. Both male and female flowers have six perianth segments, and open at the surface of the water, carried up by the slender perianth-tube. The male flowers open suddenly, the water-repellent perianth-segments reflexing against the surface-film and holding erect the nine stamens. These dehisce explosively, scattering the pollen over the surrounding water. The rather densely-set spines of the exine hold back the surface-film and prevent the pollen-grains from being wetted, so they can be freely moved over the water-surface by wind and other disturbances. The female flower lies in a shallow depression in the surface-film as in *Vallisneria*, usually resting on two of the three water-repellent stigmas which project beyond the perianth (Fig. 97). Male plants are exceedingly rare in Britain, and are apparently uncommon within the species' native range in North America.

There are several other examples of surface hydrophily in the British flora. In some species of *Potamogeton* (Pondweeds, Fig. 98) the protogynous flowers are wind-pollinated. However, in *P. filiformis* and *P. pectinatus* the lax interrupted flower-spikes float at the water-surface, and the pollen-grains are carried on the surface-film to the stigmas.* Probably all

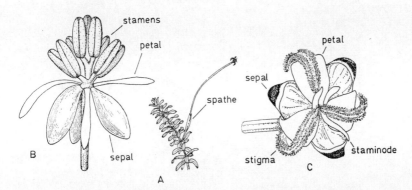

FIG. 97. Canadian Water-weed (*Elodea canadensis*). A, habit of plant. B, male flower. C, female flower. B, after H. St John (1965); A, C, Exminster, Devon.

gradations between regular anemophily and regular hydrophily are to be found in the genus – with the dense spikes of *P. natans*, standing stiffly above the floating leaves at one extreme, and the lax floating inflor-escences of *P. pectinatus* at the other. Daumann (1963) has shown that the pollen of *P. natans* rapidly loses its power of germination on contact with water; viability fell to only 10 per cent. after four hours of wetting. On the other hand, the pollen of *P. lucens* – a large submerged species, of which only the flower-spikes appear above the surface – remained 45 per cent. viable after a day in water. Daumann noticed quantities of pollen floating in the neighbourhood of the inflorescences, and showed in aquarium experiments that pollen grains could in this way reach the stigmas and germinate. The small grass-leaved species such as *P. pusillus* and *P. berchtoldii* have small few-flowered spikes, which project only a little above the surface. Their flowers differ little from those of *P. pectinatus*, and their pollination biology would be worth careful study.

In the species of *Ruppia* (Tassel Pondweeds), which grow in brackish pools and ditches, the pollen is similarly liberated at or just above the water surface, upon which it floats to the flat, shield-like stigmas (Fig. 99).

* Mr J. E. Dandy tells me that *P. pectinatus* when growing in deep water may be pollinated and set fruit without the spikes reaching the surface.

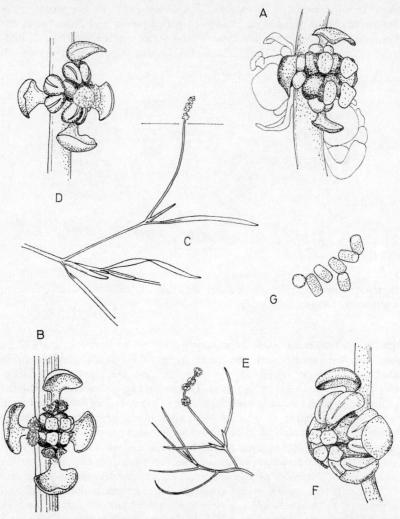

FIG. 98. Pondweeds (*Potamogeton*). A, single flower of *P. obtusifolius*. B, single flower of *P. berchtoldii*. C, habit of *P. trichoides*. D, single flower of *P. trichoides*. E, habit of *P. pectinatus*. F, single flower of *P. pectinatus*. G, pollen of *P. pectinatus*. A, C and D, Exminster, Devon; B, Shapwick, Somerset; E-G, Sturminster Newton, Dorset.

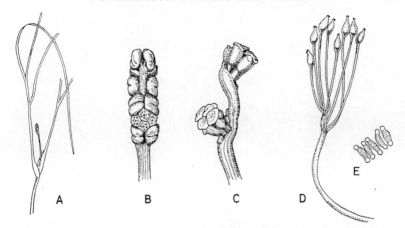

FIG. 99. Tassel Pondweed (*Ruppia maritima*). A, habit of plant. B, inflorescence at time of dehiscence of anthers (compare Plate 50a). C, inflorescence after pollination and shedding of stamens. D, shoot with ripe fruits developed from two flowers. E, pollen grains. Budleigh Salterton, Devon.

According to Gamerro (1968), who has given a detailed account of the pollination of *R. cirrhosa* (*R. spiralis*), the pollen is released into bubbles of gas expelled from the dehiscing anthers, and is scattered over the surface-film as the bubbles burst. The flowers are protandrous. After dehiscence, the anthers are shed, and the inflorescence remains floating for a time, exposing the stigmas at the surface-film, before being withdrawn beneath the surface by coiling of the peduncle as in *Vallisneria*. Pollination of our other species, *R. maritima*, apparently takes place in essentially the same way, but there is no spiral coiling of the peduncle after flowering. Some authors have suggested that *Ruppia* is normally pollinated under water. My own limited observations on *R. maritima* suggest that, although the anthers may dehisce under water, the pollen is brought to the surface by bubbles and little or no pollination takes place if the water-level is high enough to keep the stigmas submerged. However, whether pollination takes place at or beneath the surface may perhaps depend to some extent on environmental conditions, and more observation is clearly needed.

Submerged hydrophily is found in only a few genera. *Callitriche* (the Water Starworts) must be counted among these. The terrestrial and amphibious species are probably normally wind-pollinated, and flowers are borne only in the upper axils that are exposed to air (Fig. 100). By contrast, the submerged species of the section *Pseudocallitriche* (*C. hermaphroditica*, *C. truncata*) flower and fruit freely under water, in spite of the fact that they show few evident adaptations to hydrophily beyond the

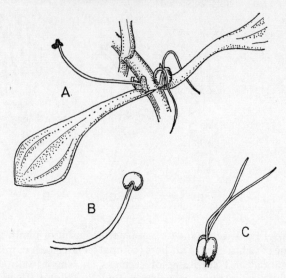

FIG. 100. Water Starwort (*Callitriche obtusangula*). A, part of flowering shoot, showing a male and a female flower. B, undehisced stamen. C, ovary and stigmas. Newton Abbot, Devon.

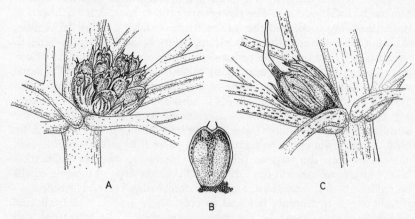

FIG. 101. Hornwort (*Ceratophyllum demersum*). A, immature male flower. B, mature stamen releasing pollen grains. C, female flower. Exminster, Devon.

reduced exine and the common presence of oil-globules in the pollen-grains. The Hornworts (*Ceratophyllum*) are more specialised and are regularly pollinated under water. The flowers are borne in the axils of the finely divided leaves (Fig. 101). The female flower consists of a one-celled ovary, containing a single ovule, with a slender, oblique style, surrounded by a small cup-shaped perianth divided into ten to fifteen lobes. In the male flower a similar perianth surrounds ten to twenty stamens. Each consists of a large anther with an expansion of the connective at the tip which acts as a float. As they mature, the stamens break off and float to the surface where they dehisce, releasing the pollen grains which sink slowly through the water as they are wafted among the submerged stems.

The genera which remain to be considered all belong to the Mono-cotyledons. *Zannichellia* (Horned Pondweed) (Fig. 102) looks at first sight very like *Ruppia* or one of the narrow-leaved *Potamogeton* species. The flowers are borne in small axillary clusters, each cluster usually com-prising one male flower, and a few female flowers surrounded at the base by a cup-shaped spathe. The male flower consists of a single stamen, with a rather long filament raising it well above the female flowers. The female flowers each consist of a single carpel with one ovule and a more-or-less funnel-shaped stigma. The pollen grains are released into the water and sink slowly on to the stigmas; it is said that they slide down the stylar canal to bring about pollination. The two British species of *Najas* are both rare and local, *N. flexilis* in the north and west from Kerry and the Lake District to the Hebrides, and *N. marina* (one of the few dioecious members of the genus) in a few of the Norfolk Broads. Their flowers are borne in the leaf axils. The male flower consists of a single stamen and a small perianth enclosed in a short tubular spathe; the female flower consists of a single carpel with one ovule, and three long stigmas at the apex. The pollen grains are rich in starch and have a greatly reduced exine; they have often begun to germinate before they escape from the envelopes of the male flower, and the developing pollen-tubes must increase their chance of being caught by the stigmas.

Among the most specialised of all hydrophilous species are the Eel-grasses (*Zostera*) of which three species occur commonly around our coasts. All three produce shoots bearing linear grass-like leaves from rhizomes rooted in sand or mud – between the tidemarks in the case of *Z. noltii* and *Z. angustifolia* (Fig. 103), and from about low water to several metres lower in the case of *Z. marina*. An admirable account of the pollination of *Zostera* is given by Clavaud (1878). Male and female flowers alternate on a short flattened axis more-or-less enclosed in a leaf sheath. The female flowers consist of an ovary, containing a solitary ovule, with two long stigmas at the apex; the male flowers, which mature 1–3 days later, consist of a single anther. The pollen 'grains' (Plate 3) are thread-like, about ¼ mm. long, and of the same density as sea-water; the exine is

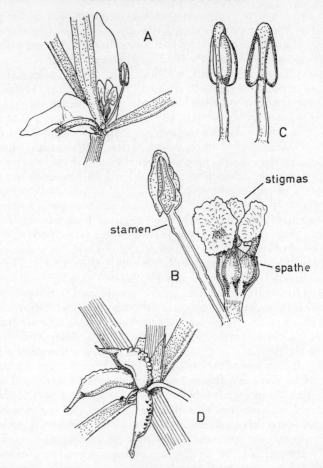

FIG. 102. Horned Pondweed (*Zannichellia palustris*). A, node showing a male flower and a group of female flowers. B, enlarged view of male and female flowers. C, two views of an undehisced stamen. D, node with ripe fruits. Newton Abbot, Devon.

extremely reduced. They are released in cloudy masses which are drifted passively across the eel-grass flats by the tides. If a pollen grain comes into contact with a narrow object it rapidly becomes curled around it; in this way pollen grains are securely anchored to any stigmas they may reach.

There are many other examples of varying degrees of specialisation for water-pollination outside the British flora; as in the case of wind-pollination it is clear that hydrophily must have arisen a number of times in

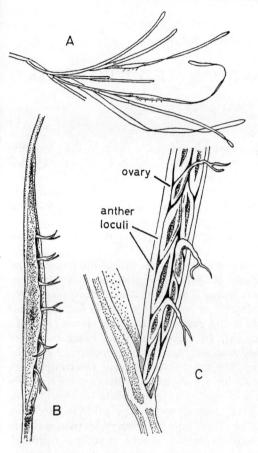

FIG. 103. Eel-grass (*Zostera angustifolia*). A, habit of plant, × 0.3. B, sheath of flowering shoot, showing projecting stigmas, × 2. C, part of flowering spike removed from sheath. Exe estuary, Devon. Compare Plate 50b.

different evolutionary lines. A useful review of the subject is given by Sculthorpe (1967). Species pollinated by water tend – not unexpectedly – to show some of the same characteristics as anemophilous plants. The perianth is much reduced, in most cases there is only a single ovule, and the flowers are commonly unisexual. To these characteristics may be added reduction in the exine of the pollen grain. The stigmas are usually large but rigid and simple; the fact that plumose stigmas like those of many wind-pollinated plants are not to be found among hydrophiles reflects the different conditions in a much denser medium. On the other hand, the pollen grains tend to be elongated (as in *Zostera*, and less strikingly in *Ruppia*) or to develop pollen-tubes precociously (as in *Zannichellia* and *Najas*); it is interesting that a similar trend towards elongation is found among fungi in the spores of some aquatic Hyphomycetes.

CHAPTER 9

SELF-POLLINATION AND APOMIXIS

SELF-POLLINATION has been mentioned many times in the preceding chapters. There are very few plants with hermaphrodite flowers in which it is completely impossible for pollen to reach the stigmas of the same flower, and if insect visits fail it is usual for at least some self-pollination to take place. Of course, self-pollination (*autogamy*) is not necessarily followed by self-fertilisation. Many plants are self-incompatible (see p. 224), and will set little or no seed even if the stigmas are well covered with their own pollen.

Every gradation exists between flowers in which mechanical and genetical adaptations effectively prevent self-fertilisation, and those that set seed freely with their own pollen if insect visitors are excluded. Even flowers apparently highly adapted to insect visits may in fact be almost entirely self-pollinated, such as the garden pea (*Pisum sativum*) and the French bean (*Phaseolus vulgaris*). In these and many other cultivated species autogamy is probably a consequence of long-continued artificial selection of plants which produce a heavy and consistent crop of seeds and breed true. Self-fertilisation has the obvious advantage of the certainty of a full seed-set, and tends to bring about homozygosity which will allow the expression of recessive genes (see Chapter 2, p. 40). On the other hand, only cross-fertilisation can maintain a degree of heterozygosity in the population, with the resulting constant source of variation through genetic recombination. In the long run, the particular balance between cross- and self-fertilisation is probably a rather constant character of a population of plants. Like any other character, this balance will be subject to natural selection and consequent change in response to the conditions of the environment (see Chapter 12, p. 381).

The cleistogamous flowers produced by many species of violets (*Viola*) may be thought of in this light. The normal flowers of the Sweet Violet (*V. odorata*) are showy, zygomorphic, and pollinated by bees. Although they are homogamous and self-compatible to their own pollen, self-fertilisation of these flowers is rare because self-pollination is prevented by structural adaptations. However, in late spring and summer the Sweet Violet produces small flowers which never open. The style and stamens remain short and poorly developed, and are still almost in contact when pollination takes place (Fig. 104). The pollen germinates within the anther-cells, and the pollen-tubes have only a short distance to traverse before reaching the stigma. Comparable, but less reduced, cleistogamous flowers are produced during the summer by the Wood Sorrel (*Oxalis acetosella*). Cleistogamous flowers occur, usually under unfavourable conditions, in many other species belonging to various families; the subject has been reviewed by Uphof (1938).

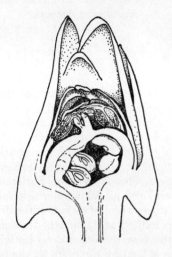

FIG. 104. Cleistogamous flower of Sweet Violet (*Viola odorata*). After A. H. Church. Compare Fig. 67.

There are a number of instances in which self-pollination has evidently arisen in response to a scarcity of suitable pollinators, or to some other breakdown in an insect pollination mechanism. The case of the Bee Orchid (*Ophrys apifera*) has already been discussed (Chapter 7, p. 250). Self-pollination among the Helleborines (*Cephalanthera* and *Epipactis*) is probably related to a more general shortage of pollinators in the dark and rather bare woods which they frequent. Similar considerations probably apply to the rather numerous self-pollinated plants of salt-marshes and wet moorland habitats. Bees are relatively scarce in the cold exposed northern and western parts of the British Isles, and few insects fly in the damp cloudy weather so common in those regions. It is scarcely surprising to find that the small zygomorphic flowers of the Water Lobelia (*Lobelia dortmanna*) and the Pale Butterwort (*Pinguicula lusitanica*) (Fig. 105) are self-pollinated, though evidently entomophilous in origin. The small white flowers of the sundews (*Drosera*) open only in hot sunshine; they are usually selfed and often cleistogamous. That entomophily is not incompatible with the insectivorous habit is shown by the large-flowered Butterworts (for example the Common Butterwort (*Pinguicula vulgaris*)) and by the African sundew (*Drosera capensis*) (often grown in glasshouses) with its showy pink flowers – though in the case of the Butterworts, at

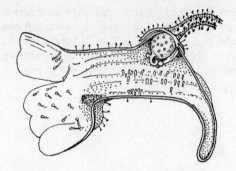

FIG. 105. Pale Butterwort (*Pinguicula lusitanica*). Side view of flower with half of calyx and corolla cut away. Aylesbeare, Devon.

least, the insects pollinating the flowers* are very much larger than those trapped and digested by the leaves. Hagerup (1951) pointed out that a good many plants generally thought of as entomophilous are apparently regularly self-pollinated in the inhospitable climate of the Faroes and in other parts of northern Europe. Probably his most surprising suggestion (Hagerup, 1950) is that rain may play an important part in bringing about self-pollination of certain species. The flowers of Buttercups (*Ranunculus* spp.) remain open during rain, and fill with water to a level which allows pollen to be swirled to the stigmas on the surface of the drop (Fig. 106). A similar mechanism operates in the Bog Asphodel (*Narthecium ossifragum*). According to Hagerup, in this case the water-drop bridging the gap between the anthers and stigma is held by the hairs on the filaments of the stamens, but my own experience in the south of England, both in rainy weather and with artificial 'rain' produced with a fine spray, is that there the water usually rests on the perianth segments (Plate 51b). If rain-pollination of this kind is as effective in the Faroes as

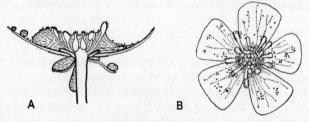

A   B

FIG. 106. Lesser Spearwort (*Ranunculus flammula*). A, vertical section of flower, showing drops of water accumulated on the flower during rain; pollen (black dots) is floating on the water surface. B, face view of a flower after rain; pollen is scattered over all parts of the flower. After Hagerup.

* See p. 212.

Hagerup's observations suggest, it will no doubt be found to be of considerable importance in our own northern and western districts.

Self-pollination and cleistogamy are also found not uncommonly in plants which may be inundated at flowering time, such as Awlwort (*Subularia aquatica*) and Mudwort (*Limosella aquatica*).

There remains a very large and important category of self-pollinated plants which certainly cannot be entirely accounted for in terms of scarcity of pollinators or their physical exclusion. These are the small annual plants, often of intermittent or unstable habitats. Such species

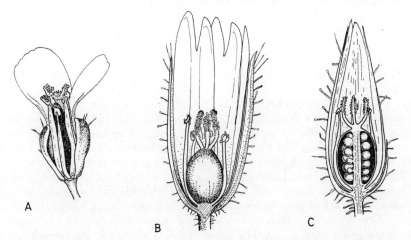

FIG. 107. Self-pollinated flowers. A, Thale Cress (*Arabidopsis thaliana*); a flower with a sepal and two petals removed to show the close proximity of the anthers and stigma. B, Sticky Mouse-ear Chickweed (*Cerastium glomeratum*); side view of flower with half of calyx and corolla removed to show anthers and stigmas. C, half-section of a flower of a cleistogamous form of the same species, lacking petals and with only five stamens. Exeter, Devon.

occur in many families of flowering plants; in the British flora they are particularly numerous among the Crucifers (*Cruciferae*, for example Thale Cress (*Arabidopsis thaliana*) (Fig. 107A)) and in the Pink family (*Caryophyllaceae*, for example 'Sticky Mouse-ear Chickweed' (*Cerastium glomeratum*) (Fig. 107B–C)). Often small-flowered, predominantly autogamous species and large-flowered, predominantly insect-pollinated species are to be found in the same genus, or in closely rated genera. Examples are the small-flowered and large-flowered cranesbills (for example Dove's-foot Cranesbill (*Geranium molle*) (Plate 51a)), and Meadow Cranesbill

(*Geranium pratense*) (Plate 6a)) or forget-me-nots (for example Early Forget-me-not (*Myosotis ramosissima*) and Wood Forget-me-not (*M. sylvatica*)) or the small-flowered cornsalads (*Valerianella*) and the larger-flowered valerians (*Valeriana*). The clovers are an interesting case. The common Red and White Clovers (*Trifolium pratense* and *T. repens*) have showy flower-heads, are self-incompatible, and are pollinated by bees (Plate 24B and c). The small annual species of dry habitats (for example *T. glomeratum*, *T. scabrum*, *T. subterraneum*, *T. suffocatum*) have small inconspicuous flowers and are probably usually selfed (Plate 51c). The recently described species *Trifolium occidentale* (Coombe, 1961) appears as an exception to prove the rule. It is a perennial, closely related to the common White Clover, but is confined to exposed, desiccated habitats in Cornwall, the Channel Islands and Brittany, where it often grows in association with a number of the annual species. The flowers are self-compatible and scentless, and experiments suggest that populations of *T. occidentale* are highly homozygous, and hence probably regularly self-fertilised.

The Crucifers show a continuous gradation from plants like the Wall-flower (*Cheiranthus cheiri*) (Fig. 35) and the Brassicas, with showy flowers producing plentiful nectar and abundantly visited by bees, to small self-pollinated annuals. The common Jack-by-the-hedge (*Alliaria petiolata*) has smaller white flowers (about 6 mm. diameter) in early spring. It is rather freely visited by small hoverflies (especially *Melanostoma* spp.) and other insects, but the flowers are automatically self-pollinated and the plant always sets a full crop of seed. Shepherd's Purse (*Capsella bursa-pastoris*) has flowers smaller again, still with nectar and sometimes visited by small insects, but usually selfed. The effect of this is obvious in the remarkable similarity of leaf-shape among the individuals of a single population, commonly differing noticeably from individuals of other uniform populations elsewhere. This is seen even more strikingly in the Whitlow-grass (*Erophila verna*), a tiny annual of thin dry stony and sandy soils. The small white flowers are produced in the earliest days of spring; they are rarely visited by insects, and almost always selfed. Local populations of *Erophila* are almost always highly inbred, homozygous, true-breeding pure lines (Fig. 108). The nineteenth-century French botanist Jordan set out to distinguish and describe these local populations, and he was able to recognise up to 200 'species' in the genus as a result of his cultivation experiments.

Self-pollination among annuals appears to cut directly across Darwin's principle that 'Nature . . . abhors perpetual self-fertilisation', and it is so common, and so regularly associated with a particular mode of life that a rational explanation is demanded. Probably it is favoured by two main factors. First, annual plants commonly grow in restricted and rather precisely defined habitats, and a breeding system with predominant self-

PLATE 39. Lords and Ladies *(Arum maculatum)*: **a**, clump of plants in flower. **b**, inflorescence with half of spathe removed to show details of lower part of spadix.

PLATE 40. Orchids: **a**, Lady's
Slipper *(Cypripedium calceolus)*;
close-up of lip to show the
large opening by which a
visiting insect enters the flower,
and one of the smaller openings
beneath the anthers by which
it leaves; compare Fig. 75.
b, Marsh Helleborine
(Epipactis palustris). **c**, Common
Helleborine *(Epipactis
helleborine)*.

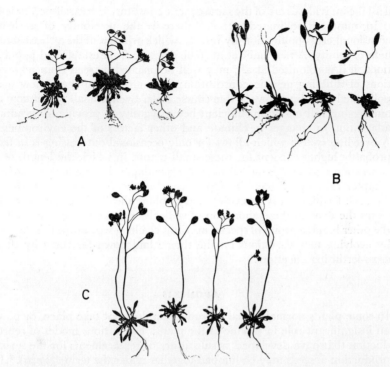

FIG. 108. Silhouettes of specimens from three populations of Common Whitlow-grass (*Erophila verna*, sensu lato). A, Oxford. B, Bishopsteignton, Devon. C, Rattery, Devon. From material in the Herbarium of the Department of Biological Sciences, University of Exeter (EXR).

fertilisation is well suited to evolving and maintaining precise adaptation to habitats of this kind. Secondly, annual plants are liable to great fluctuations in numbers from year to year; clearly, a self-pollinated plant will be able to recover more readily and quickly from a year in which the population is reduced to small numbers, or even to a single individual.* Moore and Harlan Lewis (1965) describe an interesting case of the origin of a small-flowered self-compatible form following catastrophic reduction in numbers in a population of the American *Clarkia xantiana*. Many annuals are extremely plastic in size, and Salisbury's observations have shown that, even in a normal year, the bulk of the seed in a population may be produced by a few of the largest individuals. This does not matter

* As H. G. Baker (1955) has pointed out, self-compatible species are at a great advantage in long-distance dispersal, because a single individual can give rise to a breeding population.

if all the individuals are of the same genetic constitution and self-pollinated.

In most self-pollinated annuals there is the possibility of at least occasional cross-pollination by insects, which will maintain at least some heterozygosity. A new habitat at some distance from established populations, if it is colonised at all, may well be reached by seed from populations of differing genetic constitution; from crossing of these, new self-pollinated pure lines may in due course arise. Even a small percentage of outcrossing will maintain sufficient heterozygosity to provide for gradual adaptation to changes in climate and other factors of the environment. A breeding system which allows for only occasional outcrossing is in fact probably highly efficient for these small plants, in which the length of a generation may be a hundred times shorter than it is for a forest tree (see Chapter 12, p. 382). Stebbins (1957) concluded '. . . In one sense, therefore, self-fertilisation is an evolutionary "blind alley", since it apparently closes the door to the elaboration of radically new adaptive devices. On the other hand, a group of species may travel a long way down this "alley" by evolving new variations on the theme laid down for them by their cross-fertilising ancestors.'

APOMIXIS

In some plants normal sexual reproduction does not take place, or plays an insignificant role in the life of the plant. The various modes of reproduction that have developed as substitutes or replacements for the sexual production of seeds may be lumped together under the term *apomixis*.* In a broad sense, apomixis includes the production of bulbils (for example by garlics (*Allium* spp.) and Lesser Celandine (*Ranunculus ficaria*)) and the so-called vivipary of grasses (for example *Poa alpina*, *Festuca vivipara*). However, the phenomenon that comes first to mind when apomixis is mentioned, and most relevant to our present subject, is *agamospermy* – the non-sexual formation of embryos and seeds. This may come about in various ways; an unreduced (diploid) embryo-sac may be formed or, more rarely, the embryo may be formed from a diploid cell of the nucellus or the inner integument of the ovule (see Chapter 2, p. 42). In either case the meiosis that precedes sexual reproduction is side-stepped, so development of the embryo can take place without fertilisation. Agamospermy may be complete, all of a plant's seed being produced apomictically, or it may be partial, with the same plant producing both unreduced and normal embryo-sacs.

Apomixis has some of the same consequences as constant self-fertilisation. The progeny are genetically identical to their parents, so that apomictic genera are commonly divided into large numbers of 'microspecies', composed of almost identical individuals, and differing rather

* A comprehensive account of apomixis is given by Gustafsson (1946–7).

slightly from related microspecies, which are difficult to classify satis-factorily. In obligate apomictics like the Lady's mantles (*Alchemillav ulgaris* agg.), Dandelions (*Taraxacum*) and the hawkweeds (*Hieracium*), the micro-species are generally clear-cut and stable, though they may be incon-veniently numerous. Where apomixis is only partial, as in the Brambles (*Rubus fruticosus* agg.) and the cinquefoils of the *Potentilla verna* and *P. argentea* groups, normal sexual reproduction may produce new genotypes which are then perpetuated apomictically. As a consequence, certain microspecies may be widespread and readily recognisable, but a complete and final classification virtually unattainable. Even obligate apomicts are not completely uniform genetically. Thus Dr M. E. Bradshaw (1963) was able to show by cultivation experiments that the height of the inflorescence and other characters of *Alchemilla filicaulis* varies with altitude in Co. Durham. Presumably, variation of this kind arises through the occurrence and selection of mutations over a long period of time.

The flowers of apomictic plants are normally still functional for pollin-ation. Often, however, development of the embryo in agamospermous species is autonomous. That is, it takes place without any outside stimulus. In these cases, the failure of the pollination system to degenerate, as commonly happens to useless organs in the course of evolution, evidently reflects the relatively recent origin of these plants and the genetic fixity imposed by apomixis. In other cases (for instance in the *Potentilla* species mentioned above) pollination is necessary before development will begin, though fertilisation of the egg-cell does not take place. This phenomenon is known as *pseudogamy*. The stimulus provided by the pollen tubes is probably often fertilisation of the endosperm nucleus, but pseudogamy is a reminder of the secondary effects of pollination in normal sexual species in initiating the development of parts of the fruit other than the seeds.

Apomixis differs in important respects from self-pollination. It is found most frequently in perennial plants, often growing in closed vegetation at high latitudes or high altitudes. The progeny of apomicts, like the progeny of autogamous species, are genetically identical to their parents, but apomicts may be highly heterozygous. Indeed, there is evidence that many of them may have arisen as hybrids, as in the case of the apomictic whitebeams (*Sorbus* spp.). Agamospermy has evidently arisen repeatedly in different groups of plants, and has allowed the survival of otherwise well-adapted genotypes in which normal sexual reproduction has broken down – perhaps sometimes for lack of suitable pollinators, sometimes for internal, genetic reasons. The exuberance and ubiquity of the brambles leaves little doubt that in partial agamospermy they have developed a very flexible and successful breeding system, perhaps particularly well adapted to certain kinds of instability in the habitat (see p. 382). Possibly they provide a model of the past history of other genera which are more completely apomictic at the present day.

CHAPTER 10

POLLINATION IN SOME EXOTIC PLANTS

THERE are certain types of pollination arrangements that are not found in the British Isles but are so important elsewhere that we have decided to devote a chapter to them.

DIMORPHISM OF STAMENS

One fairly ordinary device is the production in the same flower of two types of stamen, one of which provides food while the other type dusts the visiting insect with pollen. This is a general feature of the family *Melastomaceae* (Fig. 109). In the *Melastomaceae* the flowers are usually purple or pink, and the pollination-anthers are similar in colour while the food-anthers are yellow. The pollination-anthers are carried on jointed filaments on the lower half of the flower, serving as a support for an insect collecting pollen from the conspicuous food-stamens while being rather inaccessible themselves. In many species of the Spiderwort family (*Commelinaceae*) some of the stamens bear modified anthers of conspicuous colour; in some cases these produce no pollen, though it is then possible that the anthers themselves are eaten. Differentiation of stamens also occurs in *Cassia* (family *Leguminosae*); here the pollen is dry and is shed in clouds from pores at the tips of the pollination-anthers when these are vibrated by the insect, or as they spring up when the departing visitor takes its weight off them. In some few species pollinated by large *Xylocopa* bees, pollination takes place via the back or sides of the insect. In *Cassia alata* this is possible because the curved style and pollination-anthers pass round the tail of the insect and up over its back; in other species, however, the style describes an arc at one side of the flower and touches the visitor's side or back (*C. bacillaris, C. multijuga*) (Van der Pijl, 1954).

FOOD OTHER THAN POLLEN AND NECTAR

Species of *Commelinaceae* are not the only plants that offer food other than the usual nectar and pollen. Thus in *Freycinetia* (family *Pandanaceae*) sweet fleshy bracts surround the flowers and are eaten by the pollinators, which in this case are birds or bats. Two species of the family *Araceae* are pollinated by small beetles, which in the case of *Amorphophallus variabilis*

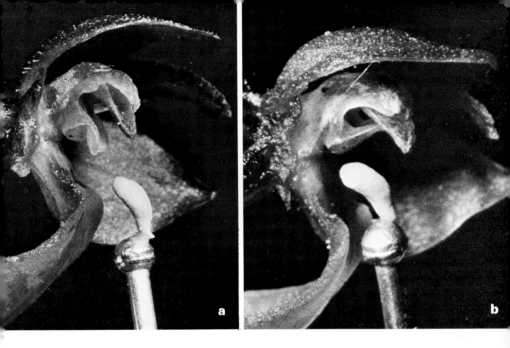

PLATE 41. Orchids: **a**, newly-opened flower of Twayblade *(Listera ovata)*; the pollinia have been removed on the head of a pin. **b**, older flower; column has curved upwards so that pollinia can now come into contact with the stigmas. **c-d**, ichneumon visiting flowers of Twayblade.

PLATE 42. Orchids. Pollination of the Twayblade *(Listera ovata)* by the skipjack beetle *Athous haemorrhoidalis*: **a**, beetle with a freshly acquired pair of pollinia on its head. **b**, beetle has arrived at another flower, and is beginning to suck nectar from the groove on the lip. **c**, pollinia have come into contact with the stigma to which pollen is firmly adhering.

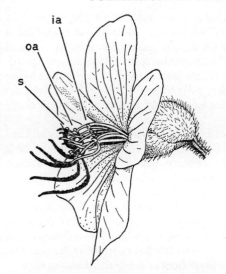

ia
oa
s

FIG. 109. Flower of *Tibouchina* (family *Melastomaceae*). The large purple pollination anthers (oa) and the style (s) are shown in black, the smaller yellow food anthers (ia) in white. Pointers to anthers indicate where they join their filaments.)

remain stationary for days on the lower part of the spathe chewing away special cells filled with starch and oil, and which in the case of *Typhonium trilobatum* eat yellow hook-like projections on the spadix above the female flowers (Van der Pijl, 1937a, 1953). The herbaceous plant *Rhodea japonica* (family *Liliaceae*), pollinated by slugs and snails, has flowers with fleshy perianth segments massed together in a spike similar to the spadix of an aroid. The fawn-coloured inflorescences are hidden among the leaves and smell of bad bread. The gastropods eat away the surface layers of the flowers, apparently without damaging the stigmas, and disperse the pollen as they go (Knuth, P.F.Y.). Imitation pollen, made from cells rich in protein and/or starch, is found around the corolla throat of *Rondeletia strigosa* (family *Rubiaceae*) and on the lip of some Orchids (*Maxillaria rufescens* and other spp., and *Polystachya* spp.) (Porsch, 1909). It is interesting that a bee (*Eulaema cingulata*) that apparently eats food of this kind has been found with pollinia of *Maxillaria rufescens* on its thorax (Dodson, 1962).

ORCHIDS

The main types of construction of the European Orchid flower have been described in Chapter 7. The vast numbers of tropical and subtropical Orchid species conform to these basic types but they show an amazing plasticity of flower form, producing seemingly endless, and sometimes bizarre, variations. From these we can infer that science has barely scratched the surface of the pollination-biology of tropical Orchids even though we are here able to describe some remarkable phenomena. In addition to the tropical species, which provide most of our examples, a whole range of peculiar terrestrial Orchids of dainty construction is found in Australia (Erickson, 1965) and one species from this continent spends its whole life underground (this is *Rhizanthella gardneri*, a saprophyte living on decaying organic matter in the soil); unless it has entirely given up cross-pollination its pollination problems would appear to be particularly acute!

One group of tropical American Orchids was at one time classified among the providers of unusual food, and the information about them available at the beginning of the century was summarised by Porsch (1909). It is now known that these flowers entice their insect pollinators by quite different means (Dodson and Frymire, 1961; Dodson, 1962). Perhaps the most extraordinary of these Orchids is the genus *Coryanthes*, the first account of which was supplied by Crüger (1865) from his observations on *C. macrantha*. The massive waxy-looking flower hangs down and part of the lip forms a bucket into which fall drops of water secreted by a pair of knobs on the column. Male bees of the genus *Eulaema* eagerly visit the flowers, being attracted initially by a strong scent (see p. 249, footnote). This leads them to an area of special tissue at the base of the lip, which they scratch with their fore-legs in order to collect the liquid scent from the surface (Plate 52a and b). This fluid affects special sense organs on the bees' front tarsi, and has an intoxicating effect on them, so that they gradually lose their grip and fall into the bucket. They swim around in the water but cannot climb the sides; however, one side has a narrow tunnel which they can get through, and in so doing they deposit or pick up pollinia (Plate 52c and d). In observations on several species of *Coryanthes*, Dodson (1965) found that the first bee to enter a flower has great difficulty in escaping, usually requiring 15 to 30 minutes to get through the exit passage because the finger-like rostellum (see Chapter 7), projecting down from the roof, slips between the thorax and the abdomen and fixes the insect. After the pollinia are removed by the first bee, the rostellum offers only slight resistance, and the next bees to arrive (species of *Euglossa* and *Eulaema*) can escape much more quickly leaving their pollinia, if any, on the stigma. Allen (1950) observed that all bees visiting *Coryanthes speciosa* fell into the bucket before alighting, and as soon as their wings touched the drops of soapy liquid hanging from the secretory knobs. He also noticed that after the escape of a bee which had taken 45 minutes to get out of the flower the scent was all gone and the bee could not be induced to re-enter the flower. However, next morning the flower again produced scent. Thus there appears to be a special arrangement to prevent immediate re-entry of the same flower and consequent self-fertilisation.

The other Orchids that cause intoxication of male bees are also strongly scented (see p. 249) and bizarre in form. In one of the Swan Orchids (*Cycnoches lehmannii*) the flower is inverted and the bee alights upside down on the lip. In order to reach the scent-producing area, the bee is forced by a projection of the lip to let go with its hind-legs, whereupon the body swings down and, if the flower is a male one, touches the anther-cover; this triggers a discharge mechanism and the pollinia are deposited on the under side of the abdomen near the tip. Female flowers are slightly smaller, and there are hooks on the column which catch the pollinia carried by the bee as its abdomen swings past them. The bee concerned here is *Eulaema*, which is larger than *Euglossa* but has similar tarsal sense organs in the male. The males of both genera are long-lived, lasting for up to six months. In another Swan Orchid (*Cycnoches egertonianum*), pollinated by *Euglossa*, the male and female flowers are very dissimilar, the female functioning in the way described while the male has a slender lip which bends under the weight of the insect and brings it into contact with the anther. In two other Orchid genera (*Gongora* and *Stanhopea*) the male bee falls from the lip as he becomes intoxicated, in such a way that his back touches the column and pollinia are deposited on his thorax (Plate 56c). In the massive waxy-looking flowers of *Stanhopea*, there are usually large prongs on either side of the lip which guide the falling bee past the tip of the column. Some species of *Stanhopea* are pollinated by *Euglossa*, but others, which have a larger gap between the tip of the column and the lip, are pollinated by the larger *Eulaema*. Another genus which intoxicates male *Eulaema* is *Catasetum*, which has inverted flowers with a large hood-like lip which the bees enter to brush the scented area. The pollinia are flung on to the bee by a special discharge mechanism, but it does not seem to have been made clear why the bee has to be drugged for this mechanism to work. The general principle of these in-toxicating Orchids, however, seems to be that after the extremely active bees have been drugged they can be manipulated by the flower in such a way that the pollinia can be accurately applied to or removed from their bodies. These Orchids apparently exploit the sexual behaviour of the bees (though not by inducing pseudocopulation), for Vogel (1966) has found that the bees collect with their fore-legs large quantities of the liquid scent, storing it in a spongy tissue within the enormously swollen hind tibiae. He considers that it must play some important part in mating behaviour, and has evidence that it is discharged at a chosen spot and serves to attract a congregation of males, which are then found by un-fertilised females. Thus it now seems that the males obtain an intoxicating, but also useful, product from the Orchids; possibly intoxication is caused by compounds not essential to the bees but always supplied by the Orchids together with those that *are* essential.

Orchids are noted for the interfertility of their species, and even of their genera. However, the possibility of hybridisation is effectively excluded if a species of Orchid is pollinated only by insect species which are not attracted to other related Orchids. Two species of *Stanhopea*, pollinated by different species of *Eulaema*, are isolated from each other by this means. On the other hand, one of the four species of *Eulaema* that visit *Catasetum macroglossum* is also the sole pollinator of *Cycnoches lehmannii*. Even when a species of bee visits more than one species of Orchid there is still a good chance of avoiding hybridisation, because the pollinia are attached to the bee in a particular position so that they can only be removed by an appropriate mechanism. Isolation does not always exist even among this group of intoxicating Orchids, for *Stanhopea oculata* hybridises freely with *S. wardii* (Dodson, 1963).

It has been noticed that eulaemas will sometimes pause when passing a *Euglossa*-pollinated orchid, as if testing the scent, and then continue on their way; in fact no intoxicating Orchids are known to be pollinated by both *Euglossa* and *Eulaema*. The bees of both these genera have very long tongues and obtain nectar from various long-tubed flowers, including some nectar-producing Orchids (which are only distantly related to the intoxicating Orchids).

A very similar type of attraction seems to be provided for bees by the Australian Orchid genus *Diuris*, though it is not associated with such elaborate mechanical arrangements. Pollination in *Diuris pedunculata* is caused by the males of *Halictus lanuginosus*, which appear to obtain an attractive fluid by boring a special tissue at the base of the flower; they have to force their way to this tissue and receive the pollinia on their faces in doing so. The visits that were observed were usually short, but occasionally bees remained on the flowers for a long time, apparently in a state of intoxication. The bees, when placed in a jar with the flowers, visited them in the normal way, but ignored flowers of *Diuris sulphurea*. This Orchid is pollinated by another larger, but unidentified, bee which is extremely active and stays in the flower only momentarily, apparently not seeking any substance there. The initial attraction to both Orchids is probably olfactory, but the flower of *D. sulphurea*, having some resemblance to its pollinator, may be seen initially by the male as a female (Coleman, 1932, 1933a and b).

Another pollination system discovered in tropical American Orchids has been found in two species of *Oncidium*. These are pollinated by the males of the solitary bee *Centris* (Dodson and Frymire, 1961; Dodson, 1962). These bees usually each have a favourite perch near an *Oncidium* plant, and from time to time take off and hover near the Orchid. The flowers are borne in long racemes, and when they are moved by the wind the bee darts in and buffets one of the flowers. The *Centris* bees appear to hold territories by chasing off all insects which fly nearby, and it seems possible that the *Oncidium* flowers, when moved by the wind, appear to the bees to be flying insects. Alternatively, the hovering bee may see the flower as a female of his own species, and his behaviour may be part of his mating process. (In Britain it is indeed possible to see leafcutter-bees (*Megachile*) hovering near females and making buffeting attacks on them.) As a result of repeated buffeting flights all the flowers of an inflorescence may be pollinated in a short space of time, but more often than not the bees appear to ignore the inflorescences altogether. However, the flowers last about three weeks, which gives them a fair chance of being visited, though investigators have little chance of seeing pollination taking place. The plants are apparently self-incompatible, and often only one flower of a pollinated plant produces a fruit.

The deception of flies by the production of smells specially attractive to them is common among tropical Orchids. One which attracts flesh-flies by its colour and its smell of bad meat is *Masdevallia fractiflexa*, from Tropical America. Without offering any tangible reward, this lures the insects into a position where they can effect pollination (Dodson, 1962). In some Orchids there is a special mechanism for throwing the flies into contact with the column. For example, in *Bulbophyllum macranthum* the two lateral sepals are directed upwards and meet near their tips. Here the flies alight and spend much time licking the surface, holding on to the outside edges of the otherwise slippery sepals. Gradually they crawl towards the centre of the flower, where they find that the sepals are parted so that they can no longer straddle them. As they slip on the surface of the sepals they clutch at the solid-looking, tongue-like lip, which affords a good grip. They then transfer their weight to the lip in a head-upwards position, whereupon they are suddenly flung backwards and downwards, for the lip is pivoted, delicately balanced, and tips with the weight of the fly. Two springy arms near the tip of the column embrace the fly, while the lip returns to its former position. The fly soon escapes but in its struggles removes pollinia on its abdomen (Ridley, 1890). The sepals secrete no nectar but appear to be covered with some other exudation, and the pollinators are always present when this Orchid is in flower. Ridley describes the pollination of a second species of *Bulbophyllum*, in which small flies are thrown head-first against the column and receive the pollinia on the head. Many species of the closely related genera *Bulbophyllum*, *Cirrhopetalum* and *Megaclinium* have pivoted lips of the type described. Whereas some species, especially those of *Cirrhopetalum*, attract flesh-flies by producing a bad smell and showing a greater or less resemblance to decaying flesh, others seem to be adapted to different tastes, for the flowers of *B. macranthum* smell of cloves and attract only a single species of fly. *Bulbophyllum* and its allies are Old World genera, but the same mechanism has evolved independently in species of *Anguloa* and *Masdevallia* in America (Dodson, 1962).

THE MILKWEED FAMILY AND ITS FLY TRAPS

A number of interesting pollination mechanisms are found in the milk-weed family (*Asclepiadaceae*) (Plate IVa). This is an important family of temperate and tropical regions, which has just failed to reach the British Isles by natural means but is found as near as northern France. The radially symmetrical flowers are generally different in appearance from Orchids, but the stamens and massive style function in an exactly similar manner, being fused together to form a column in the middle of the flower, while the pollen is usually produced in the form of pollinia, connected by bands (*retinacula*) to horny clips (*corpuscula*) which attach themselves to insect visitors. The clip and pair of bands are called a translator; the details of the construction are shown in Fig. 110. There are five translators in each flower, and they alternate with the five anthers, each of which contains two pollinia; each translator is joined to two pollinia, one from each of the two adjacent anthers. In *Asclepias syriaca* and *A. curassavica* there is a large nectarial pouch (*cucullus*) external to each anther. The visitors (bees, wasps and flies in the first case, butterflies in the second) come to the flowers for nectar, and as they fly away their legs, guided by slits on the column, tend to catch in the clips. Pollinia are thus removed from the flower attached by the clips to the claws of insects. These pollinia are then brought together and twisted through a right angle (by spontaneous movements of the connecting bands) into a position which enables them to enter a slit in the next flower visited by the insect; the pollinia wedge in the slit and break off, to be deposited on the stigmatic area. In *Vincetoxicum officinale*, the commonest European member of the family, which has small greenish white flowers, there is a nectarial pouch adjacent to each slit, and it is the proboscis of the insect that gets caught, the very small clips and pollinia being carried away on the proboscis hairs. The flowers have a disagreeable smell and are pollinated by flesh-flies, but smaller flies are sometimes trapped in the flowers because they are not strong enough to dislodge the pollinia (Müller, Knuth). Some other *Asclepiadaceae* with the pollinia connected to clips are adapted to pollination by *Lepidoptera*. In *Periploca graeca* there is a different, and probably more primitive, pollination mechanism. Between each pair of anthers is a sticky band on to which the pollen is shed. The sticky band then adheres to the proboscis of a visiting fly and is carried with the pollen to another flower.

The rather strong association with dipteran pollinators found in the Milkweed family in Europe is even more marked in other regions. In *Stapelia* and allied genera, which are cactus-like plants of Africa and southern Asia, the flowers look and smell like bad meat, characteristically being flesh-coloured or dark purplish red, often covered with hairs, and

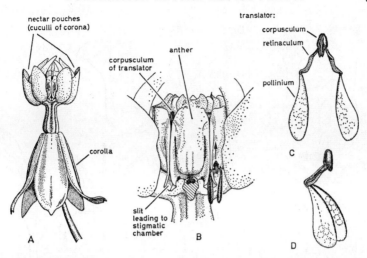

FIG. 110. *Asclepias curassavica*. A, single flower, showing the corolla, column and corona. B, enlarged view of the column, with one of the cuculli of the corona cut away at the base to show an anther and the neighbouring structures. The cut base of the cucullus is hatched; the channel at the top carries nectar to the pouch of the cucullus from the nectaries in the stigmatic cavities (see Galil & Zeroni, 1965). The pollinia to the left of the anther are still *in situ*. Those on the right have been removed, and a pollinium is shown being drawn into the stigmatic slit. C, translator and pair of pollinia, newly removed from the flower. D, side view of the same translator and pollinia a few minutes later.

often of large size – up to sixteen inches across (Plate IVa). The largest genera of this type, *Stapelia*, *Caralluma* and *Huernia*, comprise about 250 species, among which there is extensive variation in colour, patterning, type and distribution of hairs, and surface sculpture of the corolla (White and Sloane, 1937). Much of this variation suggests adaptation to pollinators with specific requirements, and there would seem to be a possibility of interesting research into the instinctive requirements and sensory discrimination of the flies that pollinate these strange flowers. When cultivated in Britain the flowers attract *Muscidae* and *Calliphoridae*, which lay their eggs in them. These are easily seen, usually near the centre of the flower where presumably the pollinia clip on to the insects.

In the genus *Ceropegia*, which occurs in Africa, Asia and Australia, the lengthened corolla tube forms a trap in which flies are imprisoned for a time. These plants are herbs, shrubs or climbers. Their flowers are often

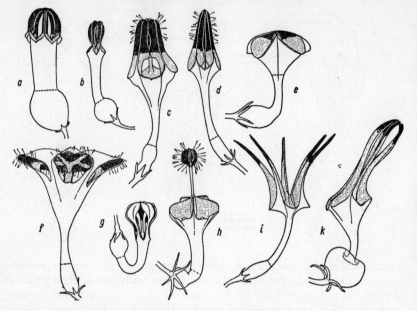

FIG. 111. Flowers of *Ceropegia*. The scent-producing areas are shown black, the slide-zones (as far as visible) stippled; shimmering hairs shown when present. Broken lines show lower limits of slide-zones. A, *C. ampliata*. B, *C. woodii*. C, *C. sandersonii* × *C. nilotica* (hybrid). D, *C. radicans*. E, *C. elegans*. F, *C. sandersonii*. G, *C. euracme*. H, *C. haygarthii*. I, *C. stapeliiformis*. K, *C. robynsiana*. Not to scale, some slightly larger than life, some slightly smaller. From Vogel (1961).

fascinatingly beautiful, the corolla most often being coloured in delicate shades of green, grey and brown, with elegantly shaped tubes and erect lobes which unite at their tips to give a lantern-like effect (Fig. 111). This lantern effect is found in some unrelated fly-trap flowers and appears to be a means of inducing insects to enter (Vogel, 1954). A study of the tiny flowers of *Ceropegia woodii* (L. Müller, 1926), published in the same year as Knoll's remarkable investigation of *Arum* (see Chapter 6), shows some striking similarities between these two unrelated plants. The method of catching insects is identical in that the interior of the flower tube is covered with minute downwardly directed conical papillae coated with a layer of oil (Fig. 112), and has downwardly pointing hairs which allow the insects to fall but hinder their efforts to climb up. A day or two after the flower has opened, the tube changes to a horizontal position and the hairs inside shrivel so that the insects can escape. In cultivation

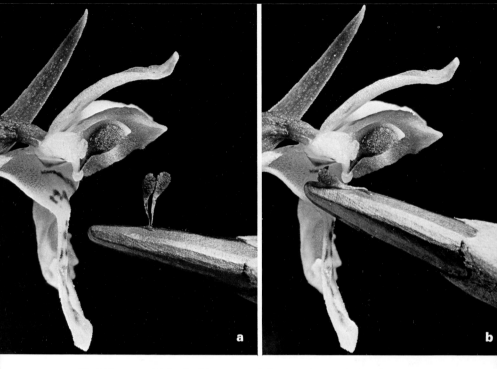

PLATE 43. Orchids: **a**, pollinia freshly removed from a flower of the Common Spotted Orchid *(Dactylorhiza fuchsii)* on the point of a pencil. **b**, pollinia about half-a-minute after removal, now in a position to strike the stigma. **c**, yellow dung-fly *(Scopeuma stercorarium)*, bearing a pair of pollinia, on Heath Spotted Orchid *(Dactylorhiza maculata)*. See also Pl. 36**a**.

PLATE 44. Orchids: **a**, five-spot burnet moth *(Zygaena* cf. *trifolii)* visiting flower of Pyramidal Orchid *(Anacamptis pyramidalis)*; the proboscis of the moth bears several pairs of pollinia. **b**, five-spot burnet on Pyramidal Orchid; the collar-like viscidium encircling the proboscis of the moth is clearly visible; in this instance the moth is sucking nectar in an inverted position, from the top of the inflorescence. **c**, Glanville fritillary butterfly *(Melitaea cinxia)*, with a pair of pollinia on its proboscis, on inflorescence of Pyramidal Orchid.

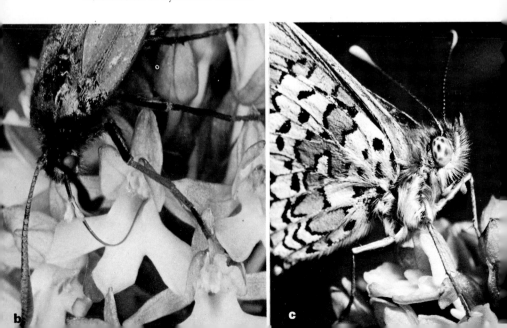

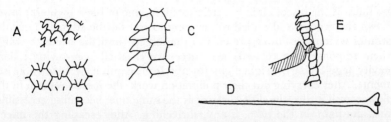

FIG. 112. A-C, papillate cells inside corolla tube of *Ceropegia woodii*. (A, hooked papillae found in entrance to tube, surface view; B, blunt papillae from within upper part of tube, surface view; C, larger, sharper papillae from lower part of tube, in section.) D-E, trap hair and papillate cells of *C. stapeliiformis*. (D, trap hair, seen from above; E, section of papillate cells lining perianth tube and base of a trap hair; one of the cells of the lining is so modified as to form the complete hair). A-C, after L. Müller (1926); D, E, after Vogel (1961).

in Vienna, this species trapped biting-midges (*Ceratopogon*), which were apparently attracted by a faint scent produced by the flowers. Several species of *Ceropegia* were studied by Vogel (1961). In each he located the scent-producing area and mapped the extent of the slide-zone (Fig. 111). He found that in cultivation in Germany five out of eight species attracted none of the available insects, while each of the others trapped female flies. *C. woodii* trapped biting-midges of the genus *Forcipomyia*, *C. stapeliiformis* trapped mainly *Madiza glabra* of the family *Milichiidae*, and a hybrid of *C. nilotica* trapped representatives of two other genera of this family. Thus the scents produced by the flowers are specific attractants to certain insects, and are presumably connected with egg-laying. The flies approach with a typical scent-orientated flight and always alight on the scent-producing area; then they investigate the slide-zone, apparently being attracted by the dark interior of the flower, and soon slip into the tube, which is darkened by red colouring on the inner surface in some species. They then pass into the prison which is often wholly or partly translucent, so that the light from it attracts the insects which have slipped into the tube. The dark part of a prison is lined with dark red which frequently does not show on the outside, and when a prison is partly translucent the bright area, or 'window pane', forms a ring round the sexual organs. The now frightened insects, on coming to this end of the prison, climb on to the pillar-like inner corona. Here they drink from the nectarial cups formed by the outer corona, there being one opposite each groove on the column. After drinking, the fly withdraws its head, and the throat membrane (in *Milichiidae*) or the base of the labellum (in *Forcipomyia*), catches in the groove, and receives the clip carrying the pollinia. If the insect already carries pollinia, one of these is caught lower down the groove and pulled off, coming to rest on the stigma. Thus the stimuli which the flower presents to the insects (and the needs to which their responses are related) are successively: smell (egg-laying); dark cavity (egg-laying); bright light (escape from captivity); taste (nourishment). After carrying out their pollination work, the insects remain in the flower until it tilts, the slide-surfaces at the same time becoming accessible, and the hairs in the tube, if any, shrivelling. After escaping the insects once again enter young flowers; the duration of imprisonment varies from less than a day to four days, according to the species of *Ceropegia*. In some species the hairs inside the tube are specially constructed with a narrow stalk and a wide asymmetric swelling just above (Fig. 112). These will thus bend downwards but not upwards or sideways. Each consists of a single cell, although it may be up to 5 mm. long.

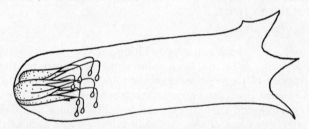

FIG. 113. *Tavaresia grandiflora*. Side view of the dangling knobs formed by the lobes of the column, with the outline of the corolla added. Based on Jaeger (1957) and on White and Sloane (1937).

THE FLY TRAPS OF BIRTHWORTS AND THE ARUM FAMILY

In the large genus *Aristolochia* (Birthwort family, *Aristolochiaceae*) all the species trap insects in their specially modified tubular perianths. In the European *A. clematitis* the greenish-yellow perianth is rather like the corolla of *Ceropegia woodii* (though having only one lobe, which is tongue-shaped), and the arrangements for trapping and releasing pollinators, and attracting them to the sexual organs, are exactly the same. Biting-midges and other small flies alight on the tongue, being attracted by the smell of the flower (Knoll, 1956; P.F.Y.). More complex structures are found in two South American species of this family, *Aristolochia lindneri* and *A. grandiflora* (Figs. 114, 115). In both species there are conspicuous perianth lobes with a long tail-like appendage, a dark antechamber or trap, a brighter prison with which the trap connects by a funnel-shaped passage, and a 'window-pane' encircling the reproductive organs (Plate

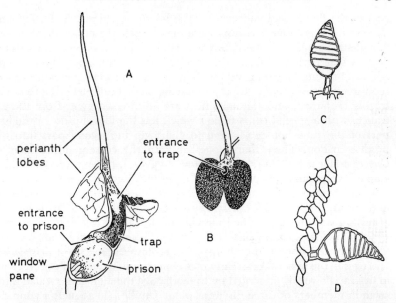

FIG. 114. Flower of *Aristolochia lindneri*. A, flower in first stage of anthesis with half the perianth cut away. The dark parts are purple-coloured; the prison is bright but an aggregation of dots sets off the 'window pane'. B, flower seen from above; tail-like lobe of perianth is foreshortened because it points up towards viewer. C, a multicellular trap hair seen from above. D, the same from the side, with adjoining papillate cells of trap wall. After Cammerloher (1933) and Lindner (1928).

54b). They also have minute downwardly pointing papillae, covered (at least in *A. lindneri*) with grease, as in *Arum*, together with larger hairs. On the day the flowers open, a bad smell is produced by the perianth lobes, and flies are attracted to them and trapped in the prison where nectar is secreted. On the second day no smell is produced, and the stigmas bend together so that they cannot receive pollen. The anthers then dehisce and the prisoners, newly dusted with pollen, are allowed to escape by the widening of the entrance and the shrivelling of the trap hairs. *A. lindneri* was studied by Lindner (1928) in Bolivia, and by Cammerloher (1933) in Vienna, in cultivation. On the afternoon of the first day, the dull purple perianth lobes of this species close over the entrance to the trap but leave spaces large enough for the insects escaping on the second day to emerge. The purple colour of the trap, which is confined to its inner surface, disappears on the second day, brightening this part of the flower and encouraging the insects to emerge. In Bolivia and Vienna the commonest visitors were members of the family *Sepsidae* (see Chapter 4 part 1). *A. grandiflora* was studied in Java by Cammerloher (1923). The perianth lobe of this species is very large, being about 12 cm. wide and 20 cm. long. It is yellowish in colour with a chequered pattern of reddish veins (see also Plate 54a), while the smell recalls decaying fish. The trap is U-shaped, and the hairs in it are all directed away from the entrance, with a special construction which lets the flies slip down the first part of the tube but helps them to climb up the second part into the brighter prison. These hairs, together with the corresponding hairs of *Ceropegia* (Figs. 112, 114, 115, pp. 303, 305, 307), present a most striking case of evolutionary convergence of differently constructed parts (multi-cellular as opposed to unicellular hairs) which have become identical in function and in external form. The fly most commonly caught by *Aristolochia grandiflora* was an unidentified member of the family *Muscidae*, measuring 4–5 mm. in length (see Chapter 4 part 1). Cammerloher found that flies trapped by this flower, and another, larger-flowered species of *Aristolochia*, laid eggs in the prison. As already mentioned, another case of convergence in fly-trap flowers is presented by the surface of the slide-zone, which is the same in members of three different plant families. That such a zone existed in *Aristolochia saccata* was already suggested as long ago as 1839 by Robert Graham, who inferred the existence of 'a particular condition of the surface in the upper part of the tube, from secretion or other cause, which prevents adhesion of the feet of the insects, though they are able to walk along it when horizontal'.

Traps for flies and beetles are also found in many members of the family *Araceae*, and the only British genus, *Arum*, has already been described in detail in Chapter 6. It was in other genera of this family that the function of 'window panes' was first discovered. In addition, in some tropical *Araceae*, insects are induced to enter the trap by the bright appearance of the interior of the spathe, viewed from the entrance. The tissue of the spathe seems to reflect and refract the light to produce this concentration of illumination, which is enhanced by dark surrounding colours. This effect has been described in *Arisaema laminatum* by Van der Pijl (1953), where the spathe has a dark green hood and the entrance has a purple border (Fig. 117B, Plate IV); the same feature has been shown in photographs of two other species of *Arisaema* published by Barnes (1934). In *Cryptocoryne griffithii* both 'window panes' in the prison and a bright entry are found (Fig. 116, p. 308). In another member of this family, *Amorphophallus titanum*, insects are trapped by being prevented from climbing to the top of the spadix by an overhanging ridge; this has such a sharp edge that the large beetles which pollinate the flowers fall off when they try to negotiate it. The species *Typhonium trilobatum*, on the other hand, is pollinated by minute beetles not more than half a millimetre

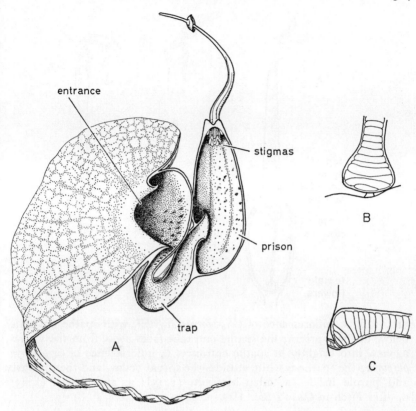

FIG. 115. *Aristolochia grandiflora.* A, flower in first stage of anthesis with half the perianth cut away. The trap is blackish purple inside and the prison lighter purple with darker freckles. B, part of a trap hair from above; swollen base limits lateral movement. C, the same, from the side; swollen base prevents hair from bending upwards but not downwards; the hairs in the right hand part of the trap are upside-down and so allow insects to climb into the prison. After Cammerloher (1923).

long. They enter the spathe in the early morning and, after they have reached the female flowers at the base, the spathe constricts just above them, making them captive. On the second day pollen is shed and collects above the constriction; after this the constriction opens slightly and the insects crawl out through the mass of pollen.

A one-way traffic system is found in this family in the Taro (*Colocasia*

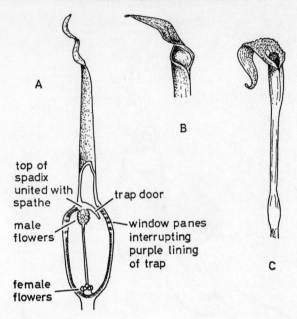

FIG. 116. A, inflorescence of *Cryptocoryne griffithii*, with spathe cut away below to show prison; the spathe entrance faces away from the viewer. B, view into brightly lit spathe entrance. C, inflorescence of *Cryptocoryne purpurea;* the spathe is white outside, the throat yellow and the tip warty and purple inside. A, after McCann (1943); B, after Vogel (1963); C, after Fitch in *Curtis's Bot. Mag.*, t. 7719.

antiquorum); flies, attracted by an unpleasant smell, enter at the base of the spathe, and a constriction prevents them from going beyond the female flowers. Later the same day the smell fades and the entrance closes up, imprisoning the insects. At night the flies are admitted to the upper part of the spathe where the pollen is shed, and on the second day this part opens, releasing the insects, now dusted with pollen (Cleghorn, 1913). Another one-way route for pollinators has been described in some Indian *Arisaema* species by Barnes (1934). In these, the spadix, unlike the spathe, is not slippery, and the insects climb down it, passing over the stamens and stigmas and out of a hole in the spathe at the bottom. An example is *A. tortuosum*, which produces both male and hermaphrodite inflorescences. *A. leschenaultii* and some other species are dioecious, and only the males have the basal opening to the spathe. In the female inflorescences the flowers are tightly packed and have stigmas that project out to the wall of the spathe. The insects – mainly tiny fungus-gnats (family *Mycetophilidae*)

– push down between the stigmas until they are jammed and die. These species of *Arisaema* have a great preponderance of male plants, which evidently reduces the danger of the female inflorescences being blocked up by flies which are not carrying pollen. Larger flies are excluded by a ring of filaments formed by sterile flowers, similar to those of *Arum*.

The large orchid genus *Paphiopedilum* is the tropical counterpart of the bee-pollinated Slipper Orchid (*Cypripedium*), described in Chapter 7, and has the same one-way pollination system. The upper sepal is often striped like the hood of the spathe of some *Arisaema* species (Plate IVd, Fig. 117), and there may be long twisted, warty petals, recalling the spathe tip of *Cryptocoryne* (Fig. 116). Further, the flower colours are prevailingly green, brown, dull red, purple and white, in various combinations, and there may be a bad smell. All these features suggest that *Paphiopedilum* species (like the trap flowers) are adapted to lure flies or beetles, and large flies were in fact long ago seen visiting the flowers of plants of this genus cultivated in Italy (Delpino, 1871).

Some Orchids trap insects by a movement of the lip, induced physiologically when the pollinator touches a sensitive area. This effect is called

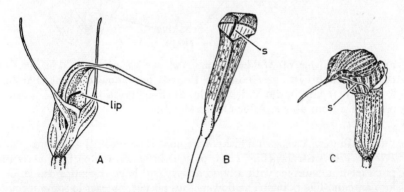

FIG. 117. A, flower of the orchid, *Pterostylis falcata*. Colour mainly greenish; the two tailed lateral sepals form a panel at the front of the flower; upper sepal interlocks with petals to form rear panel and hood. B, C, species of *Arisaema* to which *Pterostylis* shows resemblances (B, *A. laminatum*, a species with illuminated entrance; C, *A. wallichianum*, spathe pale green with purple stripes, spots and mottling; markings darkest within the hood; s – spadix appendix). A, after Coleman (1934); B, after Van der Pijl (1953) and Blume's *Rumphia*; C, after Engler, *Das Pflanzenreich*.

irritability, and an example is provided by *Masdevallia muscosa* (Oliver, 1888). This New World Orchid (Fig. 118) has been studied only in cultivation, but the likely pollinators are *Diptera*. The ridge on the distal, triangular part of the lip is the sensitive area, and touching it causes this part to rise up so that an insect settled on it will be carried into the funnel formed by the united bases of the sepals. The only escape passage for an insect is now between the lip and the column, where the pollinia and stigma are situated, and the lip remains in the trap position for about

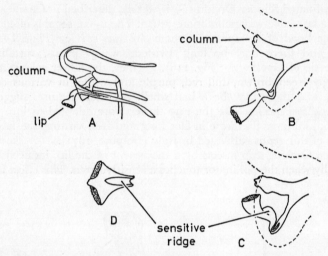

FIG. 118. Flower of *Masdevallia muscosa*. A, side view. B, side view of column and lip in open position. Dotted lines show edge of chamber formed by sepals. C, the same with lip in closed position. D, view of distal part of lip from above. After Oliver (1888).

thirty minutes. A very similar arrangement is found in the Orchid genus *Pterostylis* in Australia (Fig. 117A, p. 309) in which some of the perianth parts form a chamber with a hood above, and have their tips drawn out into antenna-like points. The flowers thus possess, at least in some species, an extraordinary resemblance to the fly-trap inflorescences of *Arisaema* (Fig. 117C, p. 309). This resemblance is heightened by dull green or reddish colouring with darker vertical striping, frequently by the bright-looking interior of the flowers (shown in coloured plates in Curtis's Botanical Magazine and in Nicholls, 1955 and 1958) and by the tip of the narrow upright lip showing itself at the entrance to the flower like a spadix. The sensitive area is at the base of the lip and often has the form of a filamentous appendage. The insect visitors are mosquitoes and other

PLATE 45. Orchids: **a**, Lesser Butterfly Orchid *(Platanthera bifolia)*. **b**, Greater Butterfly Orchid *(Platanthera chlorantha)*. **c**, Scented Orchid *(Gymnadenia conopsea)*. **d**, Frog Orchid *(Coeloglossum viride)*.

PLATE 46. Wind-pollinated trees: **a**, Alder *(Alnus glutinosa)*. **b**, Pedunculate Oak *(Quercus robur)*; **f**-female flowers. **c**, English Elm *(Ulmus procera)*. **d**, Ash *(Fraxinus excelsior)*.

gnats or midges, and if they are carrying pollen when they spring the lip they pollinate the stigma when thrown against the column with their backs towards it. The column has two wings near the tip which project towards the lip, and the only way the insect can escape is by pushing between these wings with its back towards the column, so that it picks up the pollinia on its thorax. There is evidence that in the course of repeated visits to the flowers the insects become intoxicated. A different species of insect appears to be responsible for the pollination of each species of *Pterostylis* (Coleman, 1934; Sargent, 1909, 1934).

FLORAL MOTION AS AN ATTRACTION

The attractiveness to flies of some of the Milkweed family is apparently enhanced by their possession of vibratile organs (Vogel, 1954). An example is supplied by the genus *Tavaresia* (allied to *Stapelia*), in which the corolla is tubular or bell-shaped and the base of the column is produced into ten long filaments, limp towards their tips and each terminated by a dark red knob (Fig. 113, p. 304). These knobs hang down and constantly vibrate, apparently in response to air movements; and the corolla is translucent so that the vibrations can be seen from all round the flower. Further, some *Ceropegia* species have large purple unicellular hairs, 3 mm. long and with a constricted flexible base. These hairs are found on the borders of the corolla lobes, and hang down in still air; at the slightest breeze, however, they take on a rapid oscillation, producing a strange effect to the eye. Vibratile hairs, of similar size but multicellular in construction, are found in clusters on the tips of the petals of the tropical Orchid, *Cirrhopetalum ornatissimum*. In another tropical Orchid, *Bulbophyllum medusae*, each perianth segment is drawn out into a thread several inches long, those from each cluster of flowers hanging down and forming a waving plume. The minute, pivoted lips of some species of *Megaclinium*, also a tropical Orchid, likewise oscillate in the wind. (All three of these Orchid genera have been mentioned previously on p. 299.)

OSMOPHORES

It will have been noticed that tail-like structures are widespread in fly-pollinated flowers, for example in *Aristolochia*, in the Orchids, and in the non-British members of the *Arum* family. In the latter the 'tails' may be formed by the tip of the spathe (Fig. 117B, Plate IVb), or by the appendix of the spadix, as in some *Arisaema* species. In considering the probable significance of 'tails' as alighting places for insects, Van der Pijl (1953) drew attention to the propensity of flies for alighting on suspended objects such as fly-papers and electric lights. There is, however, marked variation in the type of 'tail', some being erect, antenna-like rods, and others very long threads or ribbons which trail on the ground. An investigation by Vogel (1963) showed that most tail-like structures in fly-trapping or fly-deceiving flowers are scent-producing organs. He found that the scents are often highly specific in the attraction they exert on insects, and are responsible for the initial attraction of insects to the flowers – hence the prominent position of the scent-sources. Those scent-organs which produce a very powerful scent during a short period contain big food reserves which are dissipated during the production of scent, and this is frequently accompanied by a pronounced rise in temperature caused by the rapid respiration in the organ. In addition to the classic example of *Arum maculatum* (see Chapter 6), this kind of heat-production is now known in several other species of *Araceae*, as well as in some species of *Ceropegia* and *Aristolochia*. There are, however, some species in each of these groups which produce their scent over a period of many days, without any abnormally rapid metabolism.

THE BLOSSOM AS A BREEDING SITE

We have already seen that eggs are laid by insects in some flowers that imitate the normal breeding sites of the insects. In the aroid *Alocasia pubera*, flies are known to complete their development in the inflorescence and pupate in the spathe (Van der Pijl, 1953). Another plant that provides a breeding site for its pollinators is the source of cocoa and chocolate, *Theobroma cacao* (family *Sterculiaceae*); the pollinators are biting-midges and they breed in the decaying pods (Dessart, 1961). The breeding of pollinating flies takes place in the male inflorescences of *Artocarpus heterophylla*, a tropical tree of the family *Moraceae* (Van der Pijl, 1953). The flowers are minute and are produced in large numbers on massive receptacles. The inflorescences have a smell of overripe fruit and are pollinated by small bees and by *Diptera* of two genera. After flowering the male flower-heads drop, and it is at this stage that the pollinating flies breed in them. In this way the tree secures a population of flies constantly near to it. In the genus *Ficus* (fig-trees), which also belongs to the *Moraceae* family, a complete symbiotic interdependence between plant and insect is found; the exclusive pollinators are certain species of chalcid-wasp which develop in some of the flowers and never breed anywhere else.

In fig-trees there are numerous tiny unisexual flowers arranged on fleshy receptacles, which are hollow and bear the flowers on the inner surface; the opening to the outside is very small and is usually closed by flexible scales. The best known member of this very large genus is the edible fig (*Ficus carica*), the so-called fruit of which is an entire inflorescence (the receptacle and its contents). In the wild form of the edible fig, three types of receptacle are formed at different times of the year. The first type is formed in winter and contains many neuter flowers, and a smaller

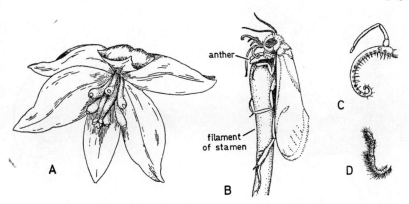

FIG. 119. Yucca and Yucca-moth. A, flower of *Yucca aloifolia*. B, *Pronuba yuccasella*, female, gathering pollen from a stamen. C, maxillary palp and tentacle (coiled) of the moth. D, labial palp. After Riley (1892).

number of male flowers which are confined to the region of the entrance. This type of receptacle is invaded by tiny female chalcid-wasps of the species *Blastophaga psenes*, which lay eggs in the neuter flowers and then die. The offspring of the wasps complete their development in the ovaries of the flowers (one wasp to each flower). The male wasps hatch first and emerge from the ovary into the interior of the receptacle, where they bore into the ovaries occupied by females, fertilise the females, and then die. The females now emerge and leave the receptacle, receiving pollen at the entrance from the male flowers which have only just opened. It is now June, and the wasps find their way to the second type of receptacle. This type contains either a mixture of neuter and female flowers, or female flowers only. The wasps lay their eggs in the flowers, but only those laid in neuter flowers develop. The ovary of the neuter flower (a modified form of the female flower) is incapable of producing seed, and the style is short with an open canal leading to the ovary. In the female flower, on the other hand, the solid style is too long to permit a wasp to reach the ovary with its ovipositor, and eggs not placed in the ovary fail to develop. The pollen which the wasps bring with them fertilises the female flowers, and so seed is set in these. The development of the wasps takes place as before, fertilised females emerging in autumn and going to the third type of receptacle which is smaller than the others. Here there are only neuter flowers, in which the insects develop that will emerge in winter and restart the cycle in the first type of receptacle (McLean and Cook, 1956; Grandi, 1961).

The male fig-wasps are highly modified, having reduced legs and eyes

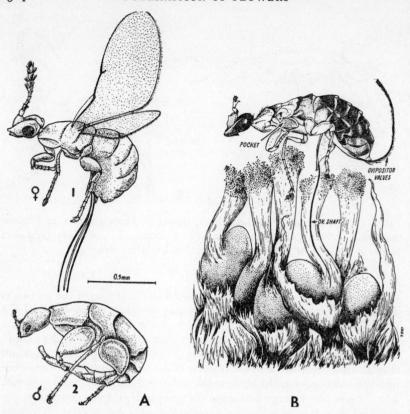

FIG. 120. Fig-wasps. A, female and male of *Blastophaga quadraticeps*. B, female of *Ceratosolen arabicus* in the act of oviposition; the fore-legs are raised in the process of extracting pollen from the thoracic pouches. From Galil and Eisikovitch (1968b and 1969).

and no wings (Fig. 120). The females are normal, though they may lose their wings and parts of their antennae in their struggles to get past the scales at the entrance to the receptacle (Grandi, 1961). This resistance offered by the scales is apparently an adaptation on the part of the plant to prevent the entry of insects lacking the instinctive persistence of the female fig-wasp. When it lays an egg, the female fig-wasp injects a drop of a special secretion into the ovary of the neuter flower, and this stimulates the development of the unfertilised ovule into a gall, which later provides the nourishment for the wasp larva. Thus the plant provides special flowers in which the pollinators breed, the winter inflorescences being devoted solely to this use. Some cultivated forms of the edible fig however, are entirely female, but in these no pollination is necessary for the fruits to develop (McLean and Cook, 1956). In other species of *Ficus*, which are mainly tropical, neuter flowers are provided for the pollinating wasps to breed in, but the life-cycles differ from that of the edible fig, there being two main types (Wiebes, 1963). In one type there are two kinds of receptacle, that containing female flowers only and that containing both neuter and male flowers. Fertilised female wasps emerging from the latter enter the next generation of either the same kind of receptacle or else the female kind. In the first case the wasps lay their eggs, and in the second they pollinate the flowers. The two kinds of receptacle are normally on different plants (dioecism). In the other type of life-cycle all three kinds of flower occur in the same receptacle (monoecism). Fertilised female wasps pollinate the female flowers and breed in the neuter flowers, while their departing female offspring are later dusted with pollen from the male flowers. Usually the neuter flowers have short styles and the female longer styles, which makes oviposition in the latter difficult or impossible. In *Ficus sycomorus* (Fig. 120) and *F. religiosa*, however, it has been found that long-styled and short-styled flowers are not physiologically different, for small percentages of the short-styled produce seeds, and of the long-styled produce galls (Galil and Eisikovitch, 1968a and b).

Another significant observation on *F. religiosa* is that female wasps newly emerged from the galls are attracted to the anthers, push their heads amongst them and even eat some of the pollen. The dependence of the wasp (*Blastophaga quadraticeps*) on the plant is emphasised by experiments on this species in which female wasps which had not been in contact with pollen were allowed to enter the inflorescences. These inflorescences dropped off without forming galls or seeds (Galil and Eisikovitch, 1968b). A further stage of specialisation in pollen transfer is seen in the special pouches on the underside of the thorax of the fig-wasp (*Ceratosolen arabicus*) (Fig. 120), which pollinates *Ficus sycomorus* in East Africa. Galil and Eisikovitch (1969), in studying this pollinator, were puzzled at its effectiveness, in view of the small amount of pollen carried on the body – it always cleans itself carefully on emerging from the inflorescence. The accidental squashing of an insect led to the discovery of the pouches, each of which can probably carry 2000–3000 pollen grains. The insects could not be seen loading these pouches but ovipositing females continued their activities despite the opening of the fig for observation, and it was found that after each egg had been laid the fore-legs were used to scratch some pollen out of the pouches and brush it on to the stigmas (Fig. 120). In this way a mixture of gall-flowers and seed-flowers develops in the area where the female has been working. After this discovery it was found that *Blastophaga quadraticeps* and some other fig-wasps had pollen pouches. One of these, *B. tonduzi*, is a New World species and was studied more fully by Ramirez (1969). The females of this wasp also search for and manipulate the anthers of the fig (*Ficus hemsleyana*, in this case), and were found to carry pollen in pouches on the thorax. However, these pouches are different in form and position from those of *Ceratosolen arabicus*, and there are two pairs on each insect. Ramirez found pouches in many other New World species of *Blastophaga* and in other genera of fig-wasps from various parts of the world; he even found cavities in the head of *B. psenes* which he thought might be used for pollen transport.

The fig-wasps (family *Agaonidae*) of south-east Asia and of Hong Kong have been correlated with their host-plants, using the results of Professor E. J. H. Corner's investigations into the taxonomy of *Ficus* (Wiebes, 1963; Hill, 1967). It was found that, with few exceptions, each species of fig-wasp confines itself to one species of fig-tree. The exceptions are of the kind that one might expect; for example, one species of wasp pollinating two very closely related species of *Ficus*, or two closely related wasps pollinating the same *Ficus* in different areas. This species-to-species relationship means that the rates of speciation of the fig-wasps and the fig-trees have been about equal. Some *Ficus* species are known to be inter-fertile, but no natural hybrids have been found and it is thought that the wasps may identify their particular host by scent. Most figs are inhabited by one or more additional species of chalcid-wasp that play no part in pollination.

The relationship between the yucca plant and the moth that pollinates it is similar to that between fig-trees and fig-wasps, and is equally famous, especially on account of the remarkable pollinating behaviour of the insects (Riley, 1892; McKelvey, 1947). All the yuccas (family *Agavaceae*) are American, and all those species which occur east of the Rockies are pollinated by a single species of moth, *Tegeticula* (*Pronuba*) *yuccasella*. The moth is very variable, but its variation does not appear to be correlated with the species of yucca in which it breeds. The moth is related to *Eriocrania*, a genus with primitive mouth-parts described in Chapter 4 part 1. The maxillae of *Tegeticula* each comprise a galea similar to that of *Eriocrania* (Fig. 119, p. 313), a palp and a special maxillary tentacle, prehensile and spinous, formed by the modification of the basal joint of the palp. These small moths are active at night and spend the day at rest in the flowers, which they resemble in colour.

The yucca produces a showy inflorescence of numerous large creamy white flowers, the perianth segments of which converge towards their tips so that the flowers are partially closed. The flowers are scented and smell most strongly at night. Nectar is sometimes secreted at the base of the ovary but it is not drunk by the yucca-moth, which apparently does not feed. The nectar may, however, keep other insects, which are attracted to the flowers, away from the stigmas, and it is believed that in fact these other insects never effect pollination. The female yucca-moth, when visiting the flowers, has a stereotyped pattern of behaviour: it descends into a flower, climbs up a stamen from the base and bends its head closely over the top of the anther. The tongue uncoils and reaches over the top of the anther, apparently steadying the moth's head. All the pollen is then scraped into a lump under the head by the maxillary palps, and held fast by the maxillary tentacles and the trochanters of the fore-legs (Fig. 119, p. 313). As many as four stamens may be climbed and the pollen collected in this way. As a rule, the moth then flies to another flower, where it closely investigates the condition of the ovary, being able to tell if the flower is of the right age and whether eggs have already been laid in it. If the flower is suitable, the moth again climbs up the stamens from the base but this time goes between them on to the ovary. The moth then reverses a little way down between the stamens and the ovary and lays an egg, boring into the ovary with its ovipositor (Fig. 119, p. 313). After this it at once climbs up to the stigmas, which are united to form a tube, and thrusts some of its pollen down into the tube, working energetically with the galeae and tentacles. The most usual behaviour of the moth is to lay one egg in each of the three cells of the ovary, and to carry out pollination after laying each egg. Sometimes, however, the moth may lay as many as twelve eggs in a flower and pollinate it the same number of times. Since an unpollinated yucca flower soon dies, this behaviour of the moth ensures that there will be food for its larvae, which is provided by the abnormal growth of one or more of the ovules in the neighbourhood of each moth's egg. The remaining ovules, which are numerous, develop into seeds and, just as they are ripening, the moth larvae emerge and pupate underground. The adult moths always emerge in the flowering season of the yuccas in their area. The emergence of any one season's brood is spread over a period of three years after pupation, thus ensuring the continuance of the moth species even if, as occasionally happens, the yuccas fail to flower in a particular year. In this relationship the moth, like the fig-wasp, ensures the seed-production of its food-plant, while the plant provides food and shelter for the young of its pollinator.

West of the Rockies, two more species of yucca-moth (*Tegeticula*) are found. One pollinates *Yucca whipplei*, which occurs in southern California, and is so distinct that it has sometimes been placed in a separate genus. It has glutinous pollen massed into two pollinia in each anther; the moths use their galeae in gathering it and carry out pollination in the daytime (Powell and Mackie, 1966). The remaining species of yucca-moth pollinates only *Yucca brevifolia* in the Mohave Desert of California. It has a hard body and scaleless wings and appears to mimic sawflies of the genus *Dolerus*. *Yucca brevifolia* has flowers with very firm perianth segments, scarcely parted at the tips, so that the moth has to force its way in as the fig-wasps do. The restriction of each of these Californian moths to one species of yucca is a curious contrast with the versatility of the eastern yucca-moth (*T. yuccasella*).

Closely allied to the yucca-moth is the bogus yucca-moth (*Prodoxus*). Superficially it resembles *T. yuccasella*, but its maxillae have no tentacles. It breeds in the ovaries or peduncles of yucca flowers and, since it does not pollinate the flowers, it depends for its existence on the true yucca-moth.

The yucca-moth and the fig-wasp both belong to groups in which feeding by the larvae on the internal parts of plants is frequent, so we may conclude that this process probably evolved first, the insects originally being at most incidental pollinators of the host-plant. It seems more likely, on the other hand, that the insects pollinating *Alocasia pubera* (a fly-trap flower, see p. 312) were pollinators before they took to breeding in the plant. In the cases of the pollinators of *Artocarpus heterophylla* and *Theobroma cacao*, there seems to be no strong evidence either way.

BIRD-POLLINATION

There are no bird-pollinated flowers in Europe, nor in Asia north of the Himalayas. Bird-pollination is, however, common in some other temperate regions and in the tropics. In the tropics, in particular, the flowers are largely adapted to bird-pollination (Porsch, 1933), there being a comparative dearth of highly developed flower-visiting insects. In the cooler parts of the temperate regions any bird-pollination that takes place is mainly by summer migrants, insects being the chief pollinators here.

Bird-pollination is known to occur up to a height of about 12,000 ft. in the mountains of East Africa and South America, the birds migrating locally to these altitudes. In latitude, bird-pollination extends from the southern tip of South America to Alaska in North America, while its nearest occurrence to Britain is in Israel.

There is a whole range of birds showing different degrees of adaptation to a floral diet (usually nectar). According to Porsch (1924), the importance of birds as pollinators was much underestimated, particularly among the less specialised birds and flowers, but flower-visiting has now been recorded for about fifty bird families. The prejudices of European observers and the difficulties of observation contributed to this earlier underestimate.

Some of the adaptations seen in bird-pollinated flowers are parallel to those found in insect-pollinated flowers. Examples are: food supply, conspicuousness, guide-marks, size, shape and positioning of flowers. Some of the bird-flowers are extremely large, while the smallest are no bigger than a Bluebell (*Endymion*). In relation to the size of the flowers, large quantities of nectar are secreted, and this is thin (with only about 5 per cent. of sugar) and sometimes slimy. Species of *Banksia* in Australia produce so much nectar that it is used as food by the aboriginals (Werth,

PLATE 47. Wind-pollinated herbs I: **a**, spikelet of False-oat *(Arrhenatherum elatius)*.
b, inflorescence of a sedge, *Carex demissa*. **c**, flowers of Woodrush *(Luzula forsteri)*.
d, inflorescence of Ribwort Plantain *(Plantago lanceolata)*.

PLATE 48. Wind-pollinated herbs II: **a**, inflorescence of Salad Burnet *(Poterium sanguisorba)*. **b**, shoots of Dog's Mercury *(Mercurialis perennis)* with male and female inflorescences; **s**, stigmas. **c**, Bur-reed *(Sparganium erectum)*. **d**, Mare's Tail *(Hippuris vulgaris)*.

1956), while showers of nectar can be brought down by shaking the branches of *Erythrina* and *Grevillea* (Swynnerton, 1916a). The nectar of most widely-open flowers is easily visible, and some tubular flowers, for example *Antholyza bicolor*, also store their nectar where it can be seen, with the result that the birds by-pass those flowers which are empty. Many other tubular flowers have a swelling at the base in which the nectar accumulates. The bird-flowers are often strongly constructed and the ovaries are usually protected from accidental pecks, either by being inferior or by being separated from the nectar-secreting zone. This separation is sometimes achieved by the storage of the nectar in a spur or by the raising of the ovary on a stalk (Grant, 1950b). The flowers occasionally have flexible pedicels, as in many *Fuchsia* species, so that if a bird probes too forcibly the flower is pushed away and escapes damage. Flexible pedicels are also a hindrance to attempts to pierce the perianth for nectar (Swynnerton, 1916a). Except in extremely large flowers, the flower itself is usually without any alighting place, but many plants provide portions of bare stem, leaf-stalks, bracts, or flower buds, in a convenient position for the birds to perch on. For example, a great many South African plants have rather short inflorescences surmounting a stout bare stem, while in the genus *Puya*, a South American member of the pineapple family (*Bromeliaceae*), the rigid inflorescence-branches are prolonged beyond the flowers. Some Australian plants are pollinated by birds standing on the ground (for example *Brachysema*, family *Leguminosae*; see Porsch, 1927).

Good examples of adaptation to different types of bird are seen in the genus *Erythrina* (family *Leguminosae*). In Indonesia, different species are adapted to birds of different sizes (Fig. 121A–C, p. 320) which feed while perched on the peduncles, and the flowers face inwards so that they can be easily reached from this position. On the other hand, an American species adapted to hummingbirds (which feed on the wing) provides no alighting place and has its flowers facing outwards to suit hovering birds (Docters van Leeuwen, 1931). American bird-flowers are generally longer-tubed than the Asiatic ones, in correspondence with the generally longer bills of the American pollinating birds (Van der Pijl, 1937b).

Bird-flowers are scentless, for birds have little or no sense of smell; they are, however, highly sensitive to colour. Red and orange are much more commonly found among bird-flowers than among flowers pollinated by insects, while reddish blue and violet, on the other hand, are rarer. There are two features of the colouring of bird-flowers by which many of them can be recognised: one is the prevalence of harsh colours (Plate III), and the other is the frequency of peculiar colour combinations ('parrot coloration', such as a mixture of green, yellow and scarlet) which are particularly common in the pineapple family and often extend to the bracts (Porsch, 1924). Guide-marks are frequent but are not usually as well defined as those of insect-pollinated flowers (Plates I, IIIa); the

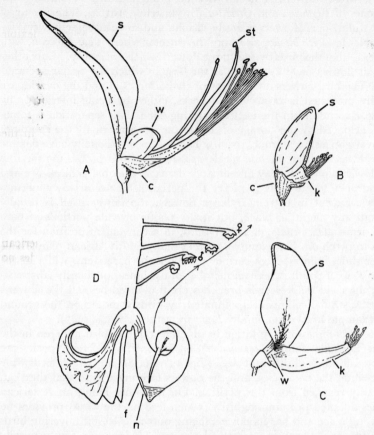

FIG. 121. A-C, flowers of *Erythrina* (A, *E. variegata* var. *orientalis*, pollinated by starlings [*Sturnopastor* sp.] and other birds of similar size [orioles, drongos, thrushes, etc.]; B, *E. subumbrans*, pollinated by bulbuls, sunbirds and white-eyes; C, *E. umbrosa*, pollinated by hummingbirds; these flowers are all red; c–calyx, k–keel, s–standard, st–stigma, w–wings). Shape of calyx suggests that it is important in keeping petals and stamens in position. D. in- florescence of a hummingbird-pollinated sp. of *Marcgravia* (incomplete); one of the flattened nectar vessels cut open. ♂ – flower in male stage, ♀ – flower in female stage, f–flower rudiment, n–nectary. Arrows show course of departing bird's head. A-C, after Docters van Leeuwen (1931); D, after Wagner (1946).

ultra-violet component in these guide-marks is very weak (Kugler, 1966).

It has been found convenient here to divide the bird-flowers very roughly into five groups according to the shape of the flower (Table 8, p. 322). The unsymmetrical, horizontal flowers of Group I resemble the dogfish, or shark; the lower perianth lobes are turned back and suggest fins, while the large upper lobe is directed forward, covering the stamens and style and giving an impression of an overhung mouth (Plate IIIa). Group 2 flowers are tubular but without an overhanging upper lip, the mouth being more or less radially symmetrical and usually slightly flared (Plate IIIe). An interesting feature of some members of this group is that the flower-tube remains closed at the top until probed by a bird, when it opens explosively, scattering a cloud of pollen over the visitor. It may also be noted that some families with a basically asymmetric flower-structure (*Acanthaceae*, *Gesneriaceae* and *Scrophulariaceae*) have produced species of plants having regular tubular flowers, as one of their forms of adaptation to bird-pollination. In both Groups 1 and 2 the flower-tube may be either straight or curved, and in Group 2 the stamens and style (or the style only) may in some cases protrude from the tube. An example is *Macranthera flammea*, which has a special arrangement for ensuring contact of the stigma and anthers with the neck of a hummingbird visitor. The narrow tubular flowers stand erect on springy pedicels, while the style and stamens in turn protrude from the tube. The hovering bird inserts its bill and then drops down so as to bring the flower into a more convenient horizontal position. In this way either the stigma or the anthers are levered into contact with a precise spot on the nape of the bird's neck (Pickens, 1927). In the brush-flowers (Group 3) the stamens protrude far beyond the perianth. The flowers are usually small and grouped into dense masses, but sometimes they are larger and more widely spaced. Brush-flowers are particularly common in the family *Myrtaceae* in Australasia, a good example being provided by the genus *Callistemon*, in which the bright red stamens are arranged like the bristles of a bottle-brush (Plate IIIf). Flowers with protruding stamens are particularly common in Australia, partly because the dry climate makes protection of the pollen from rain unnecessary. Closely allied in form to the brush-flowers are the flowers arranged in capitula (Group 4). Here the flowers of each head face one way, in line with the peduncle, and the heads are closely surrounded by bracts. They often resemble *Compositae*, particularly thistles, but in fact there are very few true *Compositae* that are bird-pollinated. The flowers of Group 5 have entirely free perianth parts and hold their nectar in spurs formed from single perianth members. This has the advantage of directing a bird's bill away from the ovaries (Grant, 1950b). These spurred flowers belong to mainly insect-pollinated families, and it seems that the possession of a spur has made it easy for some species to develop into bird-flowers.

Table 8.   Family and distribution of various bird-pollinated flowers

---

1. 'DOGFISH FLOWERS'

Columnea gloriosa	Gesneriaceae	C. America
Rechsteineria cardinalis	Gesneriaceae	Trop. America
Mimulus cardinalis	Scrophulariaceae	N. America
Salvia splendens	Labiatae	Trop. S. America
Leonurus leonotis	Labiatae	S. Africa
Antholyza spp.	Iridaceae	Africa

2. TUBULAR FLOWERS

Fuchsia spp.	Onagraceae	C. America–temp. S. America
Zauschneria spp.	Onagraceae	N. America
Manettia inflata	Rubiaceae	Trop. S. America
Lonicera sempervirens	Caprifoliaceae	N. America
Erica spp.	Ericaceae	S. Africa
Macleania spp.	Ericaceae	Trop. America
Astroloma humifusum	Epacridaceae	Australia
Russelia juncea	Scrophulariaceae	C. America
Penstemon centranthifolius	Scrophulariaceae	N. America
Macranthera flammea	Scrophulariaceae	N. America
Rechsteineria lineata	Gesneriaceae	Trop. S. America
Aeschynanthus lobbianus	Gesneriaceae	Indonesia
Odontonema schomburgkianum	Acanthaceae	Trop. S. America
Loranthus kraussianus and *dregei*	Loranthaceae	S. Africa (Explosive)
	Many Bromeliaceae	C. America–temp. S. America
Ravenala madagascariensis	Musaceae	Madagascar (Explosive)
Kniphofia spp.	Liliaceae	Africa
Aloe spp.	Liliaceae	Africa
Lachenalia spp.	Liliaceae	Africa

3. BRUSH-FLOWERS

Greyia sutherlandii	Melianthaceae	S. Africa (Flowers well spaced)
Callistemon spp.	Myrtaceae	Australia
Beaufortia sparsa	Myrtaceae	Australia
Eucalyptus spp.	Myrtaceae	Australia (Flowers well spaced)
Acacia celastrifolia	Leguminosae	Australia
Calliandra fulgens	Leguminosae	C. America
Banksia spp.	Proteaceae	Australia
Nuytsia floribunda	Loranthaceae	Australia (Flowers well spaced)

4. CAPITULA

Dryandra spp.	Proteaceae	Australia
Protea spp.	Proteaceae	S. Africa
Mutisia spp.	Compositae	S. America (Flower stalks pendent)
Haemanthus natalensis	Amaryllidaceae	S. Africa

5. SPURRED FLOWERS

Aquilegia canadensis	Ranunculaceae	N. America (5 spurs)
Delphinium cardinale	Ranunculaceae	N. America (1 spur)
Tropaeolum pentaphyllum	Tropaeolaceae	Trop. S. America (1 spur)
Impatiens niamniamensis	Balsaminaceae	Trop. Africa (1 spur)
Impatiens capensis	Balsaminaceae	N. America (1 spur)

A South African bird-pollinated plant which does not easily fall into any of these groups is *Strelitzia reginae* (family *Musaceae*) which has very unequal perianth segments. Three outer orange segments serve for display and of the three blue inner segments two conceal the stamens while the third, which is much smaller, conceals the nectary (Plate IIId). The two segments that conceal the stamens have lobes on which a bird must stand in order to reach the nectar; the weight of the bird pushes the lobes apart so that the anthers are then exposed and dust the under side of the bird with pollen (Scott-Elliot, 1890a).

The tropical American genus, *Marcgravia*, some species of which are bird-pollinated, is another very distinctive plant, having its nectar contained in modified bracts. In an unnamed species studied by Wagner (1946), these are arranged in a whorl which terminates the pendent inflorescence, and above them a series of flowers develops on stalks which lengthen with age (Fig. 121D, p. 320). The shape of the nectar vessel forces the hummingbird visitors to depart in such a way that their heads touch the flowers. A similar species (*M. sintenisii*) was observed by Howard (1970) to be visited by todies (*Todus*), hummingbirds and honeycreepers (see Table 9, p. 325); its nectaries are at first yellow but turn bright red when the flowers open.

One further unusual plant is *Freycinetia* of the Pacific region, which provides fleshy, sugary bracts surrounding the flowers. These are eaten particularly by crows, which pollinate certain species (Porsch, 1930).

Birds are so active and inquisitive that the evolution of their habit of feeding on nectar was to have been expected. However, there are two ways in which they may have been led to this source of food. One is by visiting foliage to drink raindrops, a habit common among birds of tropical forests, and the other is by going to flowers to look for small insects.

Nectar is mainly a source of carbohydrate, and all nectar-feeding birds eat other foods too, either insects or fruits. A captive sunbird was observed by Scott-Elliot (1890b) to be very expert at catching insects, and it survived six weeks, apparently on insect food alone. A captive humming-bird, on the other hand, survived for a month on nothing but sugar-water (Ridgway, 1891). It is the young birds that need most protein, and it is believed that the young of nectar-feeding birds are fed to a large extent on insects, but instances are known of their being fed on nectar as well.

The nectar-feeding birds show great rapidity and precision of movement, together with considerable constancy to flowers of a single species. They are therefore highly efficient pollinators, and individual birds probably visit many thousands of flowers in a day (Porsch, 1933; Scott-Elliot, 1890b).

In all regions where there are nectar-feeding birds there is, however, a tendency for the birds to steal nectar by piercing the sides of tubular flowers of all sizes. It was considered by Sargent (1918) that the prevalence of destructive nectar-robbing by birds in Western Australia created a strong evolutionary pressure upon insect-pollinated plants to adapt themselves to bird-pollination and so increase the chances of legitimate visitation. Swynnerton (1916a) made a special study of nectar-thieving by birds in Rhodesia. His most interesting results concerned thieving from bird-pollinated flowers by their legitimate pollinators. For example, in *Leonotis mollissima* the flowers are in dense spherical clusters placed at intervals up erect stems, and they radiate horizontally or slightly downwards, being reached legitimately by birds perched just below. Some of the birds preferred to do this and were thus valuable pollinators, only piercing the calyces or using previously made punctures, when the flowers were disarranged and so had ceased to protect one another (compare p. 219). Other birds of the same species, however, always pierced the flowers or used punctures, usually approaching from above; but even if approaching from below, when legitimate visiting would have been more convenient, they went to some trouble to steal the nectar through the tough sides of the flowers, probably owing to their having a dislike of getting pollen on their plumage. In *Leonotis* and other genera, damage depended much on whether the plants were included in the 'beat' of destructive birds. Birds which are not regular nectar-feeders may sometimes steal nectar too, and this occurs even in Europe, the main genera concerned being *Sylvia* (warblers) and *Parus* (titmice); in Britain the titmice attack American currant (*Ribes sanguineum*), gooseberry (*R. grossularia*) and cherry and almond (*Prunus* spp.) (Swynnerton, 1916b; P.F.Y.).

The eight most important families of flower-visiting birds (according to Porsch, 1924) are listed in Table 9, but there is much variation in the proportions of nectar-feeding species in the different families.

There are about three hundred species of hummingbirds (Plate 35, p. 254) and, although the area which is richest in species is the northern part of the Andes, studies of their behaviour have been made mainly in Central and North America. Most hummingbirds are very tiny – the smallest is only $2\frac{1}{4}$ inches long, and half of this length is accounted for by bill and tail – and their rapidly vibrating wings make the hum which gives them their name. The largest species (*Patagona gigas*) is $8\frac{1}{2}$ inches long and flaps its wings with a slow butterfly-like motion when hovering at flowers (Ridgway, 1891). In North America the hummingbirds are migratory, and there are notable coincidences between the movements of the birds and the flowering periods of the plants they visit (Pickens, 1936; Bené, 1946; K. A. and V. Grant, 1968).

Hummingbirds nearly always feed while in flight, and their method of feeding on nectar was observed by Moller (1931b) when studying them in Costa Rica. They paused momentarily in front of horizontally tubed flowers and seemed able to tell whether a flower was likely to be productive. If a bird wished to probe a flower, it dropped a short distance from the pausing point and then came diagonally upwards into the flower, usually remaining there for 1.4–3 seconds; on leaving, the bird again dropped a short distance before flying off. When visiting flowers with

Table 9. Scientific and English names of the main flower-visiting families of birds, with the genera mentioned in the text

Trochilidae Hummingbirds	North and South America	*Patagona gigas* *Trochilus colubris* (Ruby-throated hummingbird) *Trochilus alexandri* (Black-chinned hummingbird)
Coerebidae Honeycreepers (or sugarbirds)	Tropical America	
Nectariniidae Sunbirds	Africa, South-west Asia to Philippines	*Cinnyris osea* *Nectarinia* *Arachnothera longirostris* (Long-billed spider-hunter)
Zosteropidae White-eyes	Africa, Asia, Australasia	
Meliphagidae (a) Sugarbirds (b) Honey-eaters	South Africa Australasia	*Promerops* *Acanthorhynchus* (Spinebills) *Anthornis melanura*
Dicaeidae Flowerpeckers	Asia, Australasia	
Drepanididae Hawaiian honeycreepers	Hawaiian Islands	
Psittacidae subfam. *Loriinae* Brush-tongued parrakeets (or Lories)	Australasia	

upright tubes, however, the birds rose as they reversed out (Bené, 1946; Wagner, 1946). These authors provide ample evidence that nectar is taken in this way on a large scale by many species of hummingbird and, even when feeding on insects, these birds usually prefer to find them in the flowers which are most suited to the size and shape of their bills. However, they are obliged to dive right inside large trumpet-shaped flowers such as those of the Trumpet Creeper (*Campsis radicans*), which is generally considered, in fact, to be primarily adapted to hummingbirds. Moreover, Moller saw hummingbirds climb into the even larger trumpets of *Solandra brachycalyx*, but this species may be adapted to bat-pollination.

Aggressive behaviour in the defence of small feeding territories is very conspicuous in hummingbirds, and these territories are sometimes only a few yards across (Bené, 1946; Pickens, 1944). Indeed, a single flowering tree in Costa Rica was seen by Moller (1931b) to be parcelled out territorially between numerous hummingbirds of three species. Such behaviour evidently limits the dispersal of pollen and thus restricts the range over which out-crossing, if any, takes place.

Hummingbirds readily learn to feed from tubes and small vessels containing liquid food, even without any floral adornments (Ridgway, 1891). They are thus easily subjected to feeding experiments, which have been used in investigations of their behaviour in relation to flower-visiting. Experiments on black-chinned hummingbirds were made with sugar-syrup coloured with tasteless dyes. Various colours were visited and preferences varied from one experiment to another, but Bené (1941, 1946) concluded that, as in bees, there were no inherited colour preferences in these hummingbirds, although their preferences were capable of being modified by conditioning. Similar results were obtained with ruby-throated hummingbirds, which were presented with artificial flowers containing sugar-water (Pickens, 1941). Bené found that one part by volume of sugar in 2½ parts of water and one part of honey in seven parts of water were the minimum concentrations that the birds would take. Both concentrations have the same calorific value.

Bené noted that individual hummingbirds had differing preferences for particular flower species in the same area, and if the favourite flowers were sufficiently abundant others were rarely visited. This also recalls the behaviour of bees and indicates a degree of constancy valuable for pollination. Moreover, birds that had been feeding from glass containers investigated other objects which resembled these feeders. In Bené's list (1946) of the flowers most visited by hummingbirds, whether for nectar or for insects, and including many species not specialised to hummingbird pollination, the commonest colour is red. In addition, all the flowers in Table 10, below, are red, or red with orange or yellow. Thus it seems that there is selection in favour of red in hummingbird flowers, and this is presumably more because most insects do not perceive it than because the birds actually prefer it. K. Grant (1966) suggests that the uniform

Table 10.   The eight most important hummingbird flowers east of the Mississippi, according to James (1948)

	Family	American name
Aesculus pavia	Hippocastanaceae	Red Buckeye
Aquilegia canadensis	Ranunculaceae	Wild Columbine
Campsis radicans	Bignoniaceae	Trumpet Creeper
Impatiens capensis	Balsaminaceae	Jewelweed
Lobelia cardinalis	Campanulaceae	Cardinal Flower
Lonicera sempervirens	Caprifoliaceae	Trumpet Honeysuckle
Macranthera flammea	Scrophulariaceae	——
Monarda didyma	Labiatae	Oswego Tea

colouring of the North American hummingbird flowers is related to the migratory habits of the birds, which involve immigration from the south at breeding time followed by ascent to higher altitudes after breeding. The hummingbirds thus have only to seek red flowers on entering a new area with a different flora, since the hummingbird flowers have all evolved the same colour as those from which the birds have already learnt to feed. The North American hummingbirds are all much alike in bill-length, so that there is no question of plants specialising to suit particular types of hummingbird. In tropical America, on the other hand, the birds vary in bill-length, so there is opportunity here for specialisation, and as the birds are resident they learn to distinguish individual plant species.

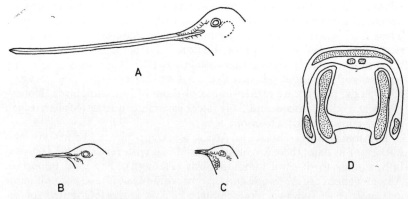

FIG. 122. A-C, heads of hummingbirds to show extremes of bill-length and a commonly occurring intermediate size. D, transverse section of bill of hummingbird showing overlap of upper and lower mandibles; unshaded area is horn, shaded is bone. Greatly magnified. A-C, after Ridgway (1891); D, after Moller (1930).

Consequently the hummingbird flowers of this area are much less uniform in form, size and colour. This theory and others are further expounded in a book on hummingbird pollination in western North America (K. A. and V. Grant, 1968), which includes many colour photographs of flower-visits.

The needle-like bills of hummingbirds are usually straight or slightly decurved (Fig. 122A–C); in rare instances they have a stronger downward curvature or are curved upwards. They are unlike those of most other birds in that there is a strong overlap between the upper and lower mandibles (Fig. 122D). During nectar-feeding, the tip of the bill is opened sufficiently to allow the tongue to move rapidly in and out (Moller, 1930, 1931b).

The tongues of birds have few internal muscles but are made up of bones and cartilages, covered with a horny sheath and extending into the head, where they are activated by muscles (Lucas, 1897) (Fig. 123). The tongue of the hummingbird is deeply bifurcated, and each lobe is a thin lamina rolled up lengthways to form a slender tube. Nectar is probably held in this double tube by capillarity until the tongue is brought back into the beak, at which stage the nectar can be sucked back into the mouth and swallowed. The end of each tube is papery and frayed giving a brush-like effect, which may increase the capillarity and helps in the capture of small insects.

The second important group of flower-visitors found in America is the family *Coerebidae*, the honeycreepers. Some of them have peculiar bills, adapted to piercing flowers and stealing their nectar. Others feed on nectar in a legitimate manner, and are tit-like in size and movements. They feed when perched, performing difficult contortions in order to reach flowers, and usually probe the flowers farther from them instead of those most conveniently situated (Moller, 1931b). Their bills are sharply pointed and commonly slightly decurved and fairly short. Their tongues (Fig. 124A, p. 331) are rather similar to those of hummingbirds. Honeycreepers feed on insects and fruit juices as well as nectar (Moller, 1931a and b; Porsch, 1930; Thomson, 1964).

The main flower-visitors in Africa are the sunbirds, which are also important in Asia; these are mostly small birds with very slender decurved bills, similar to those of hummingbirds (Plate 55). The most important African genera are *Nectarinia* and *Cinnyris*, while *Cinnyris osea* is a pollinator in Israel. A well-known representative in Asia is the long-billed spider-hunter, in which the bill is enormous in relation to the size of the body; this bird is a regular flower-visitor as well as specialising in catching spiders while hovering. Sunbirds may hover while feeding at flowers, but usually they perch. Van der Pijl's account (1937b) of their acrobatic habits while drinking from flowers is almost identical with Moller's description of the honeycreepers. The tongue of a sunbird is channelled, but divides into two separate rods near the tip; when pressed against the upper mandible, the channel forms a sucking tube (Fig. 124c).

The white-eyes are small warbler-like birds with very slender, but usually short, bills (Plate 55). They resemble honeycreepers in their gutter-shaped tongues with brush-like lobes, and also in their habit of sucking the juices of fruits as well as the nectar of flowers (Moller, 1931a).

The sugarbirds of southern Africa (Plate 55) belong to the genus *Promerops* and breed only in *Protea* vegetation, being almost confined to it. Their peak of breeding coincides with the flowering of certain species of *Protea* and as well as feeding on the nectar of the flowers they use the fluff from the inflorescences for their nests (Thomson, 1964). The honey-eaters of Australia and the Pacific islands belong to the same family of birds.

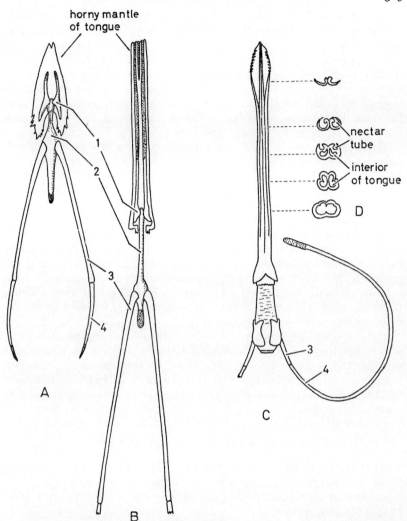

horny mantle
of tongue

1

2

3

4

A

B

nectar
tube

interior
of tongue

D

3

4

C

FIG. 123. A, bones and mantle (outlined) of an unspecialised avian tongue. B, the same parts of the Rufous Hummingbird, incomplete, tongue longitudinally sectioned. C, view from above of tongue of another hummingbird (*Eulampus holosericeus*). The protruding bones are coiled round the outside of the cranium; the tongue is bifurcated but the lobes do not diverge in life. D, transverse sections of c; mantle only. Corresponding bones bear the same numbers; those of the hummingbirds are greatly elongated; cartilage shaded. Based on Lucas (1897), Moller (1930) and Ridgway (1891).

Together with the parrakeets, they are the main pollinators of three important tree and shrub families of Australia (*Myrtaceae, Proteaceae* and *Epacridaceae*). They are numerous in species, and their size ranges from about the same as a goldcrest to as large as a jay. Most of them feed on nectar, pollen, and insects (Thomson, 1964). Some, such as the spinebills, have long slender bills like those of the sunbirds and can hover while feeding; others have short slender bills or even stout ones. *Anthornis melanura*, for example, has a beak similar to that of a blackbird, and its tongue is shown in Fig. 124D.

The flowerpeckers of Asia and Australasia frequently have bills well adapted to flower-visiting, and are particularly associated with the Mistletoe family (*Loranthaceae*) both as pollinators and as distributors of seed. In this family of plants bird-pollination is very common, and is also carried out by a variety of other birds including the sunbirds in Asia and Africa (Evans, 1895).

The various species of Hawaiian honeycreeper have bills of different lengths, adapted to the various species of tree belonging to the subfamily *Lobelioideae*, which they pollinate. However, they also visit for nectar the open flowers of a tree of the family *Myrtaceae*, which is believed not to have reached Hawaii until long after the tubular-flowered tree lobelias (Perkins, 1903). The tongue of these honeycreepers is rolled to form a tube, and has a fringed tip (Fig. 124B).

The last birds to be mentioned here are the brush-tongued parrakeets, which have the usual short hooked bill of the parrots and a tongue that is very short for a flower-visiting bird. The tip is three-lobed and provided with dense tufts of hair. These birds move in flocks and visit *Eucalyptus* flowers, feeding on pollen and nectar and frequently becoming thoroughly dusted with pollen.

MAMMAL-POLLINATION

Visits to flowers by bats are known to have been observed as early as 1772, but it was only towards the end of the nineteenth century that bats were recognised as significant pollinators (Baker and Harris, 1957; Jaeger, 1954b; Knuth).

Bat-pollination is confined to the tropics, but in the Andes it occurs up to 3400 m. above sea level, where frosts are frequent at night. Flower-visiting bats are found mainly where there is a succession of suitable flowers for them all the year round, but those that pollinate cacti in southern Arizona are probably migratory (Vogel, 1958). Most bat-pollinated plants are trees or woody climbers, but some are dwarf shrubs, herbs or herbaceous climbers. Observation of bat-pollination is even more difficult than that of bird-pollination, for not only does it take place mainly in high trees but also it is confined to dusk and darkness. How-

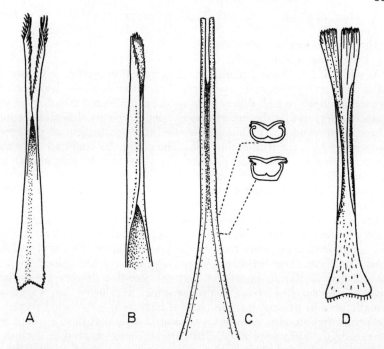

A   B   C   D

FIG. 124. A, tongue of a honeycreeper (*Coerebidae*), from above; each lobe is twisted to form an incomplete tube, open on its outer edge. B, tongue of a Hawaiian honeycreeper (*Drepanididae*), from above and slightly to one side. C, tongue of a sunbird (*Nectariniidae*), from above; internal muscles can draw the upper surface downwards, so increasing the space between the tongue and the upper mandible and producing suction; the construction of the horny mantle shown in the sections facilitates this action. D, tongue of a honey-eater (*Meliphagidae*), from above. A, B, after Lucas (1897); C, D, after Moller (1930).

ever, bat-pollination can often be deduced from the presence of claw-marks on the flowers or the fallen corollas.

The chief characteristics of bat-pollinated flowers are as follows (Van der Pijl, 1936):

(1) they first open in the evening and their pollen and nectar are available at night;

(2) they have a disagreeable, sour and musty scent;

(3) they or their inflorescences are strong enough to bear a bat;

(4) their colour is often dingy;

(5) they produce much mucillaginous nectar, accessible to a bat's tongue;

(6) they produce much pollen;

(7) their position is exposed.

As with bird-pollinated plants, the flower-food offered may be nectar, pollen or, rarely, fleshy sugary bracts. The latter occur in Indonesia in *Freycinetia insignis*, which differs from the bird-pollinated members of this genus (p. 323) in that the inflorescences open in the evening and produce a musty fruity odour. Pollination is effected by members of the genus *Cynopterus*, which are fruit-eating bats. The pollen (or stamens) of certain brush-like flowers (which also provide nectar) is eaten by bats in both the Old and the New World. Examples are found in the families *Myrtaceae*, *Sapotaceae*, *Leguminosae* and *Cactaceae* (Van der Pijl, 1936, 1956; Vogel, 1958).

The more characteristic bat-pollinated flowers have been investigated mainly in America and Africa. Vogel (1958) found that in Colombia the two commonest types were the bell- or jaw-like form (Fig. 125) and the brush-like form. The bell-like and jaw-like flowers are visited for nectar and are similar in operation, but the jaw-like flowers are unsymmetrical, the corollas having dissimilar upper and lower lips. Vogel was actually able to photograph the visits of bats to a bell-shaped flower of the genus *Symbolanthus* (family *Gentianaceae*). Although the bats clutched the corolla the visits were so brief that there seemed to be time for only a single extension of the tongue in each flower. The corolla of this plant forms a wide bell of very firm consistency, sloping slightly upwards and with short, somewhat recurved lobes. The stamens and style lie on the lower side of the bell and touch the undersides of the visitors. The flowers are odorous and open for the first time in the evening, lasting several days. A bat-pollinated plant with a rather similar flower is *Purpurella grossa* (family *Melastomaceae*), in which the petals are separate from one another but overlap to form a bell. The nectar is taken by bats while hovering, and the flowers have the 'usual cabbagy smell', but the dark red colour is unusual. The bell-shaped flowers of the genus *Cobaea* (family *Polemoniaceae*) may be known to British readers, as *C. scandens* is easily cultivated under glass or out of doors in the summer. It is a climber with pendent shoots, from which hang long pedicels supporting the flowers in a nearly horizontal position. The flowers are greenish when they open and dull purple later; their scent is somewhat sweet but also includes a cabbagy element. They are probably bat-pollinated in their native habitat, for the claw marks of bats were found by Vogel on the flowers of an allied species, *C. trianae*, in which the corolla is green with violet veins. Other families in which Vogel found bat-pollinated flowers of this general type are *Leguminosae* (represented by *Mucuna*, in which there is an explosive mechanism for the release of pollen), *Solanaceae*, *Gesneriaceae* (in both of

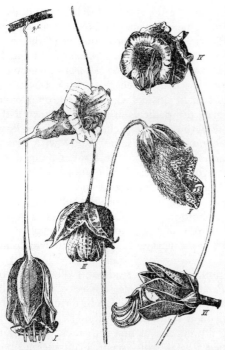

FIG. 125. Flowers of known or suspected bat-pollinated plants of South America, including two cases of 'flagelliflory' (see below). I. *Trianaea speciosa* (family *Solanaceae*). II. *Symbolanthus latifolius* (*Gentianaceae*). III. *Cayaponia* sp. (*Cucurbitaceae*). IV. *Cobaea scandens* (*Polemoniaceae*). V. *Campanaea grandiflora* (*Gesneriaceae*). VI. *Cheirostemon platanoides* (*Malvaceae*). From Vogel (1958).

which there are bird-pollinated forms closely allied to the bat-pollinated ones), *Bignoniaceae* and *Bombacaceae* (represented by *Ochroma lagopus*, the balsa-wood tree, with one of the largest of bat-pollinated flowers, 12 cm. long and 8 cm. across the mouth).

The brush-like flower forms are varied in construction and are characterised by a flower, or an inflorescence, with numerous protruding stamens. These tend to dust the bodies of the pollinators extensively, rather than in a particular place. Vogel described flowers of this type in the families *Lythraceae*, *Capparidaceae* and *Marcgraviaceae*. In one species of *Marcgravia* the inflorescence is an erect umbel and each floret has a nectary of its own formed by a bract near the base of the pedicel. All the nectaries are therefore near the centre of the umbel, where there is an open space giving easy access to them. Vogel also mentioned other bat-pollinated species of *Marcgravia* in which the inflorescences hang down on rope-like peduncles ('flagelliflory') or on leafy stems (Plate 56) and are similar to those pollinated by hummingbirds (Fig. 121, p. 320).

The two types of bat-pollinated flower so far described also occur in the Old World. A good example of the jaw-type of flower in Africa is *Kigelia africana*, the sausage tree (family *Bignoniaceae*). The flowers are dark-coloured, fleshy, sour-smelling and nocturnal, and they hang down on long stalks. The corolla is 7 cm. long and 12 cm. wide, and has a lower lip with a wrinkled surface, affording a grip for alighting bats which crawl inside.

An Asiatic example of the brush-type of flower is provided by *Gossam-*

pinus valetonii (family *Bombacaceae*), where the stamens sticking out of each corolla form a ring round a funnel-shaped space; the flowers are odorous and last one night (Van der Pijl, 1936). Two more species of the same family with brush-flowers are the Durian (*Durio zibethinus*), studied in Indonesia by Van der Pijl, and the kapok tree (*Ceiba pentandra*), studied in West Africa by Baker and Harris (1959) and Jaeger (1954a). In both of these the creamy white flowers occur in clusters where there are no leaves (Plates 32, 33). In the kapok tree the pedicels are 8 cm. long and the flowers 5 cm. in diameter, so that the spherical clusters are quite large objects. Yet another African member of the same family with brush-flowers is the baobab tree (*Adansonia digitata*). The flowers hang down and their large white petals are strongly reflexed. The numerous stamens form a tube coming down well below the petals and end in a large tuft covered with purple anthers. The style emerges from the tube and hangs down to one side (Fig. 126A). A photograph published by Jaeger (1954b) shows numerous claw marks on one of these flowers, and Harris and Baker saw bats alight on the tuft of stamens and lap up nectar which had run out on to the petals.

Another plant whose pollination by bats was studied in West Africa by Baker and Harris (1957) is *Parkia clappertoniana*, a tree of the family *Leguminosae*. Here the inflorescence is very distinctive in form, consisting of tiny flowers densely packed on a solid body of the shape shown in Fig. 126B and Plate 53. These flowers are at first red, then purple and finally salmon pink, and produce a weak fruity scent. Fruit-sucking bats cling to the inflorescences when drinking the nectar, which is produced by sterile flowers near the peduncle and collects in the trough just below. Another species of *Parkia* with similar inflorescences has been found to be bat-pollinated in Java (Docters van Leeuwen, 1938).

The bats are classified into two sub-orders, the *Megachiroptera* and the *Microchiroptera*. The first of these contains only one family, the *Pteropidae*, which are almost entirely vegetarian and are found in Africa, Asia and Australia. These bats tend to be large in size and have good eyesight, but their powers of navigation in the dark are inferior to those of the *Microchiroptera*, which have small eyes and depend on 'sonar' to detect objects in their path.

In Indonesia, the *Megachiroptera* fall into two groups, one adapted to eating fruit (for example genera *Pteropus* (flying foxes) and *Cynopterus*) and the other specialised to nectar-feeding (for example *Eonycteris* and *Macroglossus*). Both *Cynopterus* and *Pteropus* also feed on the flowers of *Freycinetia* (see p. 332) but *Pteropus* is not thought to effect pollination. The exotic fruits eaten by the fruit-bats are specially adapted to them by their peculiar smells and by growing in such a position (often on the trunk or main branches) that the bats can easily get at them. The peculiar scents of these fruits are reproduced by the flowers that attract the nectar-licking

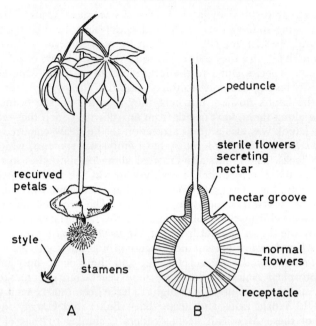

FIG. 126. Bat-pollinated flowers of the Old World. A, the Baobab (*Adansonia digitata*). B, diagrammatic section through the inflorescence of *Parkia clappertoniana*. A, after Jaeger (1954b); B, after Baker and Harris (1957).

bats. These nectar-eaters have reduced teeth, a well-developed snout, and a long slender tongue with a brush of backwardly directed hairs at the tip (the tongue of one species, which has a body length of 7.5 cm., can be protruded to a length of five to six centimetres), and they take substantial quantities of pollen in addition to nectar. Claw marks indicate that *Eonycteris* clings to the outside of flowers, whereas *Macroglossus* crawls inside when possible (Van der Pijl, 1936, 1956).

Four species of *Megachiroptera* were seen visiting flowers in West Africa by Baker and Harris (1957, 1959). The Gambian fruit bat (*Epomophorus gambianus*) (Plate 53a) and the pygmy fruit bat (*Nanonycteris veldkampii*) (Plate 53b) were seen to feed by chewing fruits to extract the juice, as well as by taking nectar. The Gambian fruit bat (wing-span 60 cm.) visited for nectar the flowers of *Parkia clappertoniana* for a period of twenty minutes just after dusk; each visit to an inflorescence lasted from 15 to 45 seconds, and the bats clung on by their hind feet, flapping their wings slowly all the time. These bats also visited the kapok tree and behaved in the same way, except that the visits lasted only a few seconds. Visits of the pygmy fruit bat (wing-span 34 cm.) to *Parkia* began about the time the other bat was leaving (in fact, the Gambian fruit bat was sometimes driven off by the new arrival) and they continued until the early hours of the morning. This smaller species clung to the inflorescences with the wing-claws and made visits lasting from one to thirty seconds. The pygmy fruit bats also visited the kapok flowers, but here they alighted on the branches and, crawling along them, took nectar from any flowers which they encountered. The kapok was also a great attraction to the straw-coloured fruit bat (*Eidolon helvum*), a large migratory bat (wing-span 100 cm.) which travels in huge flocks. These bats also crawled along the branches to reach the flowers, arriving immediately after dusk just as the flowers opened, and leaving soon after midnight. Up to forty bats were found in a single tree, and groups moved from one tree to another. The straw-coloured fruit bat was also seen visiting the baobab, and the flowers of the sausage tree were visited by the dwarf epaulet fruit bat (*Micropteropus pusillus*).

The *Microchiroptera* are mainly insectivorous, but some genera of the family *Phyllostomatidae* are vegetarians, and these occur only in tropical and subtropical America. Some of these vegetarian genera are nectar-feeders, and their evolution is thought to have been more recent than that of the Old World fruit- and nectar-bats (Baker and Harris, 1957). The tongue of these American nectar-feeders is similar to that of the Old World nectar-bats, having a brush of bristles at the tip which holds nectar by capillarity. The lower incisors are greatly reduced, allowing the tongue to move in and out without the jaws being opened, and in certain cases the tongue extension is known to be as much as 3 cm. (Vogel, 1958). Vogel found that these bats began to visit flowers at sunset but did not necessarily arrive at a particular feeding place at the same time on successive nights. They always fed in flocks and moved from tree to tree, any one tree usually being invaded repeatedly during the course of an evening. Visits to flowers lasted only a fraction of a second but, by counting the number of wing-claw marks on the flowers of the sausage tree (introduced into South America), he estimated that a flower might receive up to fifty visits in one night. Although they normally alighted, feeding from *Marcgravia* and *Purpurella* took place almost entirely while the bats were hovering. They were not active in high winds, but their feeding appeared to be unaffected by rain. Some bats were shot while visiting flowers and were found to be well dusted with pollen, their stomachs containing nectar but no pollen.

The scents of many Old and New World bat-pollinated flowers are similar, which seems rather surprising since the relationship between bats and flowers has evolved independently in the Old World and the New. However, Van der Pijl (1936) has suggested that the unpleasant smell of the glandular secretion which most bats produce, and by means of which they find each other when they gather in flocks, is similar to the scent of bat-pollinated flowers, which may have achieved adaptation to bat-pollination by imitating the smell of the bats (see p. 117). This would also explain Vogel's discovery that the sausage tree attracts the American bats.

Floral food is sometimes taken by mammals other than bats. For example, *Freycinetia arborea*, of Hawaii, which is adapted to bird-pollination, is visited by rats which have been introduced into this area. These eat the sugary bracts and apparently cause pollination (Van der Pijl, 1956). Two other bird-pollinated flowers were seen to be visited repeatedly for nectar by squirrels in Indonesia (Docters van Leeuwen, 1938). Apart from such casual occurrences there are a number of small arboreal marsupials in Australia, which feed on a mixed diet of vegetable matter, insects and nectar, or even chiefly on nectar, which they find in flowers generally regarded as mainly bird-pollinated. Some of them have long slender tongues which appear to be specially suited to nectar-feeding. These animals are remarkably skilled climbers, and some have a membrane on either side of the body, stretching between the front and back legs and giving them some ability to glide, so that they have good possibilities of carrying out cross-pollination. Although these mammals appear to be able to act as effective pollinators, it seems that up to the present no plants primarily adapted to their visits have been recognised, nor have any structures that appear to be adaptations to them (Kugler).

CHAPTER II

POLLINATION IN PLANT-BREEDING
AND COMMERCE

POLLINATION, as well as being an object of study in its own right as a biological phenomenon, is an incidental but very important process in plant-breeding and in agriculture and horticulture.

PLANT BREEDING

Plant-breeding may be carried out either as pure research into the relationships of plant species etc., or as applied research with a view to the improvement of crops and horticultural plants. In the first case it may be desired to measure degrees of interfertility of plants as well as to obtain progeny, whereas in the latter the production of progeny is usually of overriding importance. Plant-breeding requires artificial pollination, and the techniques now to be described are equally available for both pure and applied research. It is usually a simple matter to perform the transfer of pollen from anthers to stigma artificially, using, for example, a needle or a small brush. The basic requirements for normal fertilisation are that the plants should be growing within a certain temperature range which suits them and that the pollen should be viable, which it normally is if transferred direct from flower to flower. However, some plants produce a daily flush of short-lived pollen, and then pollination must be carried out at a particular time of day when the pollen is fresh. It is normally necessary to prevent both self-fertilisation and chance pollination from unspecified plants. Chance pollination is usually prevented by enclosing each flower or inflorescence in a paper bag, a method suitable for both wind-pollinated and insect-pollinated plants. Sometimes, however, whole plants are placed in insect-proof cages, which usually have walls of muslin or gauze; this method is, of course, not suitable for wind-pollinated plants. The large flowers of the vegetable marrow (*Cucurbita pepo*) may be isolated by placing a rubber band over the bud to prevent the corolla from opening. Self-pollination is prevented naturally in a few plants by self-incompatibility or by dioecism, but in other cases its prevention requires emasculation (elimination of the flower's own pollen). This is usually carried out by cutting away the anthers, an operation which is sometimes easy and sometimes not so easy, according to the size and construction of the flowers. The difficulty is much increased with those flowers which are apt to drop or wither if the perianth is damaged. In some of the cereal crops and grasses certain high or low temperatures can be used which destroy the pollen but do not injure any other parts of the flower. For Lucerne (*Medicago sativa*) the explosive mechanism is first sprung, or tripped, by cutting off the keel petals. The pollen thus discharged is then either destroyed by immersing the flowers in 57 per cent. alcohol or removed by suction through a fine glass tube, the operator wearing a binocular magnifier.

The preservation of viable pollen may make crosses possible which would otherwise be prevented by differences of flowering time, and it enables the pollen to be sent from one part of the world to another without the trouble of transporting living plants and meeting the international plant health requirements. Summaries of research results on the storage of pollen are given by Lawrence (1939), Maheshwari (1950), and by Johri and Vasil (1961). Grass pollens have the shortest life under natural conditions, in some cases requiring direct transfer to the stigma if fertilisation is to result, and in others lasting up to five hours. Many pollens, however, will last for several days or weeks, or even for many months, without any special treatment, but adjustment of the relative humidity will extend the life of all pollens. The very short-lived pollens of grasses can be made to last for several days if they are stored at high relative humidity (50–100 per cent.). On the other hand, the longer-lived pollens retain their viability best at low R.H. (0–40 per cent.), storage in a desiccator over calcium chloride (which absorbs water) prolonging life two to tenfold. Low temperature is also an important factor in prolonging the life of pollen, and other measures that in some cases improve its longevity are storage under vacuum, storage in an atmosphere of nitrogen, or in an atmosphere with a high concentration of carbon dioxide. Strong light is harmful to the survival of pollen (see p. 263).

Pollination normally leads to seed-production only if it is followed by fertilisation and by the adequate development of the embryo and the surrounding endosperm (if there is one). Usually there are no obstacles to these processes in crosses within species, and often there are none in crosses between closely related species. In other crosses, however, growth of pollen tubes towards the ovary may be retarded or seed development may be imperfect. Sometimes embryos begin to develop but then abort; in such cases they can sometimes be saved at an early stage by removal from the seed followed by culture on a sterilised nutritive jelly. Slow growth of pollen tubes can often be countered merely by early pollination (as soon as the flower opens or even before). This is recommended by Lawrence (1939) as generally worth trying provided the pollen can be made to adhere to the stigma, which may not always be possible before the stigmatic secretion has been produced. Slow growth of pollen tubes can also be countered by cutting away most of the style and placing the pollen on the stump. Germination may then be assisted by the addition to the stump of a drop of a suitable germination medium. A solution containing sugar, agar and gelatin has been used successfully for this purpose with species of *Solanum*. If this method fails, the style may be shortened by cutting away a length in the middle and sticking back into position the portion bearing the stigma. In the thorn-apple (*Datura*) the rejoining of the two pieces of a style cut in this way can be carried out by inserting them both into opposite ends of a tiny piece of grass straw until contact is made.

These methods for overcoming the slow growth of pollen tubes can also be used for the self-pollination of plants which are normally self-incompatible. For example, in breeding brassicas this process is sometimes necessary and early pollination is used 24 to 48 hours before the flowers open (Hayes, Immer and Smith, 1955). Pollination of the truncated style is the method employed for the hollyhock (*Althaea officinalis*), and no germination medium is required. Various inorganic and organic substances are known to retard the withering of the style, or to increase the percentage germination of pollen or the rate and extent of pollen tube growth. The effect of the different substances varies from species to species, but some of them have been used to overcome self-incompatibility, and they include amino-acids, vitamins, hormones, or other compounds with a hormone-like action.

Another possible technique is intra-ovarian pollination in which pollen is introduced directly into the ovary, the borings being sealed afterwards with petroleum jelly. This technique had not been tried extensively up to the time of writing, but it has been successful in limited experiments (Maheshwari and Kanta, 1964).

BEE-KEEPING

The pollination of crops leads on the one hand to the production of fruit or seed and on the other, in the case of bee-pollinated crops, to the production of honey. The behaviour of the honey-bee (*Apis mellifera*) has been dealt with in detail in Chapter 5, and many of the discoveries in that field are of great importance economically. Honey-bees on the whole behave in a manner which is for the good of the hive, and this means that they tend to seek out and exploit those crops from which they can most easily obtain large quantities of pollen and nectar. However, von Frisch (1954) found that, although the corollas of certain thistles were too long for honey-bees to work easily, they could be artificially induced to visit these plants; they then obtained a worthwhile crop of nectar from them and so produced more honey. The method of inducement was to place near the hive sugar-water in which thistle flowers had been soaked to give their scent to the liquid. When the pollination of an agricultural crop is the main concern and the flowers of the crop are relatively unattractive to the bees, the procedure just described assumes a greater importance. It has been tried on many occasions (particularly with red clover in Russia) and is sometimes successful, but unfortunately it fails quite frequently and no clear cases of success in Britain are known. But despite the occasional difficulties bee-keeping is a necessary adjunct of much crop production, for wild insects can rarely cope adequately with the huge area of plants grown. Honey-bees, however, can be brought to the crop in enormous numbers in their portable homes, and this practice is responsible for a large proportion of the honey that is produced.

SELF-POLLINATED CROPS

Problems of pollination do not arise in the raising of crop plants that are freely self-pollinated. In fact one possible method of dealing with pollination problems is to select self-fertilising cultivars. Several cultivated plants such as the sweet pea (*Lathyrus odoratus*) and the culinary pea (*Pisum sativum*) are regularly self-fertilised, although the structure of the flowers indicates that their ancestors were insect-pollinated. However, in some other cultivated plants the deterioration which normally occurs when an outbreeding species is forced to inbreed is so serious that self-fertilising cultivars are useless. This is true, for example, of lucerne (*Medicago sativa*) (Bohart, 1957).

WIND-POLLINATED SEED CROPS

Another pollination-arrangement that is independent of insects is wind-pollination, which unfortunately has the accidental consequence of giving some people hay-fever (see Chapter 8). This has economic significance in being widespread and expensive to alleviate.

An example of a wind-pollinated crop is maize or corn (*Zea mays*), a member of the grass family (*Gramineae*). The male flowers are produced at the top of the plant in panicles, referred to as tassels, while the female inflorescences, or ears, are produced lower down and are enclosed in sheaths, from each of which emerge many enormously long drooping stigmas, called silks. The tassels produce flowers for a period of about two weeks, and the silks receive pollen which has been scattered by the wind. Maize is grown for its seed, which may be required either for food or to provide the following year's crops. In the latter case, attention has to be paid to the possibility of contamination from different cultivars grown in the same area, as it is essential that named cultivars should come reasonably true from seed. However, experiments show that contamination may drop by 99 per cent. over a distance of 40 to 50 feet from the contaminant crop (Bateman, 1947b). Such results provide a basis for deciding the adequate separation between maize crops which are liable to contaminate one another.

CONTAMINATION

The same problem of contamination arises with insect-pollinated plants that are grown as seed crops – many annual flowers and vegetables, for example – and because of this problem appeals are regularly made to private gardeners not to let their brassica crops run to seed. In experiments with turnip (*Brassica napus*) and radish (*Raphanus sativus*) Bateman (1947a) found that contamination dropped by 99 per cent. over a distance of 160 feet, and Lawrence (1939) stated that it was advisable to keep cultivars of nasturtium (*Tropaeolum majus*) very well separated.

The danger of contamination has almost entirely disappeared in the double forms of China aster (*Callistephus chinensis*) because the doubling of the flowers makes them difficult for the bees to visit, so that seed-production for these forms of the aster depends on self-pollination (Lawrence, 1939).

HYBRID SEED FOR CROPS AND FLOWERS

By avoiding contamination, growers are able to produce uniform seed, which is very desirable in crop plants because it ensures similarity of reaction to particular conditions and also ensures a uniform product. It has now been found, however, that better crops can be produced by the crossing of two different uniform cultivars, for by this means the phenomenon of hybrid vigour (or positive heterosis) manifests itself, the progeny often being both extremely vigorous and at the same time very uniform. Unfortunately, if the seed of these progeny is raised, the next generation will be found to be highly variable, so the new cultivar has to be produced every year by crossing the same two chosen parents, different hybrid cultivars being obtained by using different pairs of parents. Maize was one of the first crops in which this method was used, the two chosen parent cultivars being interplanted in the same field. The tassels are removed from one cultivar so that when the cobs of this are gathered at a later stage they will have been pollinated from the tassels of the other. The female parent is usually made to outnumber the pollen source by two plants to one, or up to four to one. Use of hybrid maize became extensive in the United States in the 1930s, and by 1951 it accounted for 81 per cent. of their total crop (Hayes, Immer and Smith, 1955). Subsequently this method of producing new cultivars was extended to many insect-pollinated plants. Bodger (1960) described the crossing of cultivars of Petunias growing in separate greenhouses. The flowers of one cultivar are rendered female by the removal of the stamens a week before pollination

PLATE 49. **a**, flowers of Ling *(Calluna vulgaris)*. **b**, flower of Common Rock-rose *(Helianthemum chamaecistus)*. **c**, flower of Common Rock-rose after stimulation of the stamens.

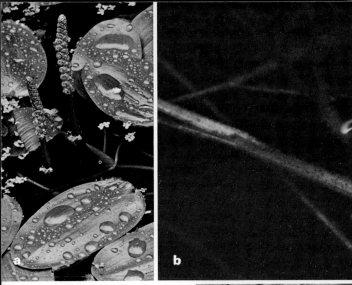

PLATE 50. **a**, inflorescences of Broad-leaved Pondweed *(Potamogeton natans)*. **b**, inflorescence of Tassel Pondweed *(Ruppia maritima)* shedding pollen just above the water surface. **c**, Eelgrass *(Zostera angustifolia)*. **d**, developing pollen of Eel-grass.

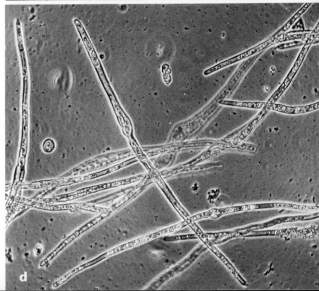

is due, and the stigmas when mature are artificially dusted with pollen collected from the other cultivar. Each female plant remains in continuous production for five months and produces about two hundred flowers, the seed capsules being collected as they ripen. If male-sterile cultivars can be bred (as in the case of onion and tomato) the laborious business of emasculation can be eliminated.

BEE-POLLINATED SEED CROPS

Lucerne

Lucerne, or alfalfa, is a field crop which requires insect-pollination to produce a full yield of seed. There is some degree of genetic self-incompatibility, and in any case the prepotent foreign pollen (i.e. from another individual of the same species) gives the better progeny, continual selfing leading rapidly to loss of vigour (see p. 384). In the flower of Lucerne the style and stamens emerge explosively from the keel when the flower is visited (see p. 200); during the explosion, referred to as 'tripping', the surface of the stigma becomes abraded and for the first time receptive to pollen. Pollination normally takes place at this stage when the stigma forcibly strikes the underside of the visiting insect, on which there is usually pollen from a flower visited previously. After the flower is tripped the stigma presses against the standard, and it is then almost impossible for it to receive any further pollen. Lucerne is a native of the steppe regions of eastern Europe, but it is a very important crop in North America where it is effectively pollinated by most of the indigenous solitary bees, especially alkali bees (*Nomia melanderi*) and leafcutter-bees (*Megachile* spp.). Unfortunately the spread of agriculture causes the number of wild bees to decrease, so that, when lucerne is grown in a newly cultivated area, yields of seed are very high at first but decrease after a number of years. Honey-bees can then be imported, but they do not always do the job required of them because, except during a short period when they are unfamiliar with the crop, they trip less than 3 per cent. of the flowers they visit for nectar. They soon learn that, by probing the keel from the side, they can get at the nectar more easily and avoid the discomfort of a sharp blow on the underside from every flower. When they are collecting pollen they are much more effective, because they succeed in tripping most of the flowers they visit, but they cannot be relied on to collect pollen from lucerne, often preferring to get their supplies from other plants. Consequently, if the honey-bee is used as a pollinator, very high densities have to be maintained. A European bee which is strongly associated with lucerne is *Melitta leporina*. This has been found at a density of 168 bees per hectare in Denmark; the tripping rate was very high and figures given by Stapel suggest that these bees were causing two or three times as much pollination in a given period as honey-bees working the same crop at a density of 7000 per hectare (see Todd, Norris and Crawford in Mittler, 1962). When honey-bees have to be relied on, pollination may sometimes be improved by the elimination of wild flowers in the neighbourhood of the crop. Heavy irrigation should be avoided when the lucerne is flowering, for the percentage of accidental tripping by honey-bees collecting nectar tends to be higher the hotter and drier the climate; if it falls below 1 per cent. it may be impossible to raise the density of honey-bees sufficiently to produce a good seed-set. It is possible that in the future lucerne may be made more attractive to honey-bees through the elimination of the buffeting from the exploding flowers. Attempts to achieve this involve searching for suitable chance floral abnormalities which can be bred into the crop (Nielsen in Mittler, 1962; Arnason in Åkerberg and Crane, 1966).

An alternative to the use of honey-bees is the restoration of the declining wild bee populations, and this has been done successfully in North America for lucerne pollination with both alkali bees and leafcutter-bees.

Alkali bees like to nest in bare, damp, rather light soil, patches of which can be prepared for them near the crops, and this is particularly successful because these bees are highly gregarious and will nest very close together (Frick, Potter and Weaver, 1960; Stephen, 1965). Bare, damp soil arises naturally where water (sometimes irrigation water) flows over an impervious layer just underground, and percolates to the surface. Evaporation then causes an accumulation of salts, which keeps down the vegetation, reducing it to, at most, a thin growth of specialised plants able to tolerate the salts. Where the natural soil is of a suitable texture, water can be supplied artificially, if necessary, to produce new nesting sites. If a gentle slope is available, a series of pits can be dug near the top and kept supplied with water up to a constant level. The seepage down the slope then produces the desired degree of moisture, and vegetation can at first be killed off artificially. On level ground long trenches can be dug, and lined in the lower part with plastic sheeting which must rise one foot up the sides. The trenches are half-filled with coarse gravel, topped by a thin layer of coarse sand, and then filled up with the natural soil. Large quantities of water must be piped to the gravel layer throughout the flight-season of the bees. The gravel acts as a reservoir, moistening the natural soil in the trench above it, and on either side for a short distance. Although this method is fairly simple, and therefore cheap, population densities rarely reach those in completely artificial beds which are in any case essential in areas where there is no suitable natural soil. For artificial beds, carefully selected soil is imported to provide optimum conditions. A bed of 30×50 feet should supply pollinators for 30 to 40 acres of lucerne, but far larger beds are sometimes made. The area is excavated

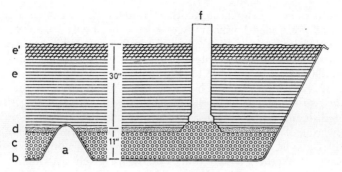

FIG. 127. Cross-section of part of artificial bed for nesting Alkali Bees. a – one of the soil ridges dividing bed into segments so that a leak cannot ruin whole bed. b – polyethylene lining. c – gravel layer forming water reservoir. d – coarse sand. e – nesting soil. e' — upper 8 inches with common salt added. f – downspout for saturating the gravel layer. These beds require little water once the reservoir is filled. From Stephen (1965).

to a depth of between three and four feet, lined with plastic sheeting, and filled with gravel, sand and the imported soil, much as described for the trenches (Fig. 127). The soil is spread and compacted by bulldozers, and salt is then mixed with it to a depth of six inches, at the rate of one to three pounds per square foot. (The salt greatly reduces summer evaporation but may cause concretion of the soil after a year or two, when calcium chloride or gypsum are added to counteract this.) Finally, the beds have to be rolled with a hand-roller. The alkali bee populations can now be quickly built up by importing, in winter or early spring, blocks of soil containing dormant prepupal bees from other areas. Owing to the gregarious habits of these bees a sufficient number must be introduced at one time to prevent them from wandering off looking for more thriving colonies. Once a colony is established, the bees tend to fly in a stream from the beds to the fields, and their flight line may cross busy roads at a height of only three or four feet. Since each bee is reckoned to be able to set about a pound of seed, notices are sometimes put up requesting motorists to slow down for alkali bees.

In Canada, leafcutter-bees have been increased by leaving strips of scrub in the lucerne fields and dumping piles of timber refuse on them for the bees to nest in. The native *Megachile* species thus encouraged, though individually capable of setting two pounds of seed per bee, are still subject to numerical fluctuations on account of the weather and parasites. The European bee, *Megachile rotundata*, which was accidentally introduced into the United States, has been increased by the provision of artificial nest sites in the form of holes drilled into blocks of wood, cans packed with drinking straws and positioned horizontally or even rolls of corrugated cardboard (Stephen, 1961; Bohart, 1962). Large populations of this bee can easily be built up because it is highly gregarious, each female choosing for its nest a hole as near as possible to an already occupied hole. Nesting units or 'hives' containing 10,000 holes or more are normal in the U.S.A. *M. rotundata* is not hardy in the Canadian winters but, being so easily managed, it can be protected in the winter and its emergence timed for the flowering of the crop in the summer. Practical details for its management in Canada, whither a million *M. rotundata* were imported in 1964 (Arnason in Åkerberg and Crane, 1966), are given by Hobbs (1965). The individual cocoons, made up of bits of leaf, are extracted from the nests in winter, and all defective cells (those which collapse when gently rolled between the thumb and forefinger) are discarded. Storage is carried out at 40° F. Fifteen days before the lucerne begins to bloom the cocoons are counted out into plastic-topped incubation trays with a few stoppered holes drilled in the side. These trays are incubated at 85° F. until about 10 per cent. of the bees have hatched; any parasites emerging during this period are disposed of. Then the trays are placed in shelters which are spread evenly through the fields. Each shelter is a box open at the side and mounted on posts, and contains a 'hive' with, say, 3000 nesting holes, made by stacking boards, grooved on each side, one on top of the other so that the grooves match to form tubes. The incubation trays, with the stoppers removed, rest on top of the 'hive' under the roof of the shelter. The bees emerge and nest in the adjoining 'hive', making their cells from lucerne leaves, each female filling two or three holes 4½ inches long. After the nesting season the grooved boards are separated, the cocoons removed to storage and the 'hives' cleaned and prepared for next year.*

Red Clover

Another field crop requiring insect-pollination is red clover (*Trifolium pratense*) which is highly self-incompatible. Its natural pollinators are bumble-bees (Plate 24c) of various species whose importance seems to vary with the tube-length of the clover cultivar being grown (see p. 200); for example, Hawkins (in Mittler, 1962) found in England that the *Bombus lapidarius* group predominated on short-tubed cultivars, the *B. hortorum* group on long-tubed cultivars in the same trial, while the *B. agrorum* group was equally frequent on all cultivars tested (Fig. 128). The workers of the short-tongued *B. terrestris* group are unfortunately persistent corolla-biters, though less troublesome on the shorter-tubed red clover flowers than on the longer. However, in Sweden, where the *B.*

* Rapid progress is shown by a picture of a shelter with 51,000 holes and a description of a mechanical cocoon-remover in the 1967 edition of Hobbs's leaflet.

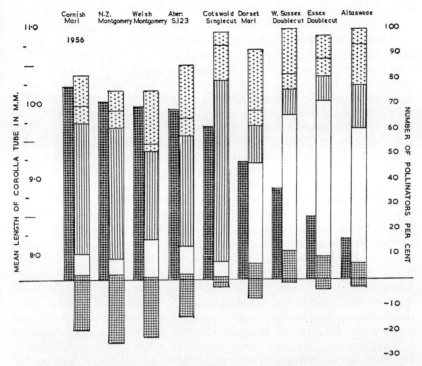

FIG. 128. Bumble-bee visitors and flower-tube length of Red Clover cultivars. Left-hand columns (black squares) show tube-length. Between 'Cotswold Singlecut' and 'Dorset Marl' the frequency of visits by the long-tongued *B. hortorum* group (vertical hatching) drops abruptly, while that of the short-tongued *B. lapidarius* group (white) increases. Near this point the proportion of negative visits (biting or feeding through holes – shown below the line) by the *B. terrestris* group also drops (white squares). Crosses represent the *B. agrorum* group. Based on figures for 1956 in Hawkins (1965). Comparable results were obtained in other years.

terrestris and *B. lapidarius* groups are usually the commonest bumble-bees on red clover, the former do not rob at the beginning of the season, so that they may be useful pollinators in years when the clover is early (Mittler, 1962, Part I). As with lucerne, pollination is dependent on the availability of nesting sites for the bees near the crop. For example, Hawkins found that one late-flowering cultivar of red clover, largely dependent on *B. agrorum* for pollination, yielded $2\frac{1}{2}$ cwt. of seed per acre when growing in open country, whereas a few miles away a field of the same cultivar, growing near a forest, yielded $6\frac{1}{2}$ cwt. per acre, apparently because the forest provided a suitable nesting habitat for the bees. Hawkins has pointed out the need for a full study of bumble-bee ecology, and has indicated the value of the white deadnettle (*Lamium album*) and also crops of the winter and spring cultivars of field bean (*Vicia faba*), which all provide nourishment for these bees before the flowering of the clover. Wild flowers and crops attractive to bumble-bees are certainly valuable outside the clover season, but in south-west Sweden the development of the cultivation of oil seed (white mustard (*Sinapis alba*) and summer rape (*Brassica napus*)) diverted the bumble-bees from the red clover and alsike clover (*Trifolium hybridum*), and also led to a reduction in the total bee population because of the use of insecticides (Wahlin in Mittler, 1962). In general, insecticides should only be used at times when there is no danger to pollinating bees.

Free and Butler (1959) have discussed the reduction of bumble-bee populations caused by modern agriculture, which involves the elimination both of hedges and of patches of neglected land. This may be offset either by the restoration of suitable ground for nesting sites, in the manner suggested by these authors, or by the provision of artificial nest boxes. In North America, Hobbs set out artificial nest boxes in wild places and, when the bumble-bee queens had established colonies in them, he moved them to the crops. It was found necessary to capture any workers not in the boxes at the time of removal, and also to use variously coloured boxes with the entrances facing in different directions to avoid confusion of neighbouring nests by the queens. In a promising experiment in Denmark the nests from the wild were placed initially in greenhouses; all the young queens raised by each colony then hibernated the following winter, and founded many more new colonies in captivity in the spring (Hobbs in Mittler, 1962; Holm in Mittler, 1962 and in Åkerberg and Crane, 1966). After surmounting the difficulties of finding a suitable material for hibernation and providing favourable conditions for the emerging queens, very good percentages of survival and nest establishment were achieved, using substantial numbers of bees. However, Holm reported that further research was required to overcome the serious losses which occurred when the established nests were moved from the greenhouses to the open fields.

Although bumble-bees are essential for the pollination of the longer-tubed cultivars of red clover, and particularly for the valuable tetraploid cultivars which have larger flowers than the others, honey-bees may be used in some countries as pollinators of the remaining cultivars, and growers sometimes hire them for this purpose. They are more effective on the second flowering of the red clover crop than on the first, possibly because the later flowers tend to be shorter-tubed or because the nectar is both sweeter and more abundant in warm dry weather (Bohart, 1957). In England, Free (1965a) found that honey-bees visiting red clover caused the same percentage of florets visited to set seed as did bumble-bees; in both cases the set by pollen-gatherers was much higher than by nectar-gatherers.

Introduction of Pollinating Bees

The introduction of new species of wild bees from foreign countries has been urged by Bohart (in Mittler, 1962). Not only would this increase the number of species available in the receiving countries, but the introduced bees, if brought in free from disease and parasites, would probably flourish better than the native bees. The success of the introduction of bumble-bees to New Zealand to fertilise the introduced red clover is well known. Moreover, several other species of wild bees, introduced into various countries by chance, have, like *Megachile rotundata*, become useful pollinators in their new areas. Both the alkali bee and *M. rotundata* have been introduced into Canada, but both are limited to certain areas with sufficiently high summer temperatures, the alkali bee the more severely (Arnason in Åkerberg and Crane, 1966).

OTHER INSECT-POLLINATED SEED CROPS

A largely self-incompatible field crop, which is pollinated mainly by short-tongued insects (bees, wasps and flies), is the carrot (*Daucus carota*). In an investigation carried out at Logan, Utah, all the insects visiting carrot flowers over a period of four years were identified, and a total of 334 species, belonging to 37 families, was recorded (Hawthorn, Bohart and Toole, 1956). The abundance of particular species varied from year to year and from one part of the flowering season to another. Among the genera commonest on average the following were judged, from their behaviour at the flowers, to be the most efficient: *Apis* and *Halictus* (bees), *Tachytes* (a solitary wasp, family *Sphecidae*), *Eristalis* and *Syritta* (both large hoverflies), and *Stratiomys* (a soldier-fly). It was suggested that growers might improve the pollination of carrots by maintaining a good supply of honey-bees, avoiding the presence of competing bloom, and providing decaying vegetation nearby for *Diptera* to breed in (Bohart and Nye, 1960).

The pollination of the self-incompatible cabbage and brussels sprout is carried out in isolation-greenhouses in some areas, owing to the great danger of contamination (see p. 342). Insect-pollinators are introduced into the houses and, while honey-bees are the easiest to procure, bumble-bees and blow-flies (*Calliphora*) give a greater seed yield. The blow-fly pupae can be obtained from factories which produce maggots for anglers (Faulkner, 1962).

FRUIT CROPS

Tomato

Although the tomato (*Lycopersicon esculentum*) is self-fertile, it cannot be relied upon to pollinate itself as it usually requires some disturbance to the flowers. Without this there may be too few flowers pollinated to give the maximum fruit-yield, and too few ovules fertilised in each flower to give well-formed fruits. In Britain, and other cool temperate climates, the tomato is an important crop, but it ripens best in greenhouses. The traditional method of causing pollination under glass is to tap the stakes to which the plants are tied, but more recently an electrical truss-vibrator has been introduced which is carried in the hand and applied in turn to each truss. Using the vibrator is slower than tapping the stakes, but it increases yield, particularly in those cultivars in which the style protrudes from the cone of anthers (which resembles that of *Solanum* (p. 191)) surrounding it. The protruding style prevails in wild forms of the tomato (whose native country is South America), and the short style apparently represents another example of inadvertent change brought about in a flower by selection for fertility in cultivation (see peas, p. 341). (In private gardens in Britain various exotic fruits, when grown under glass, require pollination by hand with the aid of the traditional camel-hair brush.)

In warm climates, the tomato can be grown out-of-doors, where it is pollinated by insects. Rick (1950) found that, although the most active pollinator of the tomato in an area of California was a bee of the genus *Anthophora*, in the tomato's native habitat in Peru a bee related to *Halictus* pollinated the same variety at least twice as efficiently.

Avocado Pear

The sub-tropical fruiting tree, avocado pear (*Persea gratissima*, family *Lauraceae*), is also self-fertile, but requires insect pollinators because its timing mechanism prevents self-pollination. Avocado trees are normally grown commercially as vegetatively propagated clones, now known to be of two distinct types which may be called A and B. In type A, the flowers open in the female stage for a morning only, and then reopen in the male stage for the afternoon of the following day. In type B, the female stage occurs during an afternoon and the male stage the following morning. Although each flower lasts for no more than two days, fresh flowers are produced each day during the season (i.e. every morning by type A and every afternoon by type B). The synchronisation of this timing mechanism precludes pollination between flowers on the same tree or type of tree; but if A trees and B trees are interplanted, insects can transfer pollen from B to A in the mornings and from A to B in the afternoons (Fig. 129).

PLATE 51. **a**, flower of Dove's-foot Cranesbill
(Geranium molle) showing close proximity of
stamens and stigma. **b**, Bog Asphodel
(Narthecium ossifragum) in artificial 'rain'; in
the lowest flower a large drop of water is
resting on the perianth segments; in the next
flower a drop is resting partly on the stamen-
hairs. **c**, self-pollinated clovers near the
Lizard, Cornwall: *Trifolium bocconei, T. dubium,
T. scabrum, T. striatum* and *T. strictum*.
Compare Pl. 24 **b-d**.

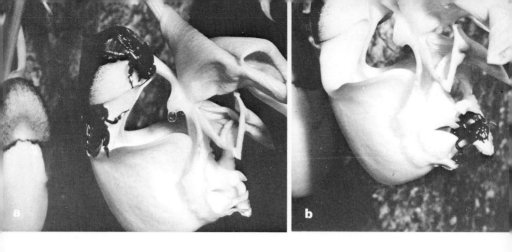

PLATE 52. Two South American orchids: **a**, pendent flower of Bucket Orchid *(Coryanthes rodriguesii)* with two bees *(Euplusia superba)* scratching at the edge of the cap-like structure at the angle of the lip; the descending arm of the lip carries the bucket; the column descends vertically and closes the right-hand side of the bucket. **b**, the same, with a bee struggling to emerge from the flower after immersion in the bucket. **c-d**, two stages in the visit of a bee *(Euglossa viridissima)* to the flower of the orchid *Gongora maculata*: **c**, bee upside-down, scratching the lip. **d**, bee loses grip and falls with back to column, down which it slides. See also Pl. 56.

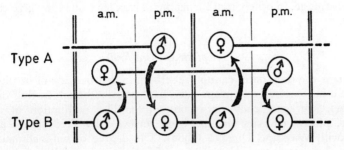

FIG. 129. Diagram to show timing of anthesis in the Avocado Pear. Horizontal lines connect the male and female stages (circles) of a single flower. Curved arrows show possibilities for pollination.

For each tree there is usually a short interval of less than an hour between the closing of the morning flowers and the opening of the afternoon ones (Robinson and Savage, 1926).

Vanilla

This tropical American Orchid is particularly successfully cultivated in Madagascar. However, its pollinating bee, *Melipona*, is not available there and workers have to pollinate the flowers artificially to ensure the ripening of the pods, which are the source of the flavouring. A small stick is used to remove the anther cover and tuck the rostellum out of the way; the anther and stigma are then pressed together, causing self-pollinatio n.

Date Palm

The dioecious date palm (*Phoenix dactylifera*) is also propagated vegetatively, the clones being either male or female. The trees are wind-pollinated, and it is only necessary to tie part of an inflorescence from a male tree into each female tree to secure pollination (see p. 19). For this purpose it is sufficient to grow one male tree for every hundred females, and the male inflorescences may be cut and tied up a little before their flowers open (Cobley, 1956). When by careful experiment the same female date palm is pollinated by male trees of several different clones, the resulting dates differ according to the source of pollen. The pollinations which produce larger seeds yield dates with more flesh (i.e. maternal tissue), the pollen having influenced the flesh indirectly ('metaxenia') through its direct genetical effect (xenia) on the seed (Nixon, 1928). By selecting suitable male clones, the size of dates can thus be increased. The larger fruits, however, ripen more slowly; and although in some climates this is an advantage, in others rapid ripening of smaller fruits may give the best yield.

Temperate Orchard Fruits

Whereas the date palm cannot be self-pollinated because of its dioecism, some other vegetatively propagated fruit-trees, although having bisexual flowers, are genetically self-incompatible. The pollination of apples, pears, plums, cherries etc. has been reviewed by Free (1962) and by Luckwill, Way and Duggan (1962). The degree of self-incompatibility varies, some fruit-tree clones (i.e. cultivars) being partially, some wholly, self-compatible, although this property may change somewhat from year to year. Since it is only necessary for a small proportion of the flowers to set fruit, a moderate degree of self-compatibility will often suffice to produce a good crop (Crane and Lawrence, 1952). However, in many cases, particularly among apples, self-incompatibility is so great that it is essential to grow two different suitable cultivars together. Only a few cultivars of apples are really well known to the general public, and the cultivars with which they have to be planted in the orchards are called pollinisers, being grown mainly for their pollen rather than for their fruit.

Bees are the most important pollinators in orchards, but other insects which probably help to pollinate the blossom from time to time are: midges and fungus-gnats (both present in stormy weather when few other insects are about), St Mark's flies and fever flies, some flies of the families *Muscidae* and *Calliphoridae*, and small flower-beetles of the family *Nitidulidae*. (All these insects are mentioned in Chapter 4, Part 1.)

Among the solitary bees, the commoner species of the genus *Andrena*, together with *Osmia rufa*, are the most important fruit-tree pollinators in Britain. They can only be significant, however, when local conditions enable good numbers to nest in or near the orchards. Even then these bees are much affected by the weather, requiring higher temperatures for their activities than honey-bees, and occurring in reduced numbers if the previous season has been unfavourable to them. It has been found that *Andrena* visit half as many flowers per minute as honey-bees but in partial compensation pollen adheres to *Andrena* more loosely and to a greater area of the body. In Nova Scotia, pollination of fruit-trees is almost entirely dependent on *Andrena* and *Halictus*, while in Norway solitary bees and bumble-bees are effective together, honey-bees being almost absent in these countries. Bumble-bees are subject to the same disadvantages of scarcity and variation from year to year as solitary bees, but they operate at a lower temperature than honey-bees and are more likely to move from tree to tree. Moreover, in certain apple and plum cultivars with long styles projecting up beyond the stamens, only bumble-bees are obliged by their size to straddle the stigmas.

Honey-bees, of course, have been carefully studied on fruit-blossom and are its most important pollinators when foraging for pollen. It has been found that they are not very effective when collecting nectar because then, besides visiting fewer flowers per minute, they often stand on the petals and probe between the stamens from the side without touching the stigmas, particularly if the stamens are long and stiff. In order to increase the proportion of pollen-gatherers, sugar-syrup is sometimes offered near the hives. The bees keep an approximate balance in the intake of the two foods and their requirement for pollen therefore increases if they can rapidly collect a large quantity of syrup from an artificial supply. However, when natural nectar supplies are very good, the bees may ignore the sugar-syrup. It has been pointed out by Free (1965b) that attempts to direct bees to a crop by feeding scented sugar-water, as described on pp. 340–1, may produce better pollination simply by increasing the collection of pollen. Honey-bees work more on the sunny side than the shady side of large fruit-trees and, when leaving one tree for another, they fly to its nearest neighbour. The bees can easily collect a full load of pollen or nectar from a single tree, but in fact they visit on average two trees per trip, though the young scout bees may visit considerably more. The bees move more often from tree to tree if small fruit-trees are used. The rule that bees always change from a less attractive crop to a more attractive one (p. 176) applies also to plants of a single species, and the significance of this for apple orchards was investigated by Free (1966). As fertilisation takes place when bees change from foraging on one cultivar to foraging on another, or when they visit two equally good cultivars indiscriminately (Free and Spencer-Booth, 1964), it is desirable that the polliniser should not be more attractive than the cropping cultivar, nor excessively inferior to it. A single honey-bee colony may forage over an area of about 120 acres, but in poor weather only those bees with foraging areas near the hive venture out. It is best to bring the bees into the orchard after flowering has begun, as they will then concentrate for a time entirely on the fruit blossom and will not give a share of their time to flowers outside the orchard until a later stage. In orchards which are grassed down, the flowering of Dandelions (*Taraxacum officinale*) should be prevented, as they produce nectar at a lower temperature than the fruit blossom and may be constantly preferred by the bees (Free, 1968).

The density of hives in orchards is usually one to the acre. They are placed in groups for convenience, and it has been found that four hives grouped in every four acres give an even density of bees throughout the orchard even in poor weather. Sometimes, at flowering time, bouquets of blossom may be cut from the pollinisers and tied into the cropping trees in order to improve the fruit set. The positioning of the polliniser trees is in any case an important matter. Ideally every other tree should be a polliniser, and all trees should be equally spaced. However, various other planting-patterns have been devised, taking into account the habits of the bees and using a smaller proportion of the pollinisers, as these are normally less valuable than the main crop. (A case in which a blackberry polliniser has been bred specially to have the same cultural and fruit qualities as the original cropping cultivar has been reported in America (Shoemaker, 1962).)

In the United States, hand-pollination has been used in areas where pollinating insects are scarce. Although laborious, it has the advantage of giving control over the distribution of fruit on the tree, and it may also eliminate the need for thinning the young fruit. In Japan, pollen has sometimes been mixed with lycopodium powder to make it go further in hand-pollination, and much effort is being put into devising economic methods for pollinating apples mechanically (Sadamori et al., 1958; Ohno, 1963).

It has been found in Canada that some virus and fungal diseases of fruit trees may be spread in pollen or by pollinating insects. One disease concerned is fire-blight of apple and pear and here an attempt has been made to turn the tables, so to speak, by supplying streptomycin to bee-hives (in pollen-inserts) and so getting the bees to distribute the fungicide (Arnason, in Åkerberg and Crane, 1966).

ECONOMIC IMPORTANCE AND IMPACT ON NATURE CONSERVATION

Perhaps the most striking evidence of the economic importance of adequate pollination is the scale of operations involved in preparing alkali bee beds, but the vast amount of research on the pollination of crops (going on all over the world and representing an enormous investment) also indicates the size of the possible economic rewards.

It may be of interest to conclude this chapter with a summary of the ways in which the demands of crop-pollination affect nature conservation. In so far as crop-pollination depends on the honey-bee, there is nothing to stop the spread on suitable land of 'total agriculture', with its elimination of natural and semi-natural habitats. However, the destruction of wild insects by insecticides is suspended at certain times of the day or year for the sake of the honey-bees. When wild flowers are considered to be competing with the crops for the attention of the bees, they may be totally eliminated, or the blooms may simply be mown at the appropriate time. Where bumble-bees and other wild bees are useful pollinators, or where ichneumon-wasps are important for pest-control (p. 136), habitats of wild flowers may be left to give them food when the crops are not in flower. The value of wild insects for pollination may also lead to the conservation of small wild areas in which they can nest. But even here the provision of artificial nests and the growing of early-flowering crops

PLATE 53. Bat-pollination: **a**, inflorescence of *Parkia clappertoniana* being visited by the Gambian fruit bat (which has a wing-span of 60 cm.). **b**, flower-clusters of the Kapok Tree being visited by a pygmy fruit bat (wing-span 34 cm.). **c**, inflorescence of Durian *(Durio zibethinus)*, another Old-World bat-pollinated tree. The petals recurve in the evening to expose a dense cluster of stamens (this has happened in one of the flowers shown).

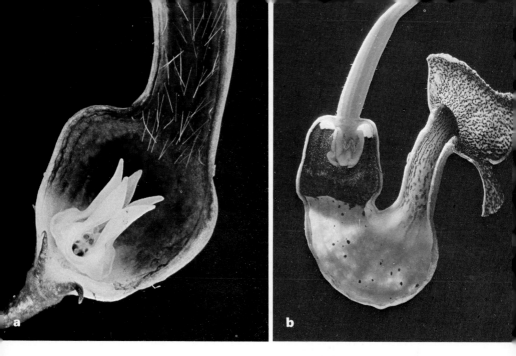

PLATE 54. **a**, base of flower of *Ceropegia woodii* cut open to show trap hairs, corona with upstanding lobes, a stamen visible between corona-lobes, and the 'window-pane' in the base of the prison. **b**, flower of Dutchman's Pipe *(Aristolochia sipho)* cut open to show colouring of interior, with window-pane, and form of reproductive organs. **c**, bumble-bee *(Bombus agrorum)* sucking nectar from the cuculli of the corona of *Asclepias curassavica* (in cultivation). Note the pollinia attached to the bee's legs.

specially for wild bees may eventually reduce the need for natural nesting and feeding sites; thus, though selected wild bees will be conserved, the wild flowers will be destroyed, and so will any birds that prey significantly on the bees. One may plausibly conclude that nature will be eliminated in areas where the terrain is suitable (other than nature reserves), except that some wild bees will be encouraged artificially; but, where unploughable areas exist alongside arable land, the wild bees and the natural habitats that support them will possibly be deliberately conserved.

*　　*　　*

Postscript. The subject of insect-pollination of crop plants is thoroughly covered in a book which appeared after this chapter was written (Free, 1970).

CHAPTER 12

THE EVOLUTION AND ECOLOGY
OF POLLINATION*

THE FOSSIL RECORD

THE rocks have yielded no more than fragmentary and tantalising glimpses of the earliest history of flowering plants. A number of fossils of supposed angiosperms have been described from the earlier parts of the Mesozoic period, but most of these are too incomplete or show too little detail for certain identification, or are attended by other doubts. It is not until the Lower Cretaceous that undoubted angiosperm fossils begin to appear in any numbers. From Aptian (Lower Greensand) times onwards, fossil leaves and pollen of angiosperms appear with increasing frequency. By Upper Cretaceous times the angiosperms had become the dominant plants in the vegetation of the Earth. The most remarkable feature of this Upper Cretaceous flora is its extraordinarily modern aspect. A large proportion of the plants can be identified with modern families and genera – though the reservation must be made that most of the identifications are of leaves, and that flowers are rarely preserved. Some of the plants identified belong to families generally held to be primitive, for instance *Magnoliaceae, Nymphaeaceae* and *Alismaceae*. However, the sixty or so families recorded from the Cenomanian† stage of the Cretaceous (mostly from North America) include for instance, *Asclepiadaceae, Boraginaceae, Caprifoliaceae, Cornaceae, Cyperaceae, Ericaceae, Fagaceae, Leguminosae, Liliaceae, Myrtaceae, Oleaceae, Palmae, Platanaceae, Rosaceae* and *Salicaceae*. Even allowing for a proportion of misidentifications, this was evidently a flora of great diversity and considerable evolutionary advancement. The most notable absences are of families with large gamopetalous zygomorphic flowers, and a few other specialised groups – *Bignoniaceae, Labiatae, Scrophulariaceae, Compositae* and *Gramineae* do not, in fact, appear in the fossil record until the Tertiary period.

The origin of flowers as we know them – and probably of the angiosperms themselves – must be closely bound up with the evolution of the flower–insect relationship. The idea that the evolution of flowers and insects proceeded hand-in-hand is borne out by a comparison of the fossil record of the angiosperms with that of the insects that pollinate them. The insects were already greatly diversified by the beginning of the Mesozoic (Fig. 130, p. 359). At the time when we may suppose that the

* Baker and Hurd (1968) discuss many of the topics considered in this chapter.
† Contemporary with the deposition of the Lower Chalk in England.

early angiosperms were evolving, most of the insect groups which include casual and more-or-less indiscriminate flower-visitors at the present day – thrips, cockroaches, bugs, beetles and the more primitive flies – were well represented. Of these, the beetles were particularly prominent, making up some 37 per cent. of the insect species known from the Mesozoic. Of the groups which now include the most specialised flower-visitors, the *Hymenoptera* were represented only by the sawflies and (from the Jurassic onwards) by ichneumons and related forms. The *Hymenoptera Aculeata* – ants, bees and wasps – do not appear until the Tertiary; bees are first recorded from the Oligocene. Among the *Diptera*, the flower-visiting hoverflies (*Syrphidae*) first appear late in the Eocene, at about the same time as the first *Lepidoptera*. However, these specialised flower-visiting insects had no doubt existed for a considerable time before we first have evidence of them, and they may well have originated during the Creta-ceous. Bats and birds must certainly have entered the field of pollination later than the insects. Bats first appear in the fossil record in the Eocene, and these are not of flower-visiting groups. The specialised nectar-feeding birds have left almost no trace in the rocks; the earliest, a hummingbird from Pleistocene deposits in Brazil, is geologically very recent.

THE ORIGIN OF THE ANGIOSPERMS

The origins of the angiosperms are to be sought among the great diversity of primitive seed-plants (*Pteridospermae*) which dominated the vegetation of the late Palaeozoic and early Mesozoic periods. There is no doubt that the ancestors of these early seed-plants were pteridophytes – spore-bearing plants having a life-cycle with an alternation of generations of the kind seen in the modern ferns, horsetails and clubmosses (Fig. 131, p. 360). The spores of a fern germinate to form a small delicate *prothallus*, bearing *antheridia* and *archegonia*. The former produce motile *antherozoids* which can swim in a film of moisture to fertilise the *egg-cell* in the flask-shaped archegonium. The fertilised egg-cell then develops into a new fern plant. In many Palaeozoic pteridophytes, as in the modern clubmoss *Selaginella*, the spores are of two kinds, *megaspores*, which produce 'female' prothalli bearing archegonia, and *microspores* which grow into smaller 'male' prothalli producing antheridia. The way in which the seed habit probably originated is illustrated by some modern species of *Selaginella*, in which the megaspore germinates to form a small prothallus bulging from a split in the spore-coat, and the archegonia are ready for fertilisation before the spore is shed. If microspores fall into the open megasporangium and complete their development there so that fertilisation can take place, the megaspore may already contain a well-developed embryo by the time it is shed from the parent plant.

In the gymnosperms the events of reproduction still show an obvious correspondence with those in a *heterosporous* pteridophyte of this kind. However, the 'megasporangium' is enclosed in the integument of the ovule, which in due course becomes detached as a true seed. The 'prothallus' is reduced to the cells and vestigial archegonia in the embryo-sac (developed from the 'megaspore'), or a few cells in the pollen grain ('microspore'). The cycads and the maidenhair tree (*Ginkgo biloba*) still produce motile antherozoids, but in the conifers the gametes are reduced to naked nuclei which pass through the pollen tube to fertilise the egg as in the angiosperms. It is generally thought that the condition in the angiosperms, described in Chapter 2, represents a further stage in reduction of a similar life-cycle.

The most important present-day group of gymnosperms, the conifers, certainly have nothing to do with the origin of the angiosperms. They are one of the few groups of plants of whose evolution we have a good fossil record, and their probable origin from the Cordaitales at the close of the Palaeozoic has been traced out by the Swedish palaeobotanist Florin (1954). Most of the present-day families of conifers existed in recognisably modern form by the Jurassic, though the *Pinaceae* did not appear until the Cretaceous. However, these gymnosperms of relatively modern type formed only a proportion of the gymnosperm flora of the Triassic and Jurassic periods. In the Triassic there were still many Pteridosperms ('seed-ferns'), remnants of a group whose remains are found in great variety and abundance in rocks of Carboniferous and Permian age. A particularly prominent part was played by another group, the Bennettitales – plants much resembling the modern cycads in habit and leaf-form which attained great diversity and world-wide distribution during the Jurassic, but became extinct before the end of the Cretaceous. These plants had reproductive shoots which give us a hint of the kind of structures from which primitive angiosperm flowers may have evolved, though it is certain that none of the known Bennettitales is close to the original angiosperm stock.

FIG. 130. The fossil record of some groups of seed-plants and insects during the Mesozoic and Cenozoic eras. In the case of the Angiosperms, the width of the stippled band indicates approximately the relative number of families recorded by the end of each period. The Lepidoptera (butterflies and moths) and Hymenoptera Aculeata (ants, bees and wasps) do not appear in the record until the Angiosperms are established as a dominant group of plants. The Diptera Brachycera are associated with the rise of the Angiosperms on the one hand, and the mammals and birds on the other; the parasitic Diptera Pupipara, which are not flower visitors, thus appear at about the same time. Dotted lines indicate uncertainty as to the precise age of the earliest occurrence. Data (somewhat simplified) from Harland *et al.* (1968).

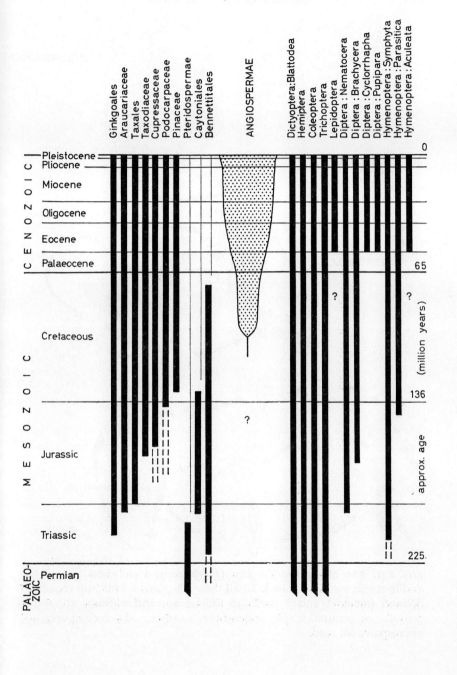

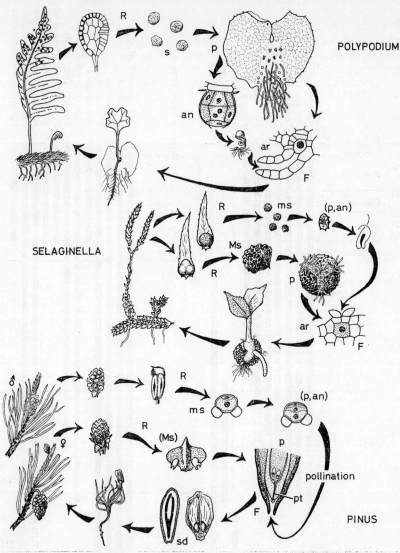

FIG. 131. The life-cycles of a fern (*Polypodium*), a clubmoss (*Selaginella*) and a coniferous tree (*Pinus*). In all three diagrams R indicates reduction division (meiosis) and F indicates fusion; an, antheridium, ar, arche-gonium, p, prothallus, pt, pollen-tube, s, spore, Ms, megaspore, ms, microspore, sd, seed.

As examples of the Bennettitales we may take two superficially very different genera, *Williamsoniella* and *Cycadeoidea*. *Williamsoniella* had a slender regularly forked stem bearing narrow entire leaves (Fig. 132, g and h). In the forks of the stem were short fertile shoots, each bearing a flower-like cluster of reproductive organs. In size and form the 'flowers' may be compared to a rather large buttercup. In place of petals there were a number of thick broadly rounded scales, each bearing pollen-sacs

FIG. 132. Some extinct seed-plants. *Lyginopteris oldhamii*, a pteridosperm common in the Coal Measures of the north of England. A, part of stem, with a leaf and pinnule. B, seed and cupule. *Caytonia*. Jurassic; Caytoniales. C, leaf. D, female inflorescence. E, vertical section of fruit, as interpreted by Harris. F, part of male inflorescence. *Williamsoniella coronata*. Jurassic; Bennettitales. G, part of plant. H, single 'flower'. *Cycadeoidea*. Jurassic; Bennettitales. I, group of plants. J, longitudinal section of hermaphrodite 'flower', as interpreted by Delevoryas. K, longitudinal section of female 'flower'.

on its inner face. The apical parts of the flower, except the extreme tip, were occupied by ovules rather like those of yew (*Taxus baccata*, p. 265), each surrounded by five or six scales, packed together forming a pattern of hexagonal areas interspersed by the tubular micropyles at the surface. *Cycadeoidea* (Fig. 132, I – K) formed a short stout trunk up to a metre or a little more in height, thickly covered with scaly leaf-bases like those of modern cycads. The stem bore numerous short-stalked flowers, each a few centimetres across, half hidden among the armour of leaf bases. Apparently some of these flowers were purely ovulate, while others were hermaphrodite. In the hermaphrodite flowers the conical axis of the flower bore at the base a whorl of complex pollen-bearing scales, fused in their lower parts. The upper parts of the floral axis were thickly covered with sterile scales amongst which the micropyles of the ovules just projected above the surface.

Because the conifers are wind-pollinated (and with their unisexual flowers probably always were so), we tend to think of wind-pollination as a characteristic feature of gymnosperms. There is no reason to suppose that this was always the case. In fact, visits of pollen-feeding beetles to the staminate cones of several genera of modern cycads are on record, and Baird (1938–9) found beetles and *Hemiptera* on cones of both sexes of the Australian cycad *Macrozamia reidlei*. The compact structure and bisexuality of the flowers of the Bennettitales makes it quite likely that they were pollinated by insects.*

The bennettitalean flowers are obviously very different in essential detail from angiosperm flowers, but they do show something of the *milieu* in which angiosperms must have evolved. Not only is the general arrangement of the parts broadly similar to the arrangement of the stamens and carpels of the angiosperm flower; the ovules receive a degree of protection from the closely packed scales surrounding them which may be compared with the protection afforded by the carpel walls of the angiosperms.

The evolution of protection for the ovules seems to have been a rather general feature of Mesozoic seed-plants. It was brought about in different ways in different groups – carpels in the angiosperms, scales between the ovules in the Bennettitales, carpel-like structures formed from the sporophyll-lobes in the Caytoniales (Fig. 132, D and E), and the woody scales of the compact cones of the conifers. Grant (1950), Van der Pijl (1960) and others have suggested that the carpels of angiosperms may have arisen as an adaptation protecting the ovules against damage by beetles. If this were so, no doubt the same factor operated in other groups of plants. Compared with the spore-bearing leaves of ferns, flowers may be thought

---

* It is interesting in this connection that Delevoryas (1968) has observed signs of damage in hermaphrodite flowers of *Cycadeoidea* of a kind which might have been caused by insects such as beetles. On the other hand, of course, such damage might be quite unrelated to pollination.

of as precocious reproductive buds. It may be (as Takhtajan has pointed out) that one of the most important features of the ancestors of the angiosperms, and the one that made possible the evolution of the characteristic angiosperm carpel, was that they had 'conduplicate vernation'. That is to say, the leaves were folded in half along the mid-vein in bud, instead of being spirally coiled like the leaves of ferns, Bennettitales and Caytoniales.

Corner (1964) has suggested that the relation between flowers and animals first arose in connection with fruit dispersal, and that in primitive seed-plants pollination and dispersal were more-or-less simultaneous. This would be consistent with the origin of seeds envisaged on p. 358, and would explain how the insects were first attracted to the ovules of seed-plants and why the development of protective structures was necessary as the seed-plants evolved flowers with smaller and more numerous ovules.

The question of when and where the angiosperms originated, and why they were so successful, remain unanswered. The simplest interpretation of the fossil evidence is that they evolved rapidly early in the Cretaceous (Hughes, 1961). If this was so, we know nothing of their immediate precursors. On the other hand, they could have originated very much earlier, diversifying initially in a part of the world where they have hitherto escaped palaeobotanical study, or in relatively dry upland regions where they escaped fossilisation (Axelrod, 1952). We simply lack the fossil evidence to decide between these possibilities, let alone to elucidate how and when the decisive steps in the evolution of closed carpels and the flower–insect relationship came about. The origin of the angiosperms is still, as Darwin described it to Hooker a century ago, 'an abominable mystery'. Ultimately, the stock from which the angiosperms evolved probably originated from one of the many groups of pteridosperms that existed in Permian times. At the present day, the distribution of the bulk of angiosperm families, including those that are believed to be the most primitive, is centred on tropical regions. There is much reason to think that the angiosperms or their precursors may have originated in the mountainous regions of the tropics, before descending to the lowlands and supplanting the existing gymnosperm flora in the less diversified topography and climate of the Cretaceous.

Whether the primary reason for the success of the angiosperms lay in the ecological and genetical advantages conferred by insect-pollination, or in the control of the breeding-system through multiple-allelomorph incompatibility made possible by the closed carpel (Whitehouse, 1950), or even outside the reproductive sphere altogether in their efficient vegetative metabolism, is a matter for conjecture. What is certain is that from the time of their first appearance the angiosperms show an extraordinary capacity for evolutionary diversification. The gymnosperms have never evolved herbaceous members. The angiosperms already had both woody and herbaceous genera during the Cretaceous, and at the present day they display a range of size, form, structure and ecological adaptation far surpassing that of any other plant group that has ever existed.

THE DEVELOPMENT AND DIVERSIFICATION OF THE ANGIOSPERMS

Many lines of evidence, including the morphology of the flowers and pollen grains, and the anatomy of the wood, lead to the conclusion that the most primitive living angiosperms are to be found among the trees and shrubs of the Magnolia family (*Magnoliaceae*) and a few related families. These include the *Winteraceae* (for example Winter's Bark, *Drimys winteri*), *Degeneriaceae* (*Degeneria*), the widespread tropical *Annonaceae*, *Lauraceae* (including the Sweet Bay (*Laurus nobilis*)) and *Calycanthaceae* (including the Wintersweet (*Chimonanthus fragrans*) and Carolina Allspice (*Calycanthus floridus*)). A notably large proportion of these plants appear to be pollinated habitually by beetles, and the same is true of the comparably primitive Water-lilies (*Nymphaeaceae*). This is itself probably a primitive feature (see pp. 357, 362).

Probably the *broad* lines of development of angiosperm flowers during their initial diversification followed the kind of pattern suggested by Leppik (1956, 1957), which has been mentioned already in Chapter 6. Leppik visualised a progressive development from primitive 'amorphic' flowers (with which the 'flowers' of *Williamsoniella* provide an analogy), through 'haplomorphic' flowers (of the *Magnolia* type) to 'actinomorphic' flowers (with the numerous parts at one level, and with radial symmetry), these in turn giving rise to 'pleomorphic' flowers (with the parts definite, and usually fewer, in number). Subsequent development to 'stereo-morphic' flowers (with significant three-dimensional form, for example corolla-tube) and 'zygomorphic' flowers (similar but with bilateral symmetry) was linked with the evolution of the structural and behavioural adaptations of the more specialised flower-visiting insects. Of course, angiosperm evolution was immeasurably more complex than this. Many different stocks have run independently through the same levels of organisation, and a particular level may be arrived at by a circuitous or even retrogressive route, as in the case of the actinomorphic 'blossoms' (capitula) of the *Compositae* (p. 220). Furthermore, these are only broad categories. In many cases close adaptation has taken place to particular pollinators. The emergence of a major type has often been followed by adaptive radiation – the evolution of a divergent range of forms adapted to different ecological niches and different pollinators. This is spectacularly demonstrated amongst the orchids, and on a less dramatic scale in many other groups, for instance among the members of the Phlox family studied by V. and K. A. Grant (1965, see p. 373). Superimposed on the pattern of adaptation to animal pollinators are the tendencies towards the development of wind-pollination, self-pollination and apomixis which have been discussed in Chapters 8 and 9. A point worth emphasising again is that adaptation takes place in relation to the whole environment of the plant, and in many cases flowers are evidently adapted to a balance between two or more pollinators or pollination mechanisms. Such situations are probably important in that they leave a range of possibilities open for subsequent evolution. Of the factors leading to small-scale proliferation of variation in flower-form more will be said later.

ADAPTATION

In writing this book we have taken for granted the idea of adaptations, or of adaptive characters, and their origin by natural selection. It is common experience that when we examine an organism we see that many of its structural and behavioural characteristics are related to its mode of life, and many instances of adaptations have been described among the animals and plants we have considered. There will always be some features the adaptive significance of which we cannot readily see. Some such features may possess undiscovered adaptive significance, while others, particularly small differences between species, are often held to be probably non-adaptive. Biologists have perhaps been too ready to make this second assumption. Those who classify, describe and identify organisms – the taxonomists – often tend to ignore the question of adaptive significance, being preoccupied with finding and listing characters purely for the purposes of classification and identification. In fact, adaptive characters relating to some constant feature of the organism's mode of life are particularly valuable in taxonomy, for they have to be constantly present if the organism is to function properly; only those adaptive features which show a variable response to the environment are unsuitable for classificatory purposes. Moreover, differences between related organisms in such characters are indicative of differences in mode of life and are therefore particularly significant indicators of species differences. The advantage to taxonomists of understanding the floral biology of the plants they work with has been emphasised by Ornduff (1969). Unfortunately, a great deal of descriptive work has been published in which investigation of structure is totally divorced from the consideration of function. This occurs not only in the description of species, but also in morphological studies aimed at investigating evolutionary relationships. Such work has been found of little use in the writing of this book, and neglect of function may well reduce the value even of the evolutionary conclusions that authors have sought to draw. Studies like Graham-Smith's superb investigation of the mouth-parts of the blow-fly are rare indeed. There would appear to be much still to learn about the functioning of the mouth-parts of bees and of their pollen-collecting apparatus (and in studying the adaptation of bees' legs for the work of pollen-collecting it has to be borne in mind that the legs are used also for nest-building). A number of morphological adaptations connected with foraging have been listed by Linsley (1958) working in California. For example, the pollen baskets of some American bees which specialise on plants with large pollen grains are made up of loosely plumose hairs, while those of bees specialising on *Compositae*, which have small grains, are densely plumose. In a number of quite unrelated bees which collect pollen mainly or exclusively from the Evening Primrose family (*Onagraceae*), in which the pollen grains are held together by elastic threads, the pollen baskets are made up of long, unbranched bristles. Bees of th American genus *Proteriades* collect pollen from the anthers of *Cryptantha* (family *Boraginaceae*), which are hidden inside the corolla-tube, by means of stiff curled hairs on the mouthparts.

This, incidentally, is an instance of one-sided adaptation – the bee is specially adapted to stealing the pollen hidden in a flower which clearly is not adapted to pollination by pollen-gatherers. The different species of *Proteriades* are not adapted to different species of *Cryptantha*, and the geographical distribution of the plant genus is much wider than that of the bee genus. A less extreme instance of one-sided specialisation is provided by *Andrena linsleyi* in the Colorado Desert. This bee is abundant wherever its principal pollen plant, *Oenothera deltoides*, occurs, but this plant is adapted to pollination by hawk-moths. The bees forage mainly in the morning, taking pollen not removed in the previous night by the moths, and can take both nectar and pollen without touching the stigmas (Linsley, MacSwain and Raven, 1963). The association of *Macropis* with *Lysimachia* (see p. 151) is also more significant for the bee than the flower, for *Macropis* is not its principal pollinator (Popov, 1958, quoted by Linsley, 1961).

There is a wealth of knowledge of the functions of floral characters, but new discoveries are still being made by those who ask 'what is it for?'. Good examples are found among the tropical orchids and insect-trapping flowers described in Chapter 10. From California again comes a rather subtle instance of floral adaptation (Linsley, MacSwain and Raven, 1964). Various species of Evening Primroses (*Oenothera*) in the Mohave Desert are adapted to bee-pollination and are visited principally by certain species of *Andrena* which obtain pollen only from *Oenothera*. Two such *Oenothera* species are the widely ranging *O. campestris*, and a geographically much more restricted species, *O. kernensis*. *O. campestris* normally opens its flowers an hour after sunrise, and is mainly pollinated by *Andrena boronensis*. However, where it grows with *O. kernensis* it follows the example of that species, both opening their flowers about sunrise. *O. kernensis* is adapted to the visits of *Andrena mohavensis*, a large, early rising species, which is found only in the area of this plant. Here, *O. campestris* appears to have modified its flowering-time locally (through natural selection) in response to the behaviour of the dominant pollinator, *Andrena mohavensis*, whose visits it receives. A similar example is provided by *Oenothera hartwegii* in Texas. This plant is adapted to pollination by hawk-moths flying in the evening and at night, but some bees are also effective pollinators. At one colony where bees were especially active it was found that the flowers were opening earlier in the evening than elsewhere, suggesting a local adaptation to bees as pollinators (Gregory, 1963, 1964).

Much work on situations of this kind has been carried out in California in connection with biosystematic studies of plants. The objects of such investigations are to study the ecology, genetics, breeding behaviour and morphological variation of plants in order to learn something of the manner of origin, biological character and evolutionary potential of species and populations of lower (and sometimes higher) rank. An interesting example is the work of Straw (1955, 1956) on a group of *Penstemon* species. Straw found that these species were capable of producing more or less fertile hybrids, and he noted that the hybrids between the two most dissimilar species of the group resembled one of the other species (Fig. 133, p. 368). One of these very dissimilar species has cylindrical scarlet flowers with no lip, borne on flexible pedicels, and is pollinated by hummingbirds probing with their bills. The other has white, pale pink or pale blue flowers with a bulging tube, borne on very strong pedicels, and is pollinated by large *Xylocopa* bees which crawl inside the corollas. The species which resembles the hybrid between these two has bluish purple flowers with a slightly bulging throat, borne on strong pedicels, and is pollinated by flower-visiting wasps (family *Masaridae*) which are smaller than *Xylocopa* and also crawl inside the flowers. These facts suggest that the wasp-pollinated species arose as a result of the crossing of the other two species, the hybrids having by chance proved to be suitable to the wasps and having been adopted by them as a food source. This would have caused the hybrids to interbreed with one another rather than to back-cross with the parent species or to remain unpollinated, natural selection among the initially very variable progeny gradually producing a new genetically uniform breeding group adapted to pollination by the wasps. An exactly similar process is postulated on similar evidence for a fourth species which is intermediate between the wasp-pollinated species and the bird-pollinated one, and is pollinated by rather small bees. What happens when the hybrids between two interfertile species do not find themselves adapted to a third pollinator is shown by two species of Columbine (*Aquilegia*) described by Grant (1952). The two species are *A. formosa*, with nodding red flowers pollinated by hummingbirds, and *A. pubescens*, with erect, long-spurred, pale yellow or white flowers pollinated by hawk-moths (Fig. 134); they grow in the mountains of California, where they have different habitats and different altitudinal ranges, but conditions occasionally allow them to meet. However, it seems that the hawk-moths never visit *A. formosa* and the hummingbirds only occasionally visit *A. pubescens* – and then only to bite the spurs from which they steal

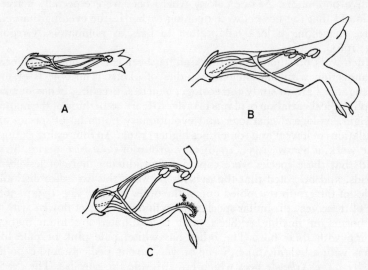

FIG. 133. Flowers of *Penstemon*; in each can be seen the gynoecium, the staminode and two of the four stamens. A, *P. centranthifolius*, pollinated by hummingbirds. B, *P. spectabilis*, pollinated by masarid wasps. C, *P. palmeri*, pollinated by *Xylocopa* bees. After Straw (1956).

the nectar without causing pollination. Hybrids, nevertheless, are found and it is believed that these are produced by occasional cross-pollinations carried out by pollen-gathering bumble-bees. The F_1 hybrids can be visited by both hummingbirds and hawk-moths, but these visitors pass from the hybrids to the parent forms or other hybrids which suit them best. Consequently the progeny of the hybrids tend to be back-crosses which approach one or other parent in form. The result, then, is not stabilisation of a new intermediate form, but introgression, that is, the passage of genes from one species into another – a process usually rigorously limited by selection, but which may on occasion provide a valuable source of evolutionary potential.

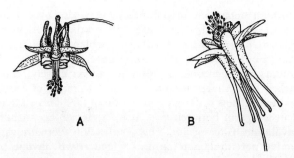

FIG. 134. Flowers of *Aquilegia*. A, *A. formosa* var. *truncata*, scarlet with yellow entrances to spurs. B, *A. pubescens*, pale yellow. After Grant (1952).

POLLINATION, REPRODUCTIVE ISOLATION AND SPECIATION

In *Penstemon* and *Aquilegia* reproductive isolation* of related species is caused by variations of form which limit the flowers to certain types of pollinator, and this is referred to as mechanical isolation. Another way in which reproductive isolation dependent on pollinator-restriction arises is through floral differences sufficient to bring into play the selective responses of flower-constant bees. This is called ethological isolation, as it depends on animal behaviour. It has been suggested by Heslop-Harrison (1958) that ethological isolation has helped in the formation of subspecies of some Orchids. In the Spotted Orchid (*Dactylorhiza fuchsii*) most populations consist mainly of plants with an evident pink or purplish flower-coloration, distinct darker striations, and hardly any scent. Within populations there is a considerable range of variation in the shape, colour and patterning of the lip of the flower, but visiting bumble-bees pass readily from flower to flower without regard to this. However, in the west of Ireland and Scotland there occur populations with white fragrant flowers, which are confined to thin highly calcareous soils; where these soils give way to deeper, neutral or slightly acid soils there is usually an abrupt change to the normal type of flower. The pale-flowered plants (*D. fuchsii* subsp. *okellyi*) possibly obtain a measure of breeding isolation through flower-constancy of the pollinators, as it is known from the experiments of Manning (see Chapter 5) that the bees can be trained to distinguish between floral differences such as are found in *D. fuchsii*, and because of the great importance of scent as a recognition mark for bees. In this way the behaviour of the pollinators may assist the establishment of ecologically differentiated forms able to coexist at close quarters.

* A general discussion of isolating mechanisms may be found in Grant (1963, Chapter 13).

Isolation through behaviour of pollinators may also play a part in the maintenance of the several ecologically differentiated sub-species in the Early Marsh Orchid (*Dactylorhiza incarnata*), which differ fairly sharply in floral characters but show only slight differences in vegetative characters.

The boundary between one species and another closely related to it is usually marked by a sharp discontinuity in the range of morphological variation, which reflects differences in mode of life. This state of adaptation is lost in a hybrid between two such species unless an intermediate habitat is available. Interspecific cross-fertilisation therefore causes waste of reproductive potential, and natural selection will favour barriers to intercrossing. Perhaps the most usual form of barrier is intersterility: the foreign pollen either fails to grow on the style, or it may grow and cause fertilisation but the seeds fail to develop.

Mechanical or ethological isolation between closely related species may bring advantages which intersterility does not confer. In some self-incompatible species of *Phlox* it has been found that there is regularly a low seed-set because the stigmas receive, on the average, fewer pollen grains than there are ovules in the flower. In this situation of pollen shortage, adaptation towards reducing waste of pollen has occurred in places where pairs of intersterile species meet and flower simultaneously. In these places, one of the species shows a preponderance of white flowers instead of the usual red and pink shades, resulting in a reduction of inter-specific cross-pollination by the butterfly visitors (Levin and Kerster, 1967). This case is unusual in that pollen wastage is important; the result is that ethologically based assortative pollination has evolved even in the presence of reproductive isolation through intersterility.

Great importance has often been attached to sterility barriers as criteria for the delimitation of species. However, there are many groups of flowering plants in which there is general agreement among taxonomists in recognising species between which sterility barriers have been shown not to exist. In some cases the related species are separated geographically, but in others geographical separation is not complete; in the latter, hybridisation is usually rare under natural conditions. There are several such groups where the flowers are highly specialised to animal pollinators and in which isolation is dependent on the behaviour of the pollinators. *Aquilegia* and *Penstemon* are notable examples, while others are provided by species of Snapdragon (*Antirrhinum*), Toadflax (*Linaria*) and *Delphinium*. The supreme example of such a group is the family *Orchidaceae*, in which even genera are frequently interfertile. In these groups it seems as though the constancy of the pollinators is sufficient to render selection in favour of sterility barriers inoperative (Grant, 1952). It is conceivable, in fact, that the interfertility which may be retained on account of the relative constancy of the pollinators of these specialised flowers has a long-term evolutionary advantage, in that it allows the origin of new species by hybridisa-

PLATE 55. Flower-visiting birds: **a**, yellow-winged sugar bird *(Cyanerpes cyaneus)*. **b**, malachite sunbird *(Nectarinia famosa)*. **c**, Indian white-eye *(Zosterops palpebrosa)*. **d**, Pucheran's emerald hummingbird *(Chlorostilbon aureoventris pucherani)*.

PLATE 56. **a**, inflorescence of
Marcgravia umbellata; the nectar-
pouches hang down in the centre,
while the unopened flowers stand
above them at an angle to their
pedicels. **b**, the cap-like corolla
drops off, and the pedicels bend
slightly downwards; the knob on
the central nectary is a rudimentary
flower – the nectaries are modified
bracts, whose associated flowers are
vestigial. **c**, single flower of the
South American orchid *Stanhopea
wardii*. The visiting bee falls down
the 'chute' between the channelled
upper surface of the lip (below the
prominent eye-spot) and the
column to its right, striking the
pollinia close to the tip of the
column in the process. Compare
Gongora (Pl. 52 **c-d.**).

tion as described for *Penstemon*, or introgression as described for *Aquilegia* (p. 368). Thus, once a certain degree of reproductive isolation is achieved through pollinator-specialisation, selection *against* intersterility may come into play.

This raises the problem of whether floral variation could, through the tendency of like to be mated with like, lead to the splitting up of a breeding group and to the development of new species, or whether splitting up by geographical separation or ecological diversification must come first, reproductive isolation either being selected for when geographically isolated relatives are brought into contact again, or having the role of reinforcing an original tendency to ecological splitting. The sequence in which flower form is secondary is the one that is believed to have occurred in most cases in which related species differ in their pollination arrangements, but it has been argued that a primary role for flower form is also theoretically possible (Grant, 1949; Heslop-Harrison, 1958).

The distinction between mechanical isolation and ethological isolation is important because it shows how absence of cross-pollination may come about between types of flowers which are equally well suited to the same type of pollinator. It thus explains why flowers of related species may differ in ways which have no mechanical significance, and are thus not overtly adaptive. The differences are there merely for the sake of variety (which *has* adaptive significance), and not so that the plant can exploit a different pollinator. Kugler has pointed out that there is much greater diversity among flowers than there is among their insect visitors and Müller (1881) noted that bee and bumble-bee flowers blooming at the same time tend to be of diverse colours, though flowers adapted to short-tongued insects are usually similar in colour to their nearest relatives. The *Labiatae* and *Leguminosae* are examples of families in which there is pronounced diversity without much basic variation in mode of functioning. Flowers isolated from their near relatives by mechanical arrangements, however, usually differ also in characters of ethological significance, chiefly scent and colour. This is true of the two *Aquilegia* species just described and of many other plants pollinated by specialised flower visitors. An example is found in the Louseworts *Pedicularis groenlandica* and *P. attolens* in North America, which dust their visitors on the underside of the abdomen and the top of the head respectively (but see p. 210), though, owing to the dusty nature of the pollen, this would not completely prevent crossing. The two Louseworts differ in colour and scent; both are visited by a species of bumble-bee (*Bombus bifarius*), the individuals of which remain quite constant to one species or the other, being capable of discriminating between them purely by scent (Sprague, 1962). The recognition marks of colour and scent act here mainly by assisting the insects to find suitable flowers.

Another role of highly specific flower characters with ethological significance is found in deceptive flowers – the trap-flowers, the intoxicating Orchids and the pseudocopulatory Orchids. The effect of the differences in this case is unlike that of the Spotted Orchids described by Heslop-Harrison because different but related species of plants in these groups attract different species of insects, and not different individuals of the same one or more species which have become temporarily conditioned to visit one flower type and not another. Moreover, these plants always have marked morphological differences restricting them to suitable visitors. Hence Grant (1949) regards pseudocopulatory orchids as examples of *mechanical* isolation. Here, however, even more than in the Spotted Orchid instance, it is likely that chance variations in, say, scent may be the initial factor in splitting a population into two different pollination-groups which become reproductively isolated. In the case of *Ophrys*, the European and North African pseudocopulatory Orchid genus (see p. 243), such initial changes (probably in some cases resulting from hybridisation) have apparently led to further floral changes making the flowers increasingly suited to different insects (and so making it convenient to us to treat them as separate species), without leading to much ecological diversification, so that several species are often found together (Heslop-Harrison, 1958). Thus we can see why Heslop-Harrison, unlike Grant, chose to regard *Ophrys* as providing an example of *ethological* isolation. The significant contrast is not simply between ethological and mechanical isolation, but between assortative pollination, induced by the offer of floral contrasts of sufficient magnitude to enable individual pollinators to develop flower-constancy, and mechanical and/or ethological restriction limiting, more or less severely, the classes of effective pollinators.*

In a genus of deceptive Orchids a newly arising floral type attractive to an insect species not previously attracted is unlikely to be cross-pollinated with individuals of the type from which it is derived, because of the specificity of the attractive odour. A floral variant in a species which offers food, however, is almost certain to cross with the parent form if it is visited at all. Such a variant would have to be sharply set off from the parent type, and to be present in fair numbers, for assortative pollination

---

* Plants with such restrictions have been termed *oligophilic* or *monophilic* (pollinated by a few or one species of visitor respectively) as opposed to *polyphilic* (plants with little restriction of pollinator) species. A related distinction may be made among flower-visiting insects. Some species of insects are *oligotropic* (visiting only a limited range of flower-types) or *monotropic* (restricted in their visits to one or a few closely related species). This situation differs fundamentally from the temporary flower-constancy shown by individuals of *polytropic* species of bees (and other insects), which may at different times visit a very wide range of flowers (see Faegri and Van der Pijl, 1966).

to occur. This state of affairs could only come into being in rather special circumstances.

Among deceptive flowers the insect-trapping species are usually pollinated by a group of several species of insects with similar habits, whereas the pseudocopulatory Orchids, and some of the intoxicating Orchids, are visited by one or a very few species of pollinator. They are thus dependent on the fortunes of these very few pollinator species, and the plant species are therefore liable to extermination if their special pollinators die out. The constant self-pollination of the Bee Orchid (*Ophrys apifera*) in northern Europe is a clear reflection of this limitation of pollination by pseudocopulation. Thus this method of pollination has evolutionary drawbacks, but this applies only to the species – for the genus, survival seems to be assured. Extreme pollinator-specialisation is also found in fig-trees (*Ficus*) where, as a rule, each species of fig has its own special pollinating fig-wasp. It is conceivable, though unlikely, that a pollinator species might die out, with consequent destruction of the plant host, but the success of the figs in general shows that their pollination system is no barrier to their evolution. In this very intimate association it seems that fig and wasp must in effect evolve and speciate together as a single biological entity.

A very good impression of pollinator-specialisation within a single family can be obtained from a beautiful and extremely interesting book on the Phlox family (*Polemoniaceae*) by V. and K. A. Grant (1965). This family is mainly North American, and its morphology, taxonomy and ecology have been extensively studied. Consequently the pollination observations can be fitted into a background of detailed knowledge of other aspects of the biology of the plants. The animals whose services in pollination are exploited by this family are large bees, small bees, hawk-moths, other moths ('noctuids'), butterflies, bee-flies, scavenger-flies (*Calliphoridae* etc.), beetles, hummingbirds and bats. But no species in the family is known which is, or could be, pollinated effectively by all of these agents. Every species, if not prevailingly self-pollinated, is specialised for pollination by some more or less restricted class of pollinators. The most restricted class is apparently the hawk-moth-pollinated group; occasionally pollination is brought about in these flowers by small bees, but by far the most regular and efficient visitors are nocturnal hawk-moths, and the flower form and behaviour are evidently fitted to these visitors. (However, casual observers are much more likely to see the visits of the bees than the hawk-moths.) Other types of flower may be pollinated by two unrelated groups of visitors, such as smallish bees and bee-flies (*Bombyliidae*), which have similar body-size, tongue-length and habits. Visits to these flowers are sometimes made by butterflies or beetles but these are ineffective in pollination, while other classes of visitor never visit this type of flower at all. The Grants liken these situations to lock-and-key situations, the keys already being in existence, and the locks being designed to fit one or a few keys, but not all of the possible keys.

In the Phlox family a phylogenetic scheme has been worked out without reference to floral biology. It is found that bee-pollination, which is the most widespread condition in the North American *Polemoniaceae*, is the only system found in the primitive members of the three tribes of the family. It seems, therefore, that in this family the other pollination systems are derived from bee-pollination; some types have arisen more than once in different evolutionary lines. Hummingbird-pollination has arisen separately in four genera in which bee-pollination is the basic condition (as it has in *Aquilegia*, *Delphinium* and *Penstemon* which we have already mentioned). In some genera of the *Polemoniaceae* there is a strong prevalence of one pollination-type (for example, bee-pollination in *Polemonium* and *Lepidoptera*-pollination in *Phlox*). In others, divergence of pollination mechanisms has occurred at a lower level in the taxonomic hierarchy – even down to the level of races within a single species. For example, in *Gilia splendens* the widespread race is pollinated mainly by bee-flies (*Bombylius*), but a high-altitude race with longer and slenderer flowers is pollinated by the very long-tongued fly *Eulonchus* (family *Cyrtidae*), another race with long stout corolla-tubes is pollinated by hummingbirds, and there is also a self-pollinating race; all the races replace each other geographically. Sometimes there may be parallels between different plants in this kind of variation, as in *Eriastrum densifolium* and *Gilia achilleifolia*, which have broad-throated flowers in coastal areas, pollinated by large bees, and small-throated flowers in the dry interior mountains, pollinated by small bees and bee-flies, the plant populations being evidently responsive in their evolution to the array of flower-visiting animals in their respective territories. Competition between plants for the visits of pollinators is considered by the Grants to be an important factor leading to specialisation in pollination.*

If we look at the plants which have highly specialised pollinators we find not only the presence of morphological features which fit them for these visitors and exclude or discourage less specialised visitors, but also that related species differ from one another particularly in floral characters. This was shown by Grant (1949) in a survey based on pollination records and taxonomic works. It was found that 40 per cent. of the taxonomic characters in plants pollinated by bees and long-tongued flies

* Competition for pollinators is a topic which merits much more study. Free (1968) concluded that competition from the flowers of dandelions in apple orchards could have a significant adverse effect on the pollination of the trees; on the other hand, Beattie (1969) considered that the presence of flowers of other species could have a beneficial effect on the pollination of Violets (*Viola* spp.) by attracting pollinators into the area.

pertained to the flower (the calyx excluded), and that the corresponding figure for bird-pollinated flowers was 37 per cent., while for plants pollinated by short-tongued insects it was 15 per cent. and for wind- and water-pollinated plants it was 4 per cent. That insects showing a high tendency to flower-constancy actually promote speciation is suggested by another survey by Grant (1949), on the flowers of southern California, an arid region where there are many species of bees. It was found that genera of bee-pollinated plants contained an average of about six species, whereas genera of plants pollinated by the less constant types of insect contained on average only about four species.

In this chapter, ethologically based diversification has been discussed at some length. However, examination of a wide range of flowers suggests that there has also been a substantial degree of ethologically based convergence, or *mimicry*. Mimicry is most spectacularly developed among animals, especially insects (Wickler, 1968). However, it is well known also among plants, where the deceptive flowers described in Chapters 7 and 10 provide some impressive instances. These are examples of *Batesian mimicry*, in which one organism obtains a one-sided advantage by imitating another organism more numerous than itself. A possible instance of Batesian mimicry in relation to pollination is provided by the Eyebright *Euphrasia micrantha*. Unlike most eyebrights, this species typically has deep pinkish or lilac flowers, so that the slender inflorescences have a noticeable resemblance to those of Ling (*Calluna vulgaris*) among which it grows. This resemblance may bring to the flowers pollinators which would otherwise remain constant to Ling, and ignore the thinly scattered and inconspicuous eyebright plants (Yeo, 1968).* The diversity to be seen among flowers is impressive, but almost equally striking is the way in which particular patterns of form and colour appear repeatedly, often in quite unrelated species. Sometimes these resemblances may arise simply because one job can be done in only a limited number of ways. However, resemblances of the kind seen between, for instance, the Buttercups (*Ranunculus*), Cinquefoils (*Potentilla*) and the Common Rockrose (*Helianthemum chamaecistus*), or in the Alps between *Ranunculus alpestris*, *Dryas octopetala* and *Chrysanthemum alpinum*, seem too numerous and striking all to be the products of mere chance convergence. It seems likely that some at least are the result of *Muellerian mimicry* – in which a number of species of similar character and behaviour, and of comparable abundance, have evolved a common 'advertising style' to their mutual advantage; obviously, mimicry of this kind is likely to be found mainly

* Schelpe (1966, p. 10, also pp. 80, 81) gives a similar instance among South African Orchids; 'An interesting case of "mimicry" is found in *Orthopenthea fasciata* in which the flowers resemble those of the rutaceous shrub *Adenandra*, amongst which it grows, but no observations are available to prove that the same insect is deluded into pollinating the flowers of two such diverse plants.'

among species with fairly unspecialised and promiscuous pollination mechanisms (but see also p. 397).

Before leaving the subject of evolution we may mention that some authors, having found physiological explanations for certain features of plants, consider that they have thereby explained them, so that there is no need to look on them as adaptations (examples are quoted by Van der Pijl, 1960). Such views ignore the fact that adaptive features are bound to have physiological causes anyway; these explain how the feature comes into existence in each individual. That each individual possesses the physiological causes of particular features can be explained through adaptation by natural selection. This is often expressed by saying that adaptive features have *proximate* and *ultimate* causes. For example, a plant may open its flowers at a particular season, or time of day, in response to climatic or meteorological conditions (the proximate factors) in such a way as to coincide with the season or hours of activity of the plant's special pollinators (the ultimate factors). The ultimate factors act through natural selection to cause the evolution of responsiveness to the proximate factors.

DIFFICULTIES OF OBSERVATION

On p. 373 we have mentioned that scavenging bees may more easily be seen on the flowers of a certain moth-pollinated plant than the legitimate pollinators. Apart from the mistaken impressions that may arise from such situations it is not always easy to determine the legitimate pollinator, not only for flowers in the canopy of tropical forests (Chapter 10), but even for terrestrial herbs. The point is well put by the Grants in their book on the Phlox family (V. and K. A. Grant, 1965). 'The activity of pollinating agents and the process of pollination often occur mainly during a few relatively brief periods during the day and during the flowering season when environmental conditions are favourable and the pollinators are on the wing in numbers. In order to understand the pollination ecology of a species it is necessary to live with its populations . . . In the present investigation a large amount of time has been spent in the field. A number of plant species have been observed fairly extensively in the wild without obtaining adequate or even any pollination records. On the other hand, we have managed by sheer luck to be in the right place at the right time to see the course of pollination in full swing in some other species.'

If we consider the observations on the Primrose (*Primula vulgaris*) summarised by Christy (1922) in the light of these remarks we can perhaps accept that the legitimate pollinators are long-tongued bees and long-tongued flies, for these have been seen, though only on rare occasions, visiting the flowers freely. More often than not little or no insect activity is seen at the flowers, and this gave rise to the supposition, unsupported by observations, that nocturnal moths might be the main pollinators (see p. 225).

POLLINATION AND THE ENVIRONMENT

The pollination relationship is inevitably influenced by the other environmental relationships of both partners. For some insects (for example, unspecialised flies and beetles) visits to flowers are an incidental among other activities. Their availability as pollinators will depend on factors which may have little to do with the occurrence of flowers. For others, such as the hoverflies and the butterflies and moths, flowers provide the principal source of food of the adults, but the occurrence and abundance of these insects will depend also on the occurrence of suitable conditions for their larvae – in the case of *Lepidoptera* on the occurrence of suitable food-plants, seldom the same species as those at whose flowers the adults feed. Even the bees, whose food requirements in both the larval and the adult state are entirely provided by flowers, are limited by their need for suitable nesting sites, and each species has its habitat preferences and can live only within a certain range of climate. Plants in their turn are limited by their particular needs with regard to light, temperature, soil-moisture, mineral nutrients, shelter and so on.

As we have already pointed out, instances of complete dependence of a plant upon a single species of pollinator are uncommon. In a few notable cases, such as those of the figs (p. 312) and *Yucca* (p. 316), the dependence is mutual and complete, and the plant and its pollinator virtually constitute a compound organism. In other instances the plant is dependent on a single pollinator but the dependence is one-sided. Stearn (1969) has given examples of close adaptation of central American *Columnea* species to particular species of hummingbirds. Within our own flora Lords and Ladies (*Arum maculatum*) appears to be largely dependent on *Psychoda phalaenoides*. In this case the pollinator is abundant and presumably does not limit the distribution of the plant, but it would be interesting to establish that this is in fact so. Several 'insect orchids' of the genus *Ophrys* quite clearly would be limited to the range of their specific pollinators; thus the Mirror Ophrys could hardly be expected to transgress far outside the range of *Campsoscolia ciliata* (p. 244). On the other hand, Kullenberg's observations on specimens in cultivation show that *O. lutea* could find suitable species of *Andrena* as pollinators far to the north of its present range in Europe. As was pointed out earlier, the detailed distribution of plant and pollinator need not coincide precisely provided there is some overlap; this is true particularly of long-lived plants with effective long-range seed-dispersal.

There are some instances of genera which are closely adapted to a narrow range of pollinators. A well-known example is provided by the monkshoods (*Aconitum*) (p. 204), whose distribution closely follows that of the bumble-bees, which are their sole pollinators over most of their range. The louseworts (*Pedicularis*) (p. 208), also specialised bumble-bee flowers, have a similar distribution.

For most plants the important factor is the general abundance of a particular category of pollinators, and correlations are often obvious between the general character of the available pollinator fauna, and the range and relative frequency of different adaptation syndromes among the plants. A classification of flower-types makes it possible to compare *pollination-spectra* from different areas.

In lowland Britain – in a region of temperate deciduous forest heavily influenced by Man – we find adaptation to a wide range of insect pollinators alongside anemophily (Chapter 8) and a good deal of self-pollination (Chapter 9). Bumble-bees are predominant among the more specialised pollinators and, correspondingly, large zygomorphic bumble-bee flowers form a conspicuous (though not predominant) part of the flora. Many flowers are visited by solitary bees, and in some cases (for example Harebell) the pollination mechanism is clearly adapted to these insects. There are a number of good examples of flowers closely adapted to pollination by butterflies and moths (see pp. 215, 241). The honey-bee, which probably originated in southern Asia, is a relatively recent arrival not native to our islands. Although it visits a wide range of plants and is an important pollinator of some, especially cultivated crops and orchards, there are no highly specialised pollination mechanisms in our flora which are at all closely adapted to this species. Pearson (1933) considered that in North America the introduction of the honey-bee had deleterious effects on the populations of the native solitary bees which could in turn have affected plant species. The same may well have happened in historic times in Europe. Certainly the activities of solitary bees are seen at their best away from the near vicinity of colonies of honey-bees.*

The proportion of wind-pollinated and self-pollinated species tends to be higher in areas where pollinating insects are scarce. This comes about not only because different species are present. It is not uncommon for a species which is generally entomophilous to be self-pollinated in outlying parts of its range, or in regions where conditions are unfavourable for insect-pollination. Hagerup (1951) has given an account of a situation of this kind in the Faroes, where virtually the only pollinating insects of any

---

* The effects of the interactions of different species and groups of insects seem to have been rather little investigated, though Kikuchi (1962, etc.) has recently studied the relative 'aggressiveness' or dominance of the diverse insects visiting *Chrysanthemum leucanthemum* in Japan. Such interactions may have significant effects on the range of pollinators visiting a particular species under different conditions, though they are likely to be most important for flowers of a moderate or rather low degree of specialisation.

consequence are *Diptera*. The flora of the islands includes a large proportion of species which are regularly anemophilous or autogamous. Insects are essential for the pollination of only a few species, and a number of species which are insect-pollinated in central and southern Europe are either wind-pollinated or selfed in the Faroes. Thus Self-heal (*Prunella vulgaris*) and Common Butterwort (*Pinguicula vulgaris*) are apparently regularly selfed in the Faroes (as they are in Denmark), although they are entomophilous in central Europe; White Clover, Bird's-foot Trefoil, Bush Vetch and Bell Heather are also apparently autogamous in the Faroes. Ling (*Calluna vulgaris*), on the other hand, appears to be pollinated largely by wind or by *Taenothrips ericae*.

Similar conditions are found in the Scottish Highlands and in the Scandinavian mountains, and these areas contrast with the Alps, which have a much richer fauna of pollinating insects and a flora much richer in large, highly coloured, regularly entomophilous flowers, especially in their warmer southern ranges.

Exposed conditions and a cold damp climate are not the only circumstances which may lead to a scarcity of pollinating insects and favour autogamy. Excessively high temperature may have something of the same effect, as Hagerup (1932) observed in the region around Timbuktu in the southern part of the Sahara. For much of the year the temperature of the ground surface reaches 70–80°C. The air temperature falls gradually to 40–45°C. at a man's height. Insects are sparse, although Timbuktu lies only 10° north of the Equator. Hagerup found that the lower-growing plants were almost all self-pollinated. On the other hand, the taller plants were mostly entomophilous. Hagerup describes a forest of *Acacia tortilis* south of Timbuktu in which '. . . the fully expanded flowers fill the *Acacia* forest with a sweet scent. The flat crowns of the acacias are covered with a profusion of small globular inflorescences, and there is a buzzing and humming of insects as in one of our native flowering lime-trees. The big [solitary wasp] *Eumenes tinctor* clumsily wends its way from flower to flower, its black body gaily decorated with light-coloured pollen, while a large green scarab feeds on the stamens of the acacias and becomes quite dusted with pollen, being thus probably also able to effect pollination. But the whole of this tropical idyll acquires a special charm by the presence of the numerous variegated birds which flit to and fro, now feeding on the buds of the flowers, now snatching insects visiting the flowers. The handsomest is the gaily coloured honeybird *Nectarinia pulchella* Bouv. which, in its adroit hunt for insects, also disperses pollen among the flowers.' Autogamy is in fact a common condition in hot arid and semi-arid areas. Hagerup also comments on the tendency of bees to visit such low-growing entomophilous species as there are only in the early hours of the morning, another common feature of such areas.

The tropical rain forests escape the extremes of temperature of the arid habitats, and present quite a different picture. These are the great regions of zoophilous pollination. Insects are abundant, and insect pollination mechanisms appear in great variety, and it is in the tropical rain forests that mechanisms depending on birds and bats are found in greatest abundance and diversity. Wind-pollination is rare for reasons (detailed by Whitehead (1969)) which are the antithesis of those which account for its prevalence in temperate forests.

These are differences between major geographical regions, but on a smaller scale similar differences exist between the plants of different habitats within any one geographical area. Thus within southern England the large zygomorphic flowers adapted to visits by bumble-bees are a particularly prominent component of the flora of hedgebanks, wood margins and the ground flora of rather open woods, while open heaths and grasslands include a higher proportion of species adapted to smaller bees and *Diptera*. Anemophily is important among the dominant trees for reasons discussed in Chapter 8, and among plants of diverse life-form in exposed habitats.

The interaction between the detailed distributions of flowers and the distribution of their pollinators may have important consequences in determining the distribution of seed-production within a plant population. A striking example in *Ophrys lutea* has been quoted already (p. 247). In the Faroes, Hagerup observed that purely entomophilous species, such as Wood Cranesbill (*Geranium sylvaticum*) and Red Campion (*Silene dioica*) were concentrated particularly in the vicinity of bird-cliffs and human habitations, which provide the breeding places of the *Diptera* which visit them.

FLORAL BIOLOGY AND GENETIC SYSTEMS

Darwin laid great stress on the importance of cross-fertilisation, and on the part played by floral mechanisms in promoting it. As we have seen, flowers display many adaptations which tend to prevent autogamy, or pollination with pollen from the same flower, but most of these mechanisms do not exclude pollination with pollen from another flower on the same plant. On the other hand, dioecism or self-incompatibility will effectively prevent anything other than cross-fertilisation from taking place. Both are very widespread among the flowering plants, but if cross-fertilisation were of paramount importance in itself we might expect them to be universal. In fact, of course, dioecious and hermaphrodite species, or self-incompatible and self-compatible species, are often found within the same genus. Species whose floral biology allows a greater or less degree of self-fertilisation are far too numerous to be mere aberrant departures from a general rule. What, then, are the factors that hold the balance of advantage between cross- and self-fertilisation?

As Darlington (1939, p. 83) pointed out, 'The efficiency of natural selection depends on the availability and potential permanence of the largest possible number of combinations of hereditary differences. These properties depend in turn on crossing over and hybridity.' New combinations of genes are produced by independent segregation of the chromosomes and crossing over within the chromosomes at meiosis, but they can only arise if there is some degree of heterozygosity in the genetic constitution of the parent plant. The fundamental importance of cross-fertilisation lies in the production of this heterozygosity. Continued self-fertilisation leads to progressive loss of heterozygosity, until ultimately an inbred population will come to consist of individuals which are homozygous for almost all of their genes. At this stage segregation and crossing over can produce virtually no new combinations of genes. The importance of cross-fertilisation for the *immediate* fitness of the individual progeny of habitually cross-fertilised species, which engaged the attention of Darwin, is a secondary consequence of a particular type of breeding system (see p. 384).

Although a highly heterozygous population is favourable for the production of new gene combinations it is not favourable for the expression of their effects. A recessive gene is almost certain to be masked by its dominant allele, and consequently selection will be slower and less effective than in a population with a lower degree of heterozygosity.

The floral biology of a species plays a crucial role by determining its *breeding system*. The rate at which new combinations of genes can be generated depends on the number of chromosomes and the frequency of chiasmata at meiosis, the degree of heterozygosity of the population, and the number of effective gametes produced in a given time. The rate at which new genotypes become established in the population depends on the intensity of selection and the average length of time between generations, and will be greatest in a population which is not too highly heterozygous. Consequently, we should expect the breeding system of a species to be closely linked with other aspects of its biology, and this is indeed found to be the case.

In long-lived perennial plants growing in closed, stable plant communities, the length of time between generations may be many years, and selection-pressure in the establishment of new individuals is intense. Under these conditions the premium is on the production of the greatest possible variety of gene-combinations, associated with a high degree of heterozygosity. Constancy of adaptation to the environment in the short run is taken care of by the longevity of the individuals. Our forest trees typically belong to this category, as do many common perennial species of woods and grassland and many aquatic plants that persist for long periods by vegetative reproduction. It is these plants which are typically self-incompatible or dioecious.

At the other extreme are the annual plants of impermanent habitats. Here the generation-time is short – there may even be several generations in a year – and selection-pressure is erratic. In a favourable year the population may expand extremely rapidly, with little elimination among the seedlings; in another season only a few individuals may survive. It is advantageous if the plants are homozygous for most genes, because this will maintain the general level of adaptation of the population through periods of low selection-pressure, and in spite of the rapid turnover of individuals. The short generation-time allows relatively rapid production of new gene-combinations even with only occasional crossing, and favourable new combinations can rapidly become established in the population (see p. 292). To the plant that suffers great fluctuations in numbers self-compatibility has the further advantage that a site can be colonised, and a population built up, by a single individual (see p. 291).

Many plants fall somewhere between these two extremes. Even in self-fertile and apparently habitually self-fertilised plants a good deal of outcrossing may take place. Thus Münzing (1930) found that the proportion of outcrossing ranged between zero and 16 per cent. in a series of progeny tests on Galeopsis tetrahit, and Watts (1958) found up to 11.5 per cent. of outcrossing in cultivated lettuce. It is seldom easy to establish the amount of outcrossing which takes place under natural conditions, but for many plants which have at least some evident adaptations favouring cross-pollination the proportion of outcrossing is likely to be greater than in the two instances quoted.

Of course, there are species which appear as exceptions to this general correlation between breeding system and ecological behaviour. An outcrossing breeding system is viable only if the agency of pollen transfer is effective enough to bring about cross-pollination of a fair proportion of the flowers. If many flowers remain unfertilised the way is open for selection in favour of alternative agencies of pollination (for example wind, see p. 275) or of self-fertility or apomixis. In some cases, such selection may be regarded as retrograde in that the future evolutionary potential of the species is sacrificed to immediate advantage. On the other hand, a species may be self-fertilised over only a part of its range, and this may allow it to occupy areas in which it would otherwise be unable to maintain itself. Kullenberg's observations suggest that this may be the situation of the Bee Orchid (Ophrys apifera).

Apomictic species are a special case. They are noticeably concentrated in open and unstable habitats in which change, though it may be drastic and repeated, is on a much longer time-scale than in the habitats of the autogamous annuals. The facultatively apomictic brambles, which were probably primarily colonisers of the gaps left in forest by the death or windthrow of old trees, are a good example. Other groups, such as the hawkweeds (Hieracium) and the Lady's mantles (Alchemilla vulgaris agg.),

probably evolved apomoxis as a response to the unstable habitats produced and maintained by the climatic upheavals of the Pleistocene
period. No doubt they originated from outcrossing ancestors as facultatively apomictic complexes like the brambles (p. 293), but under the
relatively stable conditions of the present day they persist as discrete
obligately apomictic populations in such persistent open habitats as
mountain cliffs, streamsides and meadows.

The breeding system is not the only part of the genetic system which
affects the production of new gene-combinations. There are further
factors, both internal and external. As indicated already, recombination
depends on chromosome number and crossover frequency, and perennial,
predominantly outcrossing species tend to have higher chromosome
numbers and higher chiasma frequencies than autogamous annuals. Thus,
in general, these internal factors and the characteristics of the breeding
system produce complementary effects which tend in the same direction.
However, there are cases, for instance in *Crepis* and other genera of the
Compositae (Stebbins, 1958), of annuals of impermanent habitats which
are nevertheless cross-fertilising and highly heterozygous. These plants
have evidently attained a degree of adaptive stability by evolving low
chromosome numbers and low crossover frequencies, rather than by the
more usual means of change in the breeding system.

Turning to external factors, recombination is influenced by population
size and the degree to which free gene-exchange is possible within the
population. There is always some restriction of the theoretically possible
mating and recombination, even in a self-incompatible or dioecious
species with 100 per cent. outcrossing. Mating of closely related individuals
is only to a limited degree more effective in maintaining a high level of
heterozygosity than self-fertilisation. We are very conscious of this in
relation to our own breeding system. The 'Table of Kindred and Affinity'
in the *Book of Common Prayer* is one of many cultural devices to prevent
matings between closely related individuals. But even in our own case
distance and racial and cultural preferences severely limit the possible
range of matings.

An outcrossing plant will generally be pollinated by an individual
fairly close at hand, and this individual may well be closely related to it.
Pollen may occasionally be carried for very long distances, but the great
bulk of it is deposited close to its source. For wind-pollinated forest trees
the average distance of pollen dispersal is of the order of a few hundred
feet (see p. 261); in a dense stand any given tree may have several hundred
other trees as potential pollen sources, none of which contributes more
than a few per cent. to the total pollen received (see Wang *et al.*, 1960;
Bannister, 1965). With increasing distance between trees (or with less
widely dispersed pollen) the number of potential pollen sources decreases,
and an increasing proportion of the total pollen comes from the tree's

nearest neighbours. The situation is in many ways similar if the plant is insect-pollinated. Most pollen transfer takes place on short flights from flower to flower, or from a plant to one of its nearest neighbours, but a proportion takes place on 'exploratory' and 'escape' flights which bring about relatively long-distance dispersal (Bateman, 1947c; Levin and Kerster, 1968).

The effectiveness of outbreeding depends also on seed dispersal. Most of the seeds of an oak will remain close to the parent tree, while the small winged seeds of pine or birch will be dispersed by wind over a wide area. As a consequence, under natural conditions, a more nearly random distribution of genotypes might be expected in a pine wood than in an oak wood, and (other things being equal) pine would suffer less restriction on outbreeding than oak. To a large extent, gene dispersal by pollen and by seed will have equivalent genetic effects in the population.* In some groups of plants – for instance the light-seeded orchids and the *Compositae* with pappus-borne seeds – seed-dispersal may play a major role in gene-movement within populations. Even acorns are sometimes carried considerable distances by birds, and occasional long-distance seed-dispersal of this kind may have important genetic consequences. Bannister (1965) gives an interesting account of the genetical structure of populations of the Monterey pine (*Pinus radiata*) in New Zealand in relation to seed-dispersal and establishment.

A widely outcrossing species can carry in the heterozygous state many genes which would be deleterious or even lethal if homozygous. This may be a long-term advantage by adding to the richness of the store of variation available for selection under changed circumstances; in some cases the heterozygotes themselves may be at an immediate advantage. However, in a situation of this kind, inbreeding will result in a proportion of unfit individuals – a well known fact which played a large part in the development of the Knight-Darwin 'law' that 'Nature . . . abhors perpetual self-

---

* Sewall Wright (1946) treated the problem of isolation by distance mathematically, on the assumption that the parent organisms follow a Normal distribution in space relative to their offspring. For a given individual, the genetic effects of this are equivalent to completely random mating within a population of a limited number of individuals N, for which Wright introduced the term 'neighbourhood'. For a plant growing evenly over an area at a density d, with pollen dispersed to a root-mean-square (standard deviation) distance s_p but with negligible seed-dispersal, $N = 2\pi s_p^2 d$. Seed dispersal will have twice the effect of pollen-dispersal with the same standard deviation, because it brings about dispersal of both ovules and pollen source. If the seed is dispersed with standard deviation s_s, then $N = 4\pi (s_p^2/2 + s_s^2)d$. Standard deviation of dispersal increases by a factor proportional to the square root of the number of successive generations considered. In fact, the distribution of offspring will generally not be Normal relative to the parents, but the type of distribution, whether Normal or leptokurtic (p. 261), has rather little effect on neighbourhood size (Wright, 1969, Vol. II, pp. 303–5).

fertilisation' (see p. 28). In the light of our present knowledge of genetics it is easy to see that conclusions drawn from habitually outcrossing plants cannot be generalised to plants with other kinds of breeding systems, and the fundamental importance of cross-fertilisation must be sought elsewhere (p. 381).

More detailed discussions of breeding systems in plants are given by Grant (1958) and Stebbins (1950, 1958). Fryxell (1957) gives an outline of methods by which breeding systems may be investigated, and a useful summary of the available information on breeding systems in species of flowering plants.

Compared with other organisms, the flowering plants show extraordinary variety in their genetic systems, and the diversity to be found even among closely related species suggests that features of floral structure and floral physiology are among the most accessible to natural selection of the controlling factors of the genetic system. It may well be that the evolution of the flower as a powerful and flexible mechanism for the adjustment of the genetic system has been the primary factor in the success of the flowering plants.

BIBLIOGRAPHY AND
AUTHOR INDEX

The list below includes the books and papers referred to in the text, and in addition a few general works which have not been specifically quoted. Titles of books are in *italic type*, followed by the place of publication. Titles of papers are in roman type, followed by the title of the journal or series in italic (abbreviations follow the *World List of Scientific Periodicals*), volume number and (after a colon) page-reference. A few major works have been cited in the text without dates. In such cases, the abbreviated reference (in italics) follows the full reference in this list (e.g. *'Knuth'*).

ÅKERBERG, E. and CRANE, E. (Eds.) (1966). *Proc. Second Internat. Symp. Pollination.* (*Bee World* 47; Suppl.). [343, 346, 348, 349, 354]

ALLEN, P. H. (1950). Pollination in *Coryanthes speciosa. Amer. Orch. Soc. Bull.* 19: 528–36. [296]

ANDREWS, H. W. (1953). Flies at parsley blossom. *Entomologist's Rec. J. Var.* 65: 58–9. [91]

ARBER, A. (1920). *Water Plants.* Cambridge. [277]

AXELROD, D. I. (1952). A theory of Angiosperm evolution. *Evolution* 6: 29–60. [363]

BAIRD, A. M. (1938). A contribution to the life history of *Macrozamia reidlei. J. Proc. R. Soc. West Aust.* 25: 153–75 [362]

BAKER, H. G. (1948). Stages in invasion and replacement demonstrated by species of *Melandrium. J. Ecol.* 36: 96–119. [31]

BAKER, H. G. (1949). Dimorphism and monomorphism in the Plumbaginaceae. I. A survey of the family. *Ann. Bot., Lond.* N.S. 12: 207–19. [227]

BAKER, H. G. (1955). Self-compatibility and establishment after long-distance dispersal. *Evolution* 9: 347–8. [291]

BAKER, H. G. (1957). Plant Notes: *Calystegia. Proc. Bot. Soc. Br. Isl.* 2: 241–3. [84, 86, 194]

BAKER, H. G. (1961). The adaptation of flowering plants to nocturnal and crepuscular pollinators. *Q. Rev. Biol.* 36: 64–73. [102, 194]

BAKER, H. G. (1966). The evolution, functioning and breakdown of heteromorphic incompatibility systems. *Evolution* 20: 349–68. [227]

BAKER, H. G. and HARRIS, B. J. (1957). The pollination of *Parkia* by bats and its attendant evolutionary problems. *Evolution* 11: 449–60. [30, 330, 334-6]

BAKER, H. G. and HARRIS, B. J. (1959). Bat-pollination of the Silk-Cotton Tree, *Ceiba pentandra* (L.) Gaertn. (sensu lato), in Ghana. *Jl. W. Afr. Sci. Ass.* 5: 1–9. [30, 334-5]

BAKER, H. G. and HURD, P. D. (1968). Intrafloral ecology. *A. Rev. Ent.* 13: 385–414.   [356]

BANNISTER, M. H. (1965). Variation in the breeding system of *Pinus radiata*, in Baker, H. G. and Stebbins, G. L., (Eds.), *The Genetics of Colonising Species*, pp. 353–72. Academic Press, New York and London.   [383]

BARNES, E. (1934). Some observations on the genus *Arisaema* on the Nilgiri Hills, South India. *J. Bombay Nat. Hist. Soc.* 37: 630–9.   [306, 308]

BATEMAN, A. J. (1947a). Contamination of seed crops. I. Insect pollination. *J. Genet.* 48: 257–75.   [342]

BATEMAN, A. J. (1947b). Contamination of seed crops. II. Wind pollination. *Heredity* 1: 235–46.   [341]

BATEMAN, A. J. (1947c). Contamination in seed crops. III. Relation with isolation distance. *Heredity* 1: 303–36.   [261, 384]

BEATTIE, A. J. (1969). Studies in the pollination ecology of *Viola*. I. The pollen-content of stigmatic cavities. *Watsonia* 7: 142–56.   [215, 374]

BEATTIE, A. J. (1972). Studies in the pollination ecology of *Viola*. II. Pollen loads of insect visitors. *Watsonia* 9: 13–25.   [80, 215]

BECHER, E. (1882). Zur Kenntnis der Mundtheile der Dipteren. *Denkschr. Akad. Wiss. Wien (Math.-Nat. Cl.)* 45: 123–62.   [80, 82]

BENÉ, F. (1941). Experiments on the colour preference of Black-chinned Hummingbirds. *Condor* 43: 237–42.   [326]

BENÉ, F. (1946). The feeding and related behaviour of hummingbirds, with special reference to the Black-chin, *Archilochus alexandri* (Bourcier & Mulsant). *Mem. Boston Soc. Nat. Hist.* 9: 395–480.   [324–6]

BENHAM, B. R. (1969). Insect visitors to *Chamaenerion angustifolium* and their behaviour in relation to pollination. *Entomologist* 102: 221–8.   [179]

BENNETT, A. W. (1883). On the constancy of insects in their visits to flowers. *J. Linn. Soc. (Zool.)* 17: 175–85.   [86, 99, 129]

BENSON, R. B. (1950). An introduction to the natural history of British sawflies (Hymenoptera Symphyta). *Trans. Soc. Br. Ent.* 10: 45–142.   [132, 133]

BENSON, R. B. (1959). Sawflies (Hym., Symphyta) of Sutherland and Wester Ross. *Entomologist's mon. Mag.* 95: 101–4.   [133]

BISCHOFF, H. (1927). *Biologie der Hymenopteren*. Springer, Berlin.   [133]

BLAIR, PATRICK (1720). *Botanick Essays*. London.   [24]

BLAIR, PATRICK (1721). Observations upon the generation of plants. *Phil. Trans. R. Soc.* No. 369: 216–21.   [22, 23]

BLOCK, J. M. (1962). Insect damage to flowers. *Gdnrs' Chron.* 152: 375.   [143]

BODGER, H. S. (1960). Buckets of pollen and tons of seed. *Proc. 15th Amer. Hort. Congr. and Ann. Meeting Amer. Hort. Soc.*: 18–19   [342]

BOGDAN, A. V. (1962). Grass pollination by bees in Kenya. *Proc. Linn. Soc. Lond.* 173: 57–60.   [272]

BOHART, G. E. (1957). Pollination of Alfalfa and Red Clover. *A. Rev. Ent.* 2: 355–80.   [341, 348]

BOHART, G. E. (1962). How to manage the Alfalfa Leaf-cutting Bee (*Megachile rotundata* Fabr.) for Alfalfa pollination. *Agric. Exp. Station, Utah State Univ., Logan, Circular no. 144.*   [346]

BOHART, G. E. and NYE, W. P. (1960). Insect pollinators of Carrots in Utah. *Agric. Exp. Station, Utah State Univ., Logan, Publ. no. 419.*   [349]

BOLWIG, N. (1954). The role of scent as a nectar guide for honeybees on flowers

388   BIBLIOGRAPHY AND AUTHOR INDEX

and an observation on the effect of colour on recruits. *Brit. J. Anim. Behav.* 2:
81-3   [164]
BRADLEY, RICHARD (1717). *New Improvements of Planting and Gardening.* London.
                                                                          [21]
BRADSHAW, MARGARET E. (1963). Studies on *Alchemilla filicaulis* Bus., sensu
lato, and *A. minima* Walters. Introduction, and I. Morphological variation in
A. filicaulis, sensu lato. *Watsonia* 5: 304-20.   [293]
BRAUE, A. (1913). Die Pollensammelapparate der beinsammelnden Bienen.
Jena Z. Naturw. 50: 1-96.   [162]
BRIAN, A. D. (1954). The foraging behaviour of bumble-bees. *Bee World* 35: 61-7.
                                                                          [181]
BRIAN, A. D. (1957). Differences in the flowers visited by four species of bumble-
bees and their causes. *J. Anim. Ecol.* 26: 71-98.   [151, 180, 181, 218]
BRIAN, M. V. and BRIAN, A. D. (1952). The wasp *Vespula sylvestris* Scopoli:
feeding, foraging and colony development. *Trans. R. ent. Soc. Lond.* 103: 1-26.
                                                                     [142, 143]
BRIGGS, D. and WALTERS, S. M. (1969). *Plant Variation and Evolution.* Weiden-
feld and Nicolson, London.   [38]
BUCHANAN-WHITE, F. (1898). *The Flora of Perthshire.* Edinburgh.   [29]
BURKHARDT, D. (1964). Colour discrimination in insects, in Beament, J. W. L.,
Treherne, J. E., and Wigglesworth, V. B., (Eds.), *Advances in Insect Physiology* 2.
Academic Press, London and New York.   [107, 108, 171]
BURKILL, I. H. (1897). Fertilization of spring flowers on the Yorkshire coast.
J. Bot., Lond. 35: 92-9, 138-45, 184-9.   [104]
BURTT, B. L. (1961). Compositae and the study of functional evolution. *Trans.
Bot. Soc. Edinb.* 39: 216-32.   [219]
BUTLER, C. G. (1951). The importance of perfume in the discovery of food by
the worker Honeybee (*Apis mellifera* L.). *Proc. Roy. Soc.* B 138: 403-13.
                                                                     [166, 181]
BUTLER, C. G. (1954). *The World of the Honeybee.* Collins, London.
                                                           [164, 173, 176, 179]
CALDWELL, J. and WALLACE, T. J. (1955). Biological Flora of the British Isles.
Narcissus pseudonarcissus. J. Ecol. 43: 331-41.   [196]
CAMMERLOHER, H. (1923). Zur Biologie der Blüte von *Aristolochia grandiflora*
Swartz. *Öst. bot. Z.* 72: 180-98.   [306, 307]
CAMMERLOHER, H. (1933). Die Bestäubungseinrichtungen der Blüten von
Aristolochia lindneri Berger. *Planta* 19: 351-65.   [305]
CHAMBERS, V. H. (1945). British bees and wind-borne pollen. *Nature, Lond.* 155:
145.   [151, 272]
CHAMBERS, V. H. (1946). An examination of the pollen loads of *Andrena*: the
species which visit fruit trees. *J. Anim. Ecol.* 15: 9-21.   [151, 174]
CHAMBERS, V. H. (1947). A list of sawflies (Hym., Symphyta) from Bedford-
shire. *Entomologist's mon. Mag.* 83: 91-5.   [134]
CHAMBERS, V. H. (1949). The Hymenoptera Aculeata of Bedfordshire. *Trans.
Soc. Br. Ent.* 9: 197-252.   [140]
CHRISTY, R. M. (1922). The pollination of the British primulas. *J. Linn. Soc.
(Bot.)* 46: 105-39.   [78, 225, 376]
CHURCH, A. H. (1908). *Types of Floral Mechanism.* Vol. 1. Oxford.   [29]

CLAVAUD, A. (1878). Sur la véritable mode de fécondation du *Zostera marina*. *Act. Soc. Linn. Bordeaux* 32: 109–15. [283]

CLEGHORN, M. L. (1913). Notes on the pollination of *Colocasia antiquorum*. *J. Proc. Asiat. Soc. Beng.* 9: 313–5. [308]

CLEMENTS, F. E. and LONG, F. L. (1923). *Experimental Pollination*. Washington. [31]

COBLEY, L. S. (1956). *An Introduction to the Botany of Tropical Crops*. Longmans, Green & Co., London. [351]

COLEMAN, E. (1927). Pollination of the orchid *Cryptostylis leptochila*. *Vict. Nat.* 44: 20–2. [137]

COLEMAN, E. (1928a). Pollination of *Cryptostylis leptochila* F.v.M. *Vict. Nat.* 44: 333–40. [137]

COLEMAN, E. (1928b). Pollination of an Australian orchid by the male ichneumonid *Lissopimpla semipunctata* Kirby. *Trans. ent. Soc. Lond.* 76: 533–9. [137]

COLEMAN, E. (1929a). Pollination of an Australian orchid, *Cryptostylis leptochila* F. Muell. (with a note by Col. M. J. Godfery). *J. Bot., Lond.* 67: 97–9. [137]

COLEMAN, E. (1929b). Pollination of *Cryptostylis subulata* (Labill.) Reichb. *Vict. Nat.* 46: 62–6. [137]

COLEMAN, E. (1930a). Pollination of some West Australian orchids. *Vict. Nat.* 46: 203–6. [137]

COLEMAN, E. (1930b). Pollination of *Cryptostylis erecta* R. Br. *Vict. Nat.* 46: 236–8. [137]

COLEMAN, E. (1931). Mrs Edith Coleman's further observations on the fertilisation of Australian orchids by the male ichneumonid *Lissopimpla semipunctata*, Kirb. *Proc. ent. Soc. Lond.* 6: 22–4. [137]

COLEMAN, E. (1932). Pollination of *Diuris pedunculata* R. Br. *Vict. Nat.* 49: 179–86. [298]

COLEMAN, E. (1933a). Further notes on the pollination of *Diuris pedunculata* R. Br. *Vict. Nat.* 49: 243–5. [298]

COLEMAN, E. (1933b). Pollination of *Diuris sulphurea* R. Br. *Vict. Nat.* 50: 3–8. [298]

COLEMAN, E. (1934). Pollination of *Pterostylis acuminata* R. Br. and *Pterostylis falcata* Rogers. *Vict. Nat.* 50: 248–52. [309, 311]

COLEMAN, E. (1938). Further observations on the pseudocopulation of the male *Lissopimpla semipunctata* Kirby (Hymenoptera Parasitica) with the Australian orchid *Cryptostylis leptochila* F.v.M. *Proc. R. ent. Soc. Lond.* (A) 13: 82–3. [137]

COLES, S. M. (1971). The *Ranunculus acris* L. complex in Europe. *Watsonia* 8: 237–261. [53]

COLYER, C. N. and HAMMOND, C. O. (1951). *Flies of the British Isles*. Warne, London. [84, 89, 91]

COOKE, BENJAMIN (1749). On the effects of the mixture of the farina of Apple-trees; and of Mayze or Indian Corn . . . *Phil. Trans. R. Soc.* No. 493: 205. [23]

COOMBE, D. E. (1956). Biological Flora of the British Isles. *Impatiens parviflora* DC. *J. Ecol.* 44: 701–13. [86]

COOMBE, D. E. (1961). *Trifolium occidentale*, a new species related to *T. repens Watsonia* 5: 68–87. [290

CORBET, S. A. (1970). Insects on Hogweed flowers: a suggestion for a student project. *J. biol. Educ.* 4: 133–43. [75, 76, 104]

CORNER, E. J. H. (1964). *The Life of Plants.* London. [363]

CORREVON, H. and POUYANNE, M. (1916). Un curieux cas de mimetisme chez les Ophrydacees. *J. Soc. natn. hort. Fr.* 29: 23–84. [244]

CRANE, M. B. and LAWRENCE, W. J. C. (1952). *The Genetics of Garden Plants.* Ed. 4. Macmillan, London. [50, 352]

CROSBY, J. L. (1949). Selection of an unfavourable gene-complex. *Evolution* 3: 212–30. [225]

CROSBY, J. L. (1959). Outcrossing of homostyle primroses. *Heredity* 13: 127–31. [225]

CROSBY, J. L. (1960). The use of electronic computation in the study of random fluctuations in rapidly evolving populations. *Phil. Trans. R. Soc.* Ser. B 242: 551–73.

CROWSON, R. A. (1956). Coleoptera; introduction and keys to families. *Handbooks for the identification of British insects* 4(1). Royal Entomological Society, London. [68]

CRÜGER, H. (1865). A few notes on the fecundation of orchids and their morphology. *J. Linn. Soc. (Bot.)* 8: 127–35. [296]

DARLINGTON, C. D. (1939). *The Evolution of Genetic Systems.* Cambridge. [381]

DARWIN, C. (1858). On the agency of bees in the fertilisation of papilionaceous flowers. *Gdnrs' Chron.* 824–44. [27]

DARWIN, C. (1859). *On the Origin of Species by Natural Selection.* John Murray, London. [27]

DARWIN, C, (1862a). On the two forms, or dimorphic condition, in the species of *Primula* and on their remarkable sexual relations. *J. Linn. Soc. (Bot.)* 6: 77–96. [27]

DARWIN, C. (1862b). *The Various Contrivances by which Orchids are Fertilised.* (Ed. 2, 1888). John Murray, London. [28, 230]

DARWIN, C. (1876). *The Effects of Cross and Self Fertilisation in the Vegetable Kingdom.* John Murray, London. [28]

DARWIN, C. (1877). *The Different Forms of Flowers on Plants of the same Species.* John Murray, London. [224]

DARWIN, F. (1875). On the structure of the proboscis of *Ophideres fullonica*, an orange-sucking moth. *Q. Jl. microsc. Sci.* N.S. 15: 385–9. [99]

DAUMANN, E. (1932). Über die 'Scheinnektarien' von *Parnassia palustris* und anderer Blütenarten. *Jb. wiss. Bot.* 77: 104–49. [57, 111]

DAUMANN, E. (1935). Über die Bestäubungsökologie der *Parnassia*-Blüte II. *Jb. wiss. Bot.* 81: 707–17. [111]

DAUMANN, E, (1941). Die anbohrbaren Gewebe und rudimentären Nektarien in der Blütenregion. *Beih. bot. Zbl.* 61A: 1282. [240]

DAUMANN, E. (1963). Zur Frage dem Ursprung der Hydrogamie. Zugleich ein Beitrag zur Blütenökologie von *Potamogeton. Preslia* 35: 23–30. [279]

DAUMER, K. (1956). Reizmetrische Untersuchungen des Farbensehens der Bienen. *Z. vergl. Physiol.* 38: 413–78. [170, 191]

DAUMER, K. (1958). Blumenfarben, wie sie die Bienen sehen. *Z. vergl. Physiol.* 41: 49–110. [54, 171, 173]

DELEVORYAS, T. (1968). Investigations of North American cycadeoids: structure, ontogeny and phylogenetic considerations of cones of *Cycadeoidea*. *Palaeontographica* B121: 122–33. [362]

DELPINO, FEDERICO (1868). *Ulteriori osservazioni e considerazioni sulla dicogamia nel regno vegetale*. Milano. [28]

DELPINO, F. (1871). Annotations to H. Müller, *Application of the Darwinian theory to flowers and the insects which visit them*. Trans. R. L. Packard. Naturalists' Agency, Salem. [309]

DELPINO, F. (1874). Ulteriori osservazioni e considerazioni sulla dicogamia nel regno vegetale. 2(IV). Delle piante zoidifile. *Atti Soc. Ital. Sci. nat.* 16: 151–349. [100]

DEMOLL, R. (1908). Die Mundteile der solitären Apiden. *Z. wiss Zool.* 91: 1–51. [152–4, 159]

DESSART, P. (1961). Contribution à l'étude des Ceratopogonidae (Diptera). Les *Forcipomyia* pollinisateurs du cacaoyer. *Bull. Agric. Congo* 52: 525–40. [312]

DEXTER, J. S. (1913). Mosquitoes pollinating orchids. *Science* 37: 867. [76]

DIMMOCK, G. (1881). *The Anatomy of the Mouthparts and the Sucking Apparatus of some Diptera*. Dissertation. Boston. [82]

DOBBS, ARTHUR (1750). Concerning bees and their method of gathering wax and honey. *Phil. Trans. R. Soc.* 46: 536–49. [24]

DOCTERS VAN LEEUWEN, W. M. (1931). Vogelbesuch an den Blüten von einigen *Erythrina*-Arten auf Java. *Annls. Jard. Bot. Buitenz.* 42: 57–96. [319, 320]

DOCTERS VAN LEEUWEN, W. M. (1938). Observations about the biology of tropical flowers. *Annls. Jard. Bot. Buitenz.* 48: 27–68. [334, 337]

DODSON, C. H. (1962). The importance of pollination in the evolution of the orchids of Tropical America. *Bull. Am. Orch. Soc.* 31: 525–34, 641–9, 731–5. [244, 295, 296, 298, 299]

DODSON, C. H. (1963). The Mexican Stanhopeas. *Bull. Am. Orch. Soc.* 32: 115–29. [298]

DODSON, C. H. (1965). Studies in orchid pollination; the genus *Coryanthes*. *Bull. Am. Orch. Soc.* 34: 680–7. [296]

DODSON, C. H., DRESSLER, R. L., HILLS, H. G., ADAMS, R. H. and WILLIAMS, N. H. (1969). Biologically active compounds in orchid fragrances. *Science, N. Y.* 164: 1243–1249. [249]

DODSON, C. H. and FRYMIRE, G. P. (1961). Preliminary studies in the genus *Stanhopea* (Orchidaceae). *Ann. Mo. Bot. Gdn.* 48: 137–72. [296, 298]

DOYLE, J. (1945). Developmental lines in pollination mechanisms in the Coniferales. *Scient. Proc. R. Dubl. Soc.* 24(5): 43–62. [264]

DRABBLE, E. and H. (1917). The syrphid visitors to certain flowers. *New Phytol.* 16: 105–9. [62, 85]

DRABBLE, E. and H. (1927). Some flowers and their dipteran visitors. *New Phytol.* 26: 115–23. [56, 62, 74, 84–6, 95]

DUDLEY, the Hon PAUL (1724). Observations on some of the Plants in New-England . . . *Phil. Trans. R. Soc.* 33: 194 [23]

DUNCAN, C. D. (1939). A contribution to the biology of North American Vespine wasps. *Stanford Univ. Publs: Biol. Sci.* 8(1): 1–272. [133, 142]

EASTHAM, L. E. S. and EASSA, Y. E. E. (1955). The feeding mechanism of the butterfly *Pieris brassicae* L. *Phil. Trans. R. Soc.* B 239: 1–43.   [97, 98]

EDWARDS, J. (1956). Some flower visitors in north Staffs. *Entomologist's Rec. J. Var.* 68: 275.   [53]

VAN EMDEN, H. F. (1963). Observations on the effect of flowers on the activity of parasitic Hymenoptera. *Entomologist's mon. Mag.* 98: 265–270.   [136]

ERDTMAN, G. (1952). *Pollen morphology and plant taxonomy. I. Angiosperms.* Almqvist and Wiksell, Stockholm.   [276]

ERDTMAN, G. (1957). *Pollen and spore morphology/Plant taxonomy. II. Gymnospermae, Pteridophyta, Bryophyta.* Almqvist and Wiksell, Stockholm.   [276]

ERICKSON, R. (1965). *Orchids of the West.* Ed. 2. Perth, W. Australia.   [296]

EVANS, M. S. (1895). The fertilisation of *Loranthus kraussianus* and *L. dregei.* *Nature, Lond.* 51: 235–6.   [330]

FAEGRI, K. and VAN DER PIJL, L. (1966). *The Principles of Pollination Ecology.* Pergamon, Oxford, etc. ('*Faegri & Van der Pijl*').   [63, 372]

FAULKNER, C. J. (1962). Blow-flies as pollinators of *Brassica* crops. *Commercial Grower* No. 3457: 807–9.   [349]

FISHER, R. A. and MATHER, K. (1943). The inheritance of style-length in *Lythrum salicaria.* *Ann. Eugen.* 12: 1–23.   [226]

FLORIN, R. (1954). The female reproductive organs of conifers and taxads. *Biol. Rev.* 29: 367–89.   [358]

FOGG, G. E. (1950). Biological Flora of the British Isles. *Sinapis arvensis* L. *J. Ecol.* 38: 415–29.   [219]

FORD, E. B. (1945). *Butterflies.* Collins, London.   [96]

FORD, E. B. (1955). *Moths.* Collins, London.   [96]

FORD, E. B. (1964). *Ecological Genetics.* Methuen, London.   [226]

FREE, J. B. (1962). Studies on the pollination of fruit trees by Honey-bees. *Jl. R. hort. Soc.* 87: 302–9.   [352]

FREE, J. B. (1963). The flower constancy of Honey-bees. *J. Anim. Ecol.* 32: 119–31.   [176]

FREE, J. B. (1964). Comparison of the importance of insect and wind pollination of apple trees. *Nature, Lond.* 201: 726–7.   [275]

FREE, J. B. (1965a). The ability of bumblebees and Honey-bees to pollinate Red Clover. *J. Appl. Ecol.* 2: 289–94.   [348]

FREE, J. B. (1965b). The effect on pollen collection of feeding Honey-bee colonies with sugar syrup. *J. Agric. Sci., Camb.* 64: 167–8.   [353]

FREE, J. B. (1966a). The foraging areas of Honey-bees in an orchard of standard apples. *J. Appl. Ecol.* 3: 261–8.   [353]

FREE, J. B. (1966b). The foraging behaviour of bees and its effect on the isolation and speciation of plants, in Hawkes, J. G. (Ed.), *Reproductive Biology and Taxonomy of Vascular Plants.* Botanical Society of the British Isles conference report No. 9. Pergamon Press, Oxford.

FREE, J. B. (1968). Dandelion as a competitor to fruit trees for bee visitors. *J. Appl. Ecol.* 5: 161–78.   [353, 374]

FREE, J. B. (1970). *Insect Pollination of Crops.* Academic Press London and New York.   [355]

FREE, J. B. and BUTLER, C. G. (1959). *Bumblebees.* Collins, London.   [181, 348]

FREE, J. B. and SPENCER-BOOTH, Y. (1964). The foraging behaviour of Honey-bees in an orchard of dwarf Apple trees. *J. hort. Sci.* 39: 78–83.   [353]

FRICK, K., POTTER, H. and WEAVER, H. (1960). Development and maintenance of Alkali Bee nesting sites. *Washington Agric. Exp. Stations Circular* No. 366.   [343]

FRIESE, H. (1923). *Die Europäischen Bienen (Apidae)*. De Gruyter, Berlin and Leipzig.   [149, 151]

FRISCH, K. VON (1950). *Bees, their Vision, Chemical Senses and Language*. Cornell Univ. Press, Ithaca, N.Y.   [164, 169, 173]

FRISCH, K. VON (1954. *The Dancing Bees*. Methuen, London.
                                                          [164, 169, 173, 174, 176, 340]

FRYXELL, P. A. (1957). Mode of reproduction of higher plants. *Bot. Rev.* 23: 135–233.   [385]

GALIL, J. and EISIKOVITCH, D. (1968a). On the pollination ecology of *Ficus sycomorus* in East Africa. *Ecology* 49: 259–69.   [315]

GALIL, J. and EISIKOVITCH, D. (1968b). On the pollination ecology of *Ficus religiosa* in Israel. *Phytomorphology* 18: 356–63.   [314, 315]

GALIL, J. and EISIKOVITCH, D. (1969). Further studies on the pollination ecology of *Ficus sycomorus* L. *Tijdschr. Ent.* 112: 1–13.   [314, 315]

GALIL, J. and ZERONI, M. (1965). Nectar system of *Asclepias curassavica. Bot. Gaz.* 126: 144–8.   [301]

GAMERRO, J. C. (1968). Observaciones sobre la biología floral y morfología de la Potamogetonácea *Ruppia cirrhosa* (Petag.) Grande (=*R. spiralis* L. ex Dum.). *Darwinia* 14: 575–609.   [281]

GIBSON, A. H. (1893). The phanerogamic flora of St Kilda. *Trans. Bot. Soc. Edinb.* 19: 155.   [29]

GODFERY, M. J. (1925). The fertilisation of *Ophrys speculum, O. lutea* and *O. fusca. J. Bot., Lond.* 63: 33–40.   [244]

GODFERY, M. J. (1927). The fertilisation of *Ophrys fusca* Link. *J. Bot., Lond.* 65: 350–1.   [244]

GODFERY, M. J. (1929). Recent observations on the pollination of *Ophrys. J. Bot., Lond.* 67: 298–302.   [245]

GODFERY, M. J. (1930). Further notes on the fertilisation of *Ophrys fusca* and *O. lutea. J. Bot., Lond.* 68: 237–8.   [244]

GODFERY, M. J. (1933). *Monograph and Iconograph of Native British Orchidaceae.* Cambridge University Press.   [248]

GODWIN, H. (1956). *The History of the British Flora.* Cambridge University Press.
                                                                                         [277]

GOODWIN, T. W. (Ed.) (1965). *Chemistry and Biochemistry of Plant Pigments.* Academic Press, London and New York.   [50]

GOUIN, F. (1949). Recherches sur la morphologie de l'appareil buccal des Diptères. *Mém. Mus. natn. Hist. nat.*, Paris N.S.28: 167–269.   [80, 81, 83]

GRAHAM, R. (1839). *Aristolochia saccata.* Pouch-flowered Birth-wort. *Curtis's bot. Mag.* 65: t.3640.   [306]

GRAHAM-SMITH, G. S. (1930). Further observations on the anatomy and function of the proboscis of the Blow-fly, *Calliphora erythrocephala* L. *Parasitology* 22: 47–115.   [91–3, 108, 365]

GRANDI, G. (1961). The hymenopterous insects of the superfamily Chalcidoidea developing within the receptacles of figs. *Boll. Ist. Ent. Univ. Bologna* 26: 1–3.
                                                                                   [313, 314]

GRANT, K. A. (1966). A hypothesis concerning the prevalence of red coloration in California hummingbird flowers. *Am. Nat.* 100: 85–97.   [326]

GRANT, K. A. and V. (1968). *Hummingbirds and their Flowers.* Columbia University Press, New York and London.   [324, 327]

GRANT, V. (1949). Pollination systems as isolating mechanisms in Angiosperms. *Evolution* 3: 82–97.   [371, 372, 374, 375]

GRANT, V. (1950a). The pollination of *Calycanthus occidentalis. Am. J. Bot.* 37: 294–7.   [70]

GRANT, V. (1950b). The protection of the ovules in flowering plants. *Evolution* 4: 179–201.   [70, 219, 319, 321, 362]

GRANT, V. (1950c). The flower constancy of bees. *Bot. Rev.* 16: 379–98.   [170, 174, 180]

GRANT, V. (1952). Isolation and hybridisation between *Aquilegia formosa* and *A. pubescens. Aliso* 2: 341–60.   [367, 369, 380]

GRANT, V. (1958). The regulation of recombination in plants. *Cold Spring Harb. Symp. quant. Biol.* 23: 337–63.   [385]

GRANT, V. (1963). *The Origin of Adaptations.* Columbia University Press, New York.   [369]

GRANT, V. and K. A. (1965). *Flower Pollination in the Phlox Family.* Columbia University Press, New York.   [364, 373, 376]

GREGORY, D. P. (1963, 1964). Hawkmoth pollination in the genus *Oenothera. Aliso* 5: 357–84, 385–419.   [367]

GREGORY, P. H. (1961). *The Microbiology of the Atmosphere.* Leonard Hill, London, and Interscience Publishers, New York.   [260, 261]

GRENSTED, L. W. (1946). An assemblage of Diptera on Cow-parsnip. *Entomologist's mon. Mag.* 82: 180.   [62, 75, 91]

GRENSTED, L. W. (1947). Diptera in the spathes of *Arum maculatum. Entomologist's mon. Mag.* 83: 1–3.   [89, 229]

GREW, NEHEMIAH (1671). *The Anatomy of Vegetables Begun.* See 'Account' in *Phil. Trans. R. Soc.* No. 78: 3041.   [20, 23]

GREW, NEHEMIAH (1682). *The Anatomy of Plants.* London.   [20]

GRIFFITHS, D. J. (1950). The liability of seed-crops of Perennial Rye Grass (*Lolium perenne*) to contamination by wind-borne pollen. *J. Agric. Sci., Camb.* 40: 19–38.   [261, 262]

GUSTAFSSON, Å. (1946). Apomixis in higher plants. I. The mechanism of apomixis. *Lunds Univ. Årsskr.* 42: 1–67.   [292]

GUSTAFSSON, Å. (1947). Apomixis in higher plants. II. The causal aspect of apomixis. *Lunds Univ. Årsskr.* 43: 69–179. III. Biotype and species formation. *ibid.* 43: 183–370.   [292]

HAGERUP, O. (1932). On pollination in the extremely hot air at Timbuctu. *Dansk bot. Ark.* 8(1): 1–20.   [379]

HAGERUP, O. (1950a). Thrips pollination in *Calluna. Biol. Medd.* 18(4): 1–16.   [105]

HAGERUP, O. (1950b). Rain pollination. *Biol. Medd.* 18(5): 3–18.   [288]

HAGERUP, O. (1951). Pollination in the Faroes – in spite of rain and poverty of insects. *Biol. Medd.* 18(15): 1–48.   [75, 288, 378–80

HAGERUP, O. (1952a). Bud autogamy in some northern orchids. *Phytomorphology* 2: 51–60.   [234, 237, 254]

HAGERUP, O. (1952b). The morphology and biology of some primitive orchid flowers. *Phytomorphology* 2: 134–8. [231]

HAGERUP, O. and E. (1953). Thrips pollination of *Erica tetralix*. *New Phytol.* 52: 1–7. [188]

HALDANE, J. B. S. (1936). Some natural populations of *Lythrum salicaria*. *J. Genet.* 32: 393–7. [226]

HALDANE, J. B. S. (1938). Heterostylism in natural populations of the Primrose, *Primula acaulis*. *Biometrika* 30: 196–8. [224]

HAMM, A. H. (1934). Syrphidae (Dipt.) associated with flowers. *J. Soc. Br. Ent.* 1: 8–9. [53, 84]

HAMMOND, A. (1874). The mouth of the Crane Fly. *Science Gossip* 1874: 155–60. [71–2]

HARBORNE, J. B. (1963). Distribution of anthocyanins in higher plants, in Swain, T. (Ed.), *Chemical Plant Taxonomy*. Academic Press, New York. [50]

HARLAND, W. B. *et al.* (1967). *The Fossil Record*. Geological Society of London. [358]

HARPER, J. L. (1957). Biological Flora of the British Isles. *Ranunculus acris, R. repens, R. bulbosus*. *J. Ecol.* 45: 289–342. [53, 78, 134, 135, 138]

HARPER, J. L. and WOOD, W. A. (1957). Biological Flora of the British Isles. *Senecio jacobaea* L. *J. Ecol.* 45: 617–37. [76, 88, 91, 135, 151, 221]

HARRIS, B. J. and BAKER, H. G. (1959). Pollination of flowers by bats in Ghana. *Nigerian Field* 24: 151–9. [334]

HATTON, R. H. S. (1965). Pollination of Mistletoe (*Viscum album* L.). *Proc. Linn. Soc. Lond.* 176: 67–76. [275]

HAWKINS, R. P. (1965). Factors affecting the yield of seed produced by different varieties of Red Clover. *J. Agric. Sci., Camb.* 65: 245–53. [347]

HAWTHORN, L. R., BOHART, G. E. and TOOLE, E. H. (1956). Carrot seed yield and germination as affected by different levels of insect pollination. *Proc. Am. Soc. hort. Sci.* 67: 384–9. [349]

HAYES, H. K., IMMER, F. R. and SMITH, D. C. (1955). *Methods of Plant Breeding*. Ed. 2. McGraw-Hill, New York, Toronto and London. [340, 342]

HERTZ, M. (1931). Die Organisation des optischen Feldes bei der Biene. I. *Z. vergl. Physiol.* 8: 693–748. [169]

HERTZ, M. (1935). Die Untersuchungen über den Formensinn der Honigbiene. *Naturwissenschaften* 36: 618–24. [168]

HESLOP-HARRISON, J. (1958). Ecological variation and ethological isolation, in Hedberg, O. (Ed.), Systematics of Today. *Uppsala Univ. Årsskr.* 1958(6): 150–8. [369, 371, 372]

HESSELMANN, H. (1919). Iakttagelser över skogsträdspollens spridningsförmåga. *Meddn St. Skogsforsk Inst.* 16: 27–53. [262–3]

HILDEBRAND, F. (1867). *Die Geschlechtsverteilung bei den Pflanzen*. Leipzig. [28, 58]

HILL, D. S. (1967). Figs (*Ficus* spp.) and fig-wasps (Chalcidoidea). *J. nat. Hist.* 1: 413–34. [316]

HOBBS, G. A. (1965). Importing and managing the Alfalfa Leaf-cutter Bee. *Canada Dept. Agric., Publ. No. 1209.* [356]

HOBBS, G. A. (1967). Domestication of Alfalfa Leaf-cutter Bees. *Canada Dept. Agric., Publ. No. 1313.* [346]

HOBBY, B. M. (1933). Diptera and Coleoptera visiting orchids. *J. ent. Soc. S. Engl.* 1: 105–6.   [67, 240]

HOBBY, B. M. and SMITH, K. G. V. (1961). The bionomics of *Empis tesselata* F. (Dipt. Empididae). *Entomologist's mon. Mag.* 97: 2–10.   [79]

HODGES, D. (1952). *Pollen Loads of the Honeybee.* Bee Research Association, London.
                                                                                    [151, 164–5]

HOWARD, R. A. (1970). The ecology of an elfin forest in Puerto Rico. 10. Notes on two species of *Marcgravia. J. Arnold Arbor.* 51: 41–55.   [323]

HUBBARD, C. E. (1954). *Grasses.* Penguin Books, Harmondsworth.   [271]

HUGHES, N. F. (1961). Fossil evidence and angiosperm ancestry. *Sci. Progr.* 49: 84–102.   [363]

HULKKONEN, O. (1928). Zur Biologie der südfinnischen Hummeln. *Ann. Univ. Abo,* Ser. A, 3: 1–81.   [180]

HYDE, H. A. (1950). Studies in atmospheric pollen. IV. Pollen deposition in Great Britain, 1943. *New Phytol.* 49: 398–420.   [257, 275]

HYDE, H. A. (1969). Aeropalynology in Britain – an outline. *New Phytol.* 68: 579–90.   [275]

HYDE H. A. and ADAMS, K. F. (1958). *An atlas of airborne pollen grains.* Macmillan, London.   [276]

HYDE, H. A. and WILLIAMS, D. A. (1961). Atmospheric pollen and spores as causes of allergic disease: hay-fever, asthma and the aerospora. *Advmt Sci.:* 526–33.   [259]

ILSE, D. (1928). Über den Farbensinn der Tagfalter. *Z. vergl. Physiol.* 8: 658–92.   [127]

ILSE, D. (1932). Zur 'Formwahrnehmung' der Tagfalter. I. Spontane Bevorzugung von Formmerkmalen durch Vanessen. *Z. vergl. Physiol.* 17: 537–56.
                                                                                    [128]

ILSE, D. (1941). The colour vision of insects. *Proc. R. Phil. Soc. Glasg.* 65: 98–112.   [128]

ILSE, D. (1949). Colour discrimination in the Drone Fly, *Eristalis tenax. Nature, Lond.* 163: 255–6.   [113]

IMMS, A. D. (1947). *Insect Natural History.* Collins, London.   [68, 69, 107]

IMMS, A. D. (1957). *A General Textbook of Entomology.* Ed. 9, revised by O. W. Richards and R. G. Davies. Methuen, London, and Dutton, New York.
                                                                                    [96, 105]

JAEGER, P. (1954a). Note sur l'anatomie florale, l'anthocinétique et les modes du pollinisation du Fromager (*Ceiba pentandra* Gaertn.). *Bull. Inst. Français Afr. Noire,* Sér. A, 16: 370–8.   [334]

JAEGER, P. (1954b). Les aspects actuels du problème de la chéiroptèrogamie. *Bull. Inst. Fr. Afr. Noire,* Sér. A, 16: 786–821.   [330, 334, 335]

JAEGER, P. (1957). Les aspects actuels du problème de l'entomogamie. *Bull. Soc. Bot. Fr.* 104: 179–222, 352–412.   [46, 50, 304]

JAEGER, P. (1961). *The Wonderful Life of Flowers.* Trans. J. P. M. Brenan. Harrap, London.

JAMES, R. L. (1948). Some hummingbird flowers east of the Mississippi. *Castanea* 13: 97–109.   [326]

JAMES, W. O. and CLAPHAM, A. R. (1935). *The Biology of Flowers.* Clarendon Press, Oxford.

JOHRI, B. M. and VASIL, I. K. (1961). Physiology of pollen. *Bot. Rev.* 27: 325–81. [339]

JONES, E. W. (1945). Biological Flora of the British Isles. *Acer* L. *J. Ecol.* 32: 215–52. [134]

KAUSIK, S. B. (1939). Pollination and its influences on the behaviour of the pistillate flower in *Vallisneria spiralis. Am. J. Bot.* 26: 207–11. [278]

KERNER VON MARILAUN, A. (1878). *Flowers and their Unbidden Guests.* Kegan Paul, London. [145]

KERNER VON MARILAUN, A. (1902). *The Natural History of Plants.* Trans. F. W. Oliver. Blackie, London. [29, 196]

KIKUCHI, T. (1962). Studies on the coaction among insects visiting flowers. I. Ecological groups in insects visiting the Chrysanthemum flower, *Chrysanthemum leucanthemum. Sci. Rep. Tôhoku Univ.*, Ser. 4 (Biol.), 28: 17–22. II. Dominance relationship in the so-called drone fly group. *ibid.* 47–51. [378]

KIKUCHI, T. (1963). Studies on the coaction among insects visiting flowers. III. Dominance relationship among flower-visiting flies, bees and butterflies. *Sci. Rep. Tôhoku Univ.*, Ser. 4 (Biol.), 29: 1–8. [378]

KIMMINS, D. E. (1939). Empididae (Dipt.) on the flowers of *Orchis elodes* Godf. *J. Soc. Br. Ent.* 2: 37–8. [75, 77]

KIRBY, W. and SPENCE, W. (1815–26). *Introduction to Entomology.* London. [26]

KNIGHT, G. H. (1961). Some observations on pollination. *Biology and Human Affairs* 27(1): 35–42. [189]

KNIGHT, G. H. (1963). Pollen and nectar preferences. *Ann. Rep. Warwick Nat. Hist. Soc.* 9: 12–15. [53]

KNIGHT, G. H. (1968). Observations on the behaviour of *Bombylius major* L. and *B. discolor* Mik. (Dipt., Bombyliidae) in the Midlands. *Entomologist's mon. Mag.* 103: 177–82. [78]

KNIGHT, THOMAS (1799). Experiments on the fecundation of vegetables. *Phil. Trans. R. Soc.* 89: 195–204. [26]

KNOLL, F. (1921). *Bombylius fuliginosus* und die Farbe der Blumen (Insekten und Blumen I). *Abh. zool.-bot. Ges. Wien* 12: 17–119. [80, 111, 112]

KNOLL, F. (1922). Lichtsinn und Blumenbesuch des Falters von *Macroglossum stellatarum.* (Insekten und Blumen III). *Abh. zool.-bot. Ges. Wien* 12: 121–378. [99, 120, 123, 126]

KNOLL, F. (1925). Lichtsinn und Blütenbesuch des Falters von *Deilephila livornica. Z. vergl. Physiol.* 2: 329–80. [122, 124, 125, 130]

KNOLL, F. (1926). Die Arum-Blütenstände und ihre Besucher (Insekten und Blumen IV). *Abh. zool.-bot. Ges. Wien* 12: 379–481. [29, 107, 109, 228, 302]

KNOLL, F. (1927). Über Abendschwärmer und Schwärmerblumen. *Ber. dt. bot. Ges.* 45: 510–18. [125, 130]

KNOLL, F. (1935). Über den Schwärmflug der Maskenbienen (*Prosopis*). *Biologia. Gen.* 11: 115–54. [169]

KNOLL, F. (1956). *Die Biologie der Blüte.* Springer, Berlin, etc. [307]

KNUTH, P. (1906–9). *Handbook of Flower Pollination.* Trans. J. R. Ainsworth Davis (3 vols.; I, 1906, II, 1908, III, 1909). Oxford. ('*Knuth*').
[29, 53, 54, 57, 63, 64, 73–9, 88, 89, 91, 98, 99, 102, 133–5, 137, 138, 140, 330]

KÖLREUTER, J. G. (1761–6). *Vorläufige Nachricht von einigen das Geschlecht der Pflanzen betreffenden Versuchen und Beobachten.* Leipzig.   [25]

KUGLER, H. (1932). Blütenökologische Untersuchungen mit Hummeln. III. Das Verhalten der Tiere zu Duftstoffen, Duft und Farbe. IV. Der Duft als chemischer Nahfaktor bei duftenden und 'duftlosen' Blüten. *Planta* 16: 227–76, 534–53.   [168, 180]

KUGLER, H. (1938). Sind *Veronica chamaedrys* L. und *Circaea lutetiana* L. Schwebefliegenblumen? *Bot. Arch.* 39: 147–65.   [85]

KUGLER, H. (1940). Die Bestäubung von Blumen durch Furchenbienen (*Halictus* Latr.). *Planta* 30: 780–99.   [149, 155, 166, 174]

KUGLER, H. (1943). Hummeln als Blütenbesucher. *Ergebn. Biol.* 19: 143–323.
                    [169, 179–81]

KUGLER, H. (1950). Der Blütenbesuch der Schlammfliege (*Eristalomyia tenax*). *Z. vergl. Physiol.* 32: 328–47.   [110, 114, 115]

KUGLER, H. (1951). Blütenökologische Untersuchungen mit Goldfliegen (Lucilien). *Ber. dt. bot. Ges.* 64: 327–41.   [109]

KUGLER, H. (1955a). *Einführung in die Blütenökologie.* Fischer, Stuttgart. ('*Kugler*').
          [80, 84, 90, 91, 99, 115, 131, 134, 140, 162, 170, 337, 371]

KUGLER, H. (1955b). Zur Problem der Dipterenblumen. *Öst. bot. Z.* 102: 529–41.   [85]

KUGLER, H. (1956). Über die optische Wirkung von Fliegenblumen auf Fliegen. *Ber. dt. bot. Ges.* 69: 387–98.   [110, 115]

KUGLER, H. (1963). UV-Musterungen auf Blüten und ihr Zustandekommen. *Planta* 59: 296–329.   [173]

KUGLER, H. (1966). UV-Male auf Blüten. *Ber. dt. bot. Ges.* 79: 57–70.
                    [173, 321]

KULLENBERG, B. (1950). Investigations on the pollination of *Ophrys* species. *Oikos* 2: 1–19.   [245]

KULLENBERG, B. (1953). Några iakttagelser över insektsbesoken på blomman av *Parnassia palustris* L. *Svensk bot. Tidskr.* 47: 24–9.   [57]

KULLENBERG, B. (1956a). On the scents and colours of *Ophrys* flowers and their specific pollinators among the Aculeate Hymenoptera. *Svensk bot. Tidskr.* 50: 25–46.   [140]

KULLENBERG, B. (1956b). Field experiments with chemical sexual attractants on aculeate Hymenoptera males. I. *Zool. Bidr. Upps.* 31: 253–354.   [151]

KULLENBERG, B. (1961). Studies in *Ophrys* pollination. *Zool. Bidr. Upps.* 34: 1–340.   [245, 248, 377, 382]

LAGERBERG, T., HOLMBOE, J, and NORDHAGEN, R. (1957). *Våre ville planter* 6. Oslo.   [209]

LAWRENCE, W. J. C. (1939). *Practical Plant Breeding.* Ed. 2. Allen & Unwin, London.   [339, 342]

LECLERCQ, J. (1960). Fleurs butinées par les bourdons (Hym. Apidae Bombinae). *Bull. Inst. agron. Stns Rech. Gembloux* 28: 180–98.   [180]

LEDERER, G. (1951). Biologie der Nahrungsaufnahme der Imagines von *Apatura* und *Limenitis*, sowie Versuche zur Feststellung der Gustorezeption durch die Mittel- und Hintertarsen dieser Lepidoptera. *Z. Tierpsychol.* 18: 41–61.
                    [100, 126]

LEIUS, K. (1960). Attractiveness of different foods and flowers to the adults of some hymenopterous parasites. *Can. Ent.* 92: 369–76.   [135, 136]

LEPPIK, E. (1953). The ability of insects to distinguish number. *Am. Nat.* 87: 229–36. [180]

LEPPIK, E. (1956). The form and the function of numeral pattern in flowers. *Am. J. Bot.* 43: 445–55. [364]

LEPPIK, E. (1957). Evolutionary relationships between entomophilous plants and anthopilous insects. *Evolution* 11: 466–81. [364]

LEVIN, D. A. and KERSTER, H. W. (1967). Natural selection for reproductive isolation in *Phlox*. *Evolution* 21: 679–87. [370]

LEVIN, D. A. and KERSTER, H. W. (1968). Local gene dispersal in *Phlox*. *Evolution* 22: 130–9. [384]

LEX, T. (1954). Duftmale an Blüten. *Z. vergl. Physiol.* 36: 212–34. [164]

LIEBERMANN, A. (1925). Korrelation zwischen den antennalen Geruchsorganen und der Biologie der Musciden. *Z. Morph. Ökol. Tiere* 5: 1–97. [108]

LINDNER, E. (1928). *Aristolochia lindneri* Berger und ihre Bestäubung durch Fliegen. *Biol. Zbl.* 48: 93–101. [305]

LINSLEY, E. G. (1958). The ecology of solitary bees. *Hilgardia* 27: 543–99. [366]

LINSLEY, E. G. (1961). The role of flower specificity in the evolution of solitary bees. *Verhandl. XI Internat. Kongr. Entom., Wien, 1960,* 1: 593–6. [366]

LINSLEY, E. G., MACSWAIN, J. W. and RAVEN, P. H. (1963). Comparative behaviour of bees and Onagraceae. I and II. *Univ. Calif. Publ. Entom.* 33(1). [366]

LINSLEY, E. G., MACSWAIN, J. W. and RAVEN, P. H. (1964). Comparative behaviour of bees and Onagraceae. III. *Univ. Calif. Publ. Entom.* 33(2). [366]

LOEW, E. (1895). *Einführung in die Blütenbiologie auf historischer Grundlage.* Berlin. [29, 86]

LOGAN, JAMES (1736). Experiments concerning the impregnation of the seeds of plants. *Phil. Trans. R. Soc.* 39: 192. [22]

LOGAN, JAMES (1739). *Experimenta et Meletemata de Plantarum Generatione.* Leiden. (Translated as, *Experiments and Considerations on the Generation of Plants . . . Translated from the . . . Latin by J. F*[othergill]*.* London, 1747.)

LUBBOCK, Sir JOHN (LORD AVEBURY) (1875). *On British wild flowers considered in relation to insects.* Macmillan, London. [31]

LUCAS, F. A. (1897). The tongues of birds. *Rep. U.S. Natn. Mus.* 1895: 1001–20. [328, 329, 331]

LUCKWILL, L. C., WAY, D. W. and DUGGAN, J. B. (1962). The pollination of fruit crops. II and III. *Scient. Hort.* 15: 82–122. [352]

McCANN, C. (1943). 'Light-windows' in certain flowers. *J. Bombay nat. Hist. Soc.* 44: 182–4. [302]

MACIOR, L. W. (1964). An experimental study of the pollination of *Dodecatheon meadia*. *Am. J. Bot.* 51: 96–108. [192]

MACIOR, L. W. (1966). Foraging behaviour of *Bombus* (Hymenoptera: Apidae) in relation to *Aquilegia* pollination. *Am. J. Bot.* 53: 302–9. [218]

MACIOR, L. W. (1967). Pollen-foraging behaviour of *Bombus* in relation to pollination of nototribic flowers. *Am. J. Bot.* 54: 359–64. [213]

MACIOR, L. W. (1968a). Pollination adaptation in *Pedicularis groenlandica*. *Am. J. Bot.* 55: 927–32. [209, 210]

MACIOR, L. W. (1968b). Pollination adaptation in *Pedicularis canadensis*. *Am. J. Bot.* 55: 1031–5.   [209, 210]

McKELVEY, S. D. (1947). *Yuccas of the Southwestern United States*. Vol. 2. Arnold Arboretum, Jamaica Plain, Mass.   [316]

McLEAN, R. C. and COOK, W. R. IVIMEY- (1956). *Textbook of Theoretical Botany*. Vol. 2. Longmans, Green & Co., London, etc.   [313, 315]

McNAUGHTON, I. H. and HARPER, J. L. (1960). The comparative biology of closely related species living in the same area. I. External breeding-barriers between *Papaver* species. *New Phytol.* 59: 15–26.   [54]

MAHESHWARI, P. (1950). *An Introduction to the Embryology of the Angiosperms*. McGraw-Hill, New York, etc.   [339]

MAHESHWARI, P. and KANTA, K. (1964), in Linskens, H. F. (Ed.), *Pollen Physiology and Fertilization*. North Holland Publishing Co., Amsterdam.
                                                                                         [340]

MANNING, A. (1956a). The effect of honeyguides. *Behaviour* 9: 114–39.
                                                                                    [168, 169]

MANNING, A. (1956b). Some aspects of the foraging behaviour of bumble-bees. *Behaviour* 9: 164–201.   [166, 167, 180]

MANNING, A. (1957). Some evolutionary aspects of flower-constancy of bees. *Proc. R. phys. Soc. Edinb.* 25: 67–71.   [166]

MARSDEN-JONES, E. M. (1935). *Ranunculus ficaria* Linn.: life-history and pollination. *J. Linn. Soc. (Bot.)* 50: 39–55.   [69]

MEEUSE, B. J. D. (1961). *The Story of Pollination*. Ronald Press Co., New York.

MEEUSE, B. J. D. (1966). The voodoo lily. *Scient. Am.* 215: 80–88.   [229]

MILLER, PHILIP (1724). *The Gardener's and Florist's Dictionary*. London.   [22]

MILLER, PHILIP (1731). *The Gardener's Dictionary*. London. (Ed. 6, 1752).
                                                                                          [22]

MITTLER, T. E. (Ed.) (1962). *Proc. First Internat. Symp. Pollination (Swedish Seed Growers' Assoc., Comm. No. 7)*: 1–224.   [343, 346–9]

MOLITOR, A. (1937). Zur vergleichenden Psychobiologie der akuleaten Hymenopteren auf experimenteller Grundlage. *Biologia. Gen.* 13: 294–333.   [142]

MOLLER, W. (1930). Über die Schnabel- und Zungen-mechanik blütenbesuchender Vögel. I. *Biologia. Gen.* 6: 651–726.   [327, 329, 331]

MOLLER, W. (1931a). Über die Schnabel- und Zungen-mechanik blütenbesuchender Vögel. II. *Biologia. Gen.* 7: 99–154.   [328]

MOLLER, W. (1931b). Vorläufige Mitteilung über die Ergebnisse einer Forschungsreise nach Costa Rica zu Studien über die Biologie blütenbesuchender Vögel. *Biologia. Gen.* 7: 287–312.   [324, 325, 327, 328]

MOORE, D. M. and LEWIS, H. (1965). The evolution of self-pollination in *Clarkia xantiana*. *Evolution* 19: 104–14.   [291]

MORLAND, SAMUEL (1703). Some new observations upon the parts and use of the flower in plants. *Phil. Trans. R. Soc.* No. 287: 1474.   [23]

MÜLLER, H. (1879). Weitere Beobachtungen über Befruchtung der Blumen durch Insekten. II. *Verh. naturh. Ver. preuss. Rheinl.* 36: 198–268.   [104]

MÜLLER, H. (1881). *Die Alpenblumen, ihre Befruchtung durch Insekten und ihre Anpassungen an dieselben*. Leipzig.   [28, 102, 181, 371]

MÜLLER, H. (1883). *The Fertilisation of Flowers.* Trans. D'Arcy W. Thompson. London. (*'Müller'*).
[28, 56, 63, 66, 68–70, 73, 78, 80, 82, 83, 86, 99, 117, 162]

MÜLLER, L. (1926). Zur biologischen Anatomie der Blüte von *Ceropegia woodii* Schlechter. *Biologia. Gen.* 2: 799–814. [29, 302, 303]

MÜNZING, A. (1930). Outlines to a genetic monograph of *Galeopsis. Hereditas* 13: 185–341. [382]

NICHOLLS, W. H. (1955, 1958). *Orchids of Australia*, parts 3 and 4. Georgian House, Melbourne. [310]

NIXON, R. W. (1928). The direct effect of pollen on the fruit of the Date Palm. *J. agric. Res.* 36: 97–128. [351]

OHNO, M. (1963). Studies on pollen suspensions for saving labour in the pollination of fruit trees: the effect of alcohol added to the suspension on pollen germination. *J. Jap. hort. Sci.* (*J. Hort. Ass. Japan*) 31: 360–4. [354]

OLIVER, F. W. (1888). On the sensitive labellum of *Masdevallia muscosa*, Rchb. f. *Ann. Bot.* 1: 237–53. [310]

ORNDUFF, R. (1969). Reproductive biology in relation to systematics. *Taxon* 18: 121–33. [365]

PANDEY, K. K. (1960). Evolution of gametophytic and sporophytic systems of self-incompatibility. *Evolution* 14: 98–115. [227]

PARKIN, J. (1928). The glossy petal of *Ranunculus. Ann. Bot.* 92: 739–55. [50]

PARKIN, J. (1931). The structure of the starch layer in the glossy petal of *Ranunculus. Ann. Bot.* 45: 201–5. [50]

PARKIN, J. (1935). The structure of the starch layer in the glossy petal of *Ranunculus*. II. The British species examined. *Ann. Bot.* 49: 283–9. [50]

PARMENTER, L. (1941). Diptera visiting flowers of Devilsbit Scabious, *Scabiosa succisa* L. *Entomologist's Rec. J. Var.* 53: 134. [91]

PARMENTER, L. (1942). Dolichopodidae (Dipt.) associated with flowers. *Entomologist's mon. Mag.* 78: 252. [77]

PARMENTER, L. (1948). *Rhingia campestris* Mg. (Dipt., Syrphidae), a further note. *Entomologist's Rec. J. Var.* 60: 119–20. [84, 86]

PARMENTER, L. (1949). Further notes on insect visitors to the flowers of Sea Aster, *Aster tripolium* Linn. *Entomologist's Rec. J. Var.* 61: 85–6. [77]

PARMENTER, L. (1951). Notes on the genus *Empis* (Dipt., Empididae). *Entomologist's mon. Mag.* 87: 41–4. [77]

PARMENTER, L. (1952a). Flies visiting Greater Stitchwort, *Stellaria holostea* Linn. (Caryophyllaceae). *J. Soc. Br. Ent.* 4: 88–9. [55, 78, 91]

PARMENTER, L. (1952b). Flies at Ivy-bloom. *Entomologist's Rec. J. Var.* 64: 90–1. [73–5, 91]

PARMENTER, L. (1956). Beetles visiting the flowers of Dogwood, *Cornus sanguinea* L. *Entomologist's Rec. J. Var.* 68: 243–4. [67]

PARMENTER, L. (1958). Flies (Diptera) and their relations with plants. *Lond. Nat.* 37: 115–25. [76, 114]

PEARSON, J. F. W. (1933). On the ecological relations of bees in the Chicago region. *Ecol. Monogr.* 3: 373–442. [378]

PENNINGTON, W. (1969). *The History of British Vegetation.* English Universities Press, London. [277]

PERCIVAL, M. S. (1961). Types of nectar in angiosperms. *New Phytol.* 60: 235–81. [46]

PERCIVAL, M. S. (1965). *Floral Biology.* Pergamon Press, Oxford, etc.   [222]

PERKINS, R. C. L. (1903). Vertebrata, in Sharp. D., *Fauna Hawaiiensis* 1(4): 368–465.   [330]

PETERSON, A. (1916). The head-capsule and mouth-parts of Diptera. *Illinois biol. Monogr.* 3: 173–282.   [80, 81, 83]

PICKENS, A. L. (1927). Unique method of pollination by the Ruby-throat. *Auk* 44: 24–7.   [321]

PICKENS, A. L. (1936). Steps in the development of the bird-flower. *Condor* 38: 150–4.   [324]

PICKENS, A. L. (1941). A red figwort as the ideal nearctic bird flower. *Condor* 43: 100–2.   [326]

PICKENS, A. L. (1944). Seasonal territory studies of Ruby-throats. *Auk* 61: 88–92.   [325]

PIGOTT, C. D. (1958). Biological Flora of the British Isles. *Polemonium caeruleum* L. *J. Ecol.* 46: 507–25.   [106]

VAN DER PIJL, L. (1936). Fledermäuse und Blumen. *Flora, Jena* 131: 1–40.   [332, 333, 335, 337]

VAN DER PIJL, L. (1937a). Biological and physiological observations on the inflorescence of *Amorphophallus. Recl. Trav. bot. néerl.* 34: 157–67.   [295]

VAN DER PIJL, L. (1937b). Disharmony between Asiatic flower-birds and American bird-flowers. *Annls. Jard. Bot. Buitenz.* 48: 17–26.   [319, 328]

VAN DER PIJL, L. (1953). On the flower biology of some plants from Java, with general remarks on fly-traps. *Ann. Bogor.* 1: 77–99.   [295, 306, 309, 311, 312]

VAN DER PIJL, L. (1954). *Xylocopa* and flowers in the Tropics. *Proc. K. ned. Akad. Wet.,* Ser. C, 57: 413–23, 541–62.   [146, 179, 294]

VAN DER PIJL, L. (1956). Remarks on pollination by bats in the genera *Freycinetia* etc. and on chiropterophily in general. *Acta bot. neerl.* 5: 135–44.   [332, 335, 337]

VAN DER PIJL, L. (1960). Ecological aspects of flower evolution. I. Phyletic evolution. *Evolution* 14: 403–16.   [362, 376]

VAN DER PIJL, L. (1961). Ecological aspects of flower evolution. II. Zoophilous flower classes. *Evolution* 15: 44–59.   [183]

VAN DER PIJL, L. and DODSON, C. H. (1966). *Orchid Flowers: their Pollination and Evolution.* University of Miami Press, Coral Gables, Florida.   [241]

PLATEAU, F. (1885–98, etc.). Numerous references listed by Knuth.   [31]

PLATEAU, F. (1899). Nouvelles recherches sur les rapports entre les insectes et les fleurs. II. Le choix des couleurs par les insectes. *Mém. Soc. zool. Fr.* 12: 336–70.

POHL, F. (1937). Die Pollenerzeugung der Windblutler. *Beih. bot. Zbl.* A 56: 365–470.   [256]

POPHAM, E. J. (1961). Earwigs in the British Isles. *Entomologist* 94: 308–10.   [104]

PORSCH, O. (1909). Neuere Untersuchungen über die Insektenanlockungsmittel der Orchideenblüte. *Mitt. naturwiss. Ver. Steierm.* 45: 346–70.   [295, 296]

PORSCH, O. (1924). Vogelblumenstudien. I. *Jb. wiss. Bot.* 63: 553–706.   [318, 319]

PORSCH, O. (1927). Kritische Quellenstudien über Blumenbesuch durch Vögel. III. *Biologia. Gen.* 3: 475–548.   [319]

PORSCH, O. (1930). Kritische Quellenstudien über Blumenbesuch durch Vögel. V. *Biologia. Gen.* 6: 135–246. [323, 328]

PORSCH, O. (1933). Der Vogel als Blumen bestäuber. *Biologia. Gen.* 9: 239–52. [318, 323]

PORSCH, O. (1956). Windpollen und Blumeninsekt. *Öst. bot. Z.* 103: 1–18. [272]

PORSCH, O. (1957). Alte Insektentypen als Blumenausbeuter. *Öst. bot. Z.* 104: 115–64. [105, 106]

POULTON, E. B. (1932), in Godfery, M. J., Insect carriers of orchid pollinia. *Proc. ent. Soc. Lond.* 6: 70. [134]

POUYANNE, A. (1917). La Fécondation des Ophrys pas les Insectes. *Bull. Soc. Hist. nat. Afr. N.* Vol. 8. [244]

POWELL, J. A. and MACKIE, R. A. (1966). Biological interrelationships of moths and *Yucca whipplei* (Lepidoptera: Gelechiidae, Blastobasidae, Prodoxidae). *Univ. Calif. Publ. Ent.* 42. [317]

PRIME, C. T. (1954). Biological Flora of the British Isles. *Arum neglectum* (Townsend) Ridley. *J. Ecol.* 42: 241–8. [89]

PRIME, C. T. (1960). *Lords and Ladies.* Collins, London. [222]

RAMIREZ, B. W. (1969). Fig wasps: mechanism of pollen transfer. *Science* 163: 580–1. [316]

RAY, JOHN (1696–1704). *Historia Plantarum.* (3 vols.). London. [20]

REMPE, H. (1937). Untersuchungen über die Verbreitung des Blütenstaubes durch die Luftströmungen. *Planta* 27: 93–147. [263]

RENDLE, A. B. (1930). *The classification of flowering plants.* 1 (Ed. 2). Cambridge. [271]

RIBBANDS, C. R. (1949). The foraging method of individual honeybees. *J. Anim. Ecol.* 18: 47–66. [174]

RIBBANDS, C. R. (1953). *The Behaviour and Social Life of Honeybees.* Bee Research Association, London. [164]

RIBBANDS, C. R. (1955). The scent perception of the Honeybee. *Proc. R. Soc.* B 143: 367–79. [164]

RICHARDS, O. W. and HAMM, A. H. (1939). The biology of the British Pompilidae (Hymenoptera). *Trans. Soc. Br. Ent.* 6: 51–114. [140]

RICK, C. M. (1950). Pollination relations of *Lycopersicon esculentum* in native and foreign regions. *Evolution* 4: 110–22. [350]

RIDGWAY, R. (1891) The hummingbirds. *Rep. U.S. Natn. Mus.* 1890: 253–383. [323–3, 326, 327, 329]

RIDLEY, H. N. (1890). On the method of fertilization in *Bulbophyllum macranthum*, and allied orchids. *Ann. Bot.* 4: 327–36. [299]

RILEY, C. V. (1892). The Yucca moth and Yucca pollination. *Rep. Mo. Bot. Gdn.* 3: 99–158. [313, 316]

ROBINSON, T. R. and SAVAGE, E. M. (1926). Pollination of the Avocado. *U.S. Dept. Agric., Circular No. 387.* [351]

SACHS, J. (1875). *Geschichte der Botanik vom 16 Jahrhundert bis 1860.* R. Oldenbourg, München. (English translation by H. E. F. Garnsey and I. Bayley-Balfour, *History of Botany (1530–1860)*, Oxford, 1890.) [23]

SADAMORI, S. *et al.* (1958). Studies in commercial hand pollination methods for apple flowers. I. Examination of pollen diluents, of degree of pollen distribution and pollinating methods. *Bull. Tôhoku Natn. Agric. Exp. Stn.* 14: 74–81.   [354]

St John, H. (1965). Monograph of the genus *Elodea*: Part 4 and summary. *Rhodora* 67: 1–35.   [279]

Salisbury, E. J. (1969). The reproductive biology and occasional seasonal dimorphism of *Anagallis minima* and *Lythrum hyssopifolia*. *Watsonia* 7: 25–39.   [291]

Sargent, O. H. (1909). Notes on the life-history of *Pterostylis*. *Ann. Bot.* 23: 265–74.   [311]

Sargent, O. H. (1918). Fragments on the flower biology of Westralian plants. *Ann. Bot.* 32: 215–31.   [323]

Sargent, O. H. (1934). Pollination in *Pterostylis*. *Vict. Nat.* 51: 82–4.   [311]

Saunders, E. (1878). Remarks on the hairs of some of our British Hymenoptera. *Trans. ent. Soc. Lond.* 1878: 169–72.   [160]

Saunders, E. (1890). On the tongues of the British Hymenoptera Anthophila. *J. Linn. Soc. (Zool.)* 23: 410–32.   [153]

Schelpe, E. A. C. L. E. (1966). *An introduction to the South African orchids.* Macdonald, London.   [375]

Schlegtendal, A. (1934). Beitrag zum Farbensinn der Arthropoden. *Z. vergl. Physiol.* 20: 545–81.   [108, 111]

Schmid, G. (1912). Zur Ökologie der Blüte von *Himantoglossum. Ber. dt. bot. Ges.* 30: 464–9.   [242]

Schremmer, F. (1941a). Sinnesphysiologie und Blumenbesuch des Falters von *Plusia gamma* L. *Zool. Jber. Neapel* 74: 375–434.   [117, 130]

Schremmer, F. (1941b). Versuche zum Nachweis der Rotblindheit von *Vespa rufa* L. *Z. vergl. Physiol.* 28: 457–66.   [142]

Scorer, A. G. (1913). *The Entomologist's Log-book and Dictionary of the Life-Histories and Food Plants of the British Macro-Lepidoptera.* Routledge, London.   [99]

Scott, H. (1953). Discrimination of colours by *Bombylius* (Dipt., Bombyliidae). *Entomologist's mon. Mag.* 89: 259–60.   [78]

Scott-Elliot, G. F. (1890a). Note on the fertilisation of *Musa, Strelitzia reginae,* and *Ravenala madagascariensis. Ann. Bot.* 4: 259–63.   [323]

Scott-Elliot, G. F. (1890b). Ornithophilous flowers in South Africa. *Ann. Bot.* 4: 265–80.   [323]

Scott-Elliot, G. F. (1896). *The Flora of Dumfriesshire.* Dumfries.   [29]

Sculthorpe, C. D. (1967). *The Biology of Aquatic Vascular Plants.* Arnold, London.   [278, 285]

Shaw, R. J. (1962). The biosystematics of *Scrophularia* in western North America. *Aliso* 5: 147–78.   [144]

Shoemaker, J. S. (1962). Pollination requirements of Flordagrand Blackberry. *Proc. Fla. St. Hort. Soc.* 74: 356–8.   [354]

Silén, F. (1906a). Blombiologisk iakttagelser i Kittilä Lappmark. *Meddn Scc. Fauna Flora fenn.* 31: 80–99.   [243]

Silén, F. (1906b). Blombiologiska iakttagelser i södra Finland. *Meddn Soc. Fauna Flora fenn.* 32: 120–34.   [212]

Simes, J. A. (1946). Behaviour of *Bombylius* (Dipt., Bombyliidae) while feeding. *Entomologist's mon. Mag.* 89: 234.   [80]

Sladen, F. W. L. (1911). How pollen is collected by the social bees, and the part played in the process by the auricle. *Br. Bee J.* 39: 491–5; The pollen-collecting apparatus in the social bees. *ibid.* 39: 506.   [164

SLADEN, F. W. L. (1912). How the corbicula is loaded. *Br. Bee J.* 40: 138; four other papers, *ibid.* 40: 144–5, 164–6, 196, 462–3. [164]

SMALL, J. (1915). The pollen-presentation mechanism in the Compositae. *Ann. Bot.* 29: 457–70. [220]

SMALL, J. (1917). The origin and development of the Compositae. Introduction and I–III. *New Phytol.* 16: 157–8, 159–77, 198–221, 253–76. [222]

SMALL, J. (1918). The origin and development of the Compositae. IV–IX. *New Phytol.* 17: 13–40, 69–94, 114–25, 126–32, 133–42, 200–30. [222]

SMART, J. (1943). *Simulium* feeding on Ivy flowers. *Entomologist* 76: 20–21. [75]

SMITH, K. V. G. (1959). The distribution and habits of the British Conopidae (Dipt.). *Trans. Soc. Br. Ent.* 13: 113–36. [88]

SMITH, K. V. G. (1961). Supplementary records of the distribution and habits of the British Conopidae (Diptera). *Entomologist* 94: 238–9. [88]

SNODGRASS, R. E. (1956). *The Anatomy of the Honey Bee.* London and New York. [156–7, 164–5]

SPERRING, A. H. (1933). Trichoptera at Ivy blossom. *J. ent. Soc. S. Engl.* 1: 62. [106]

SPOONER, G. M. (1930). *The Bees, Wasps and Ants of Cambridgeshire.* Cambridge Natural History Society, Cambridge. [140, 143]

SPOONER, G. M. (1941). The characters of the female and distribution in Britain of *Pompilus trivialis* Dahlb., *unguicularis* Thoms. and *wesmaeli* Thoms. (Hymenoptera: Pompilidae). *Trans. Soc. Br. Ent.* 7: 85–122. [140]

SPRAGUE, E. F. (1962). Pollination and evolution in *Pedicularis* (Scrophulariaceae). *Aliso* 5: 181–209. [167, 209, 211, 371]

SPRENGEL, C. K. (1793). *Das Entdeckte Geheimniss der Natur im Bau und in der Befruchtung der Blumen.* Berlin. [26, 27, 240]

STACE, C. A. (1961). Some studies in *Calystegia*: compatibility and hybridisation in *C. sepium* and *C. silvatica. Watsonia* 5: 88–105. [194]

STACE, C. A. (1965). Some studies in *Calystegia.* 2. Observations on the floral biology of the British inland taxa. *Proc. bot. Soc. Br. Isl.* 6: 21–31. [194]

STEARN, W. T. (1969). The Jamaican species of *Columnea* and *Alloplectus* (Gesneriaceae). *Bull. Br. Mus. nat. Hist. (Bot.)* 4(5): 181–263 [377]

STEBBINS, G. L. (1950). *Variation and Evolution in Plants.* Columbia University Press, New York. [385]

STEBBINS, G. L. (1957). Self-fertilisation and population variability in higher plants. *Am. Nat.* 91: 337–54. [292]

STEBBINS, G. L. (1958). Longevity, habitat and release of variability in higher plants. *Cold Spring Harb. Symp. quant. Biol.* 23: 365–78. [383, 385]

STEPHEN, W. P. (1961). Artificial nesting sites for the propagation of the Leafcutter Bee, *Megachile (Eutricharaea) rotundata,* for Alfalfa pollination. *J. econ. Ent.* 54: 989–93. [346]

STEPHEN, W. P. (1965). Artificial beds for Alkali Bee propagation. *Agric. Exp. Station, Oregon State Univ., Cornwallis, Bulletin No. 598.* [343, 345]

STRAW, R. M. (1955). Hybridisation, homogamy and sympatric speciation. *Evolution* 9: 441–4. [367]

STRAW, R. M. (1956). Adaptive morphology of the *Penstemon* flower. *Phytomorphology* 6: 112–9. [367–8]

SUMMERHAYES, V. S. (1951). *Wild Orchids of Britain.* Collins, London.
[230, 254]
SWYNNERTON, C. F. M. (1916a). Short cuts by birds to nectaries. *J. Linn. Soc. (Bot.)* 43: 381–416. [319, 323]
SWYNNERTON, C. F. M. (1916b). Short cuts to nectaries by Blue Tits. *J. Linn. Soc. (Bot.)* 43: 417–22. [324]
TAKHTAJAN, A. (1969). *Flowering Plants: Origin and Dispersal.* Trans. C. Jeffrey. Oliver & Boyd, Edinburgh. [363]
THIEN, L. B. (1969a). Mosquito pollination of *Habenaria obtusata* (Orchidaceae). *Am. J. Bot.* 56: 232–7. [76]
THIEN, L. B. (1969b). Mosquitoes can pollinate orchids. *Morris Arboretum Bull.* 20(2): 19–23. [76]
THOMPSON, A. LANDSBOROUGH (Ed.) (1964). *A New Dictionary of Birds.* Nelson, London. [328, 330]
TILLYARD, R. J. (1923). On the mouthparts of the Micropterygoidea (Order Lepidoptera). *Trans. ent. Soc. Lond.* 1923: 181–206. [103, 104]
TRELEASE, W. (1881). The fertilisation of *Scrophularia. Bull. Torrey bot. Club* 8: 133–40. [217]
UPHOF, J. C. T. (1938). Cleistogamic flowers. *Bot. Rev.* 4: 21–49. [287]
VAILLANT, SEBASTIEN (1718). *Discours sur la Structure des Fleurs.* Leiden. [24]
VANDER PIJL: see under PIJL.
VERDCOURT, B. (1948). Scarcity of *Rhingia campestris*, Mg. (Dipt., Syrphidae). *Entomologist's Rec. J. Var.* 60: 108. [86]
VERLAINE, L. (1932a). L'instinct et l'intelligence chez les Hymenoptères. XVIII. L'odorat et la généralisation, le relatif et l'absolu chez les guêpes. *Bull. Annls Soc. ent. Belg.* 72: 311–22. [142]
VERLAINE, L. (1932b). L'instinct et l'intelligence chez les Hymenoptères. XV. Les guêpes ont-elles un langage? *Mém. Soc. r. Sci. Liége*, Sér. 3, 17(13): 1–16. [142]
VERRALL, G. H. (1909). *British Flies* 5. Gurney and Jackson, London. [78]
VOGEL, S. (1954). *Blütenbiologische Typen als Elemente der Sippengliederung.* Fischer, Jena. [101, 124, 302, 311]
VOGEL, S. (1958). Fledermausblumen in Sudamerika. *Öst. bot. Z.* 104: 491–530. [30, 330, 332, 333, 336]
VOGEL, S. (1961). Die Bestäubung der Kesselfallen-Blüten von *Ceropegia. Beitr. Biol. Pfl.* 36: 159–237. [302, 303]
VOGEL, S. (1963). Duftdrüsen im Dienste der Bestäubung: über Bau und Funktion der Osmophoren. *Abh. Math.-Naturw. Kl. Akad. Wiss. Mainz.* 1962(10): 599–763. [308, 312]
VOGEL, S. (1966). Scent organs of orchid flowers and their relation to insect pollination. *Proc. Fifth World Orchid Conference*: 253–9. Long Beach, California. [297]
VUILLEUMIER, B. S. (1967). The origin and evolutionary development of heterostyly in the angiosperms. *Evolution* 21: 210–26. [227]
WAGNER, H. O. (1946). Food and feeding habits of Mexican hummingbirds. *Wilson Bull.* 58: 69–93. [320, 325]

WANG, C. W., PERRY, T. O. and JOHNSON, A. G. (1960). Pollen dispersion of Slash Pine (*Pinus elliottii* Engelm.) with special reference to seed orchard management. *Silvae Genet.* 9: 78–86. [261, 383]

WATTS, L. E. (1958). Natural cross-pollination in Lettuce, *Lactuca sativa* L] *Nature, Lond.* 181: 1084. [223, 382.

WERFFT, R. (1951). Über die Lebensdauer der Pollenkörner in der freien Atmosphäre. *Biol. Zbl.* 70: 354–367. [263]

WERTH, E. (1956). *Bau und Leben der Blumen.* Enke, Stuttgart. [318]

WEST, R. G. (1968). *Pleistocene Geology and Biology.* Longmans, London. [227]

WHITE, A. and SLOANE, B. L. (1937). *The Stapelieae.* 3 vols. Pasadena, California. [301, 304]

WHITEHEAD, D. R. (1969). Wind pollination in the angiosperms; evolutionary and environmental considerations. *Evolution* 23: 28–35. [380]

WHITEHOUSE, H. L. K. (1950). Multiple-allelomorph incompatibility of pollen and style in the evolution of the angiosperms. *Ann. Bot.* N.S.14: 199–216. [363]

WICKLER, W. (1968). *Mimicry in Plants and Animals.* Trans. R. D. Martin. Weidenfeld and Nicolson, London. [375]

WIEBES, J. T. (1963). Taxonomy and host preferences of Indo-Australian fig-wasps of the genus *Ceratosolen* (Agaonidae). *Tijdschr. Ent.* 106: 1–112. [315, 316]

WILLIS, J. C. and BURKILL, I. H. (1895–1908). Flowers and insects in Great Britain. I. *Ann. Bot.* 9: 227–73 (1895). II. *ibid.* 17: 313–49 (1903a). III. *ibid.* 17: 539–70 (1903b). IV. *ibid.* 22: 603–49 (1908). ('*Willis and Burkill*'). [28, 53, 56, 66, 73–9, 88–91, 93–4, 99, 101–2, 104–5, 133–6, 138, 140–1, 143, 212]

WODEHOUSE, R. P. (1945). *Hayfever Plants.* Chronica Botanica, Waltham, Mass. [275]

WOLFF, T. (1950). Pollination and fertilisation of the Fly Ophrys, *Ophrys insectifera* L. in Allindelille fredskov, Denmark. *Oikos* 2: 20–59. [245]

WRIGHT, J. W. (1953). Pollen-dispersion studies: some practical applications. *J. Forestry* 51: 114–18. [261]

WRIGHT, S. (1946). Isolation by distance under diverse systems of mating. *Genetics, Princeton* 31: 39–59. [384]

WRIGHT, S. (1969). *Evolution and the genetics of populations.* University of Chicago Press. [384]

YARROW, I. H. H. (1945). Collecting bees and wasps, in The Hymenopterist's Handbook. *Amat. Ent.* 7: 55–81. [143]

YEO, P. F. (1968). The evolutionary significance of the speciation of *Euphrasia* in Europe. *Evolution* 22: 736–47. [375]

YOUNG, D. P. (1962). Studies in the British *Epipactis*. VI. Some further notes on *E. phyllanthes*. *Watsonia* 5: 136–9. [254]

INDEX